Applications of Biopolymers in Science, Biotechnology, and Engineering

Applications of Biopolymers in Science, Biotechnology, and Engineering

Edited by

Arunima Reghunadhan
Department of Chemistry
TKM College of Engineering, Karicode
Kollam, Kerala
India

Akhina H.
International and Interuniversity Centre for
Nanoscience and Nanotechnology
Kottayam, Kerala
India

Sabu Thomas
Mahatma Gandhi University
Priyadarshini Hills P.O.
Kottayam, Kerala
India

WILEY

Published by John Wiley & Sons, Inc., Hoboken, New Jersey.

Published simultaneously in Canada.

For general information on our other products and services or for technical support, please contact our Customer Care Department within the United States at (800) 762-2974, outside the United States at (317) 572-3993 or fax (317) 572-4002.

Wiley also publishes its books in a variety of electronic formats. Some content that appears in print may not be available in electronic formats. For more information about Wiley products, visit our web site at www.wiley.com.

A catalogue record for this book is available from the Library of Congress

Hardback ISBN: 9781119783442; ePub ISBN: 9781119783466;
ePDF ISBN: 9781119783459; oBook ISBN: 9781119783473

Cover image(s): © Ali DM/Shutterstock; Roman Chernikov/Shutterstock; Aldona Griskeviciene/Shutterstock; David T Clarke/Shutterstock; ART production/Shutterstock; KEITH CHAMBERS/SCIENCE PHOTO LIBRARY/Getty Images; Javier Zayas Photography/Getty Images
Cover design: Wiley

Set in 9.5/12.5pt STIXTwoText by Integra Software Services Pvt. Ltd., Pondicherry, India

Contents

List of Contributors *vii*
Preface *xiii*

1 **Introduction to Biopolymers, Their Blend, IPNs, Gel, Composites, and Nanocomposites** *1*
Mehvish Mumtaz, Nazim Hussain, Mubeen Ashraf, Hafiz Muhammad Husnain Azam, and Anwar Iftikhar

2 **Synthetic Biopolymers: Properties, Fabrication, and Applications** *31*
Reeba Mary Cherian, Rini Thresia Varghese, Jithin Joy, Cintil Jose, and Sabu Thomas

3 **Role of Biopolymers and Their Composites in Sustainable Agriculture: Recent Developments and Future Perspectives** *57*
Abhisekh Saha, Sreedeep, and Uttam Manna

4 **Biopolymer in Bioengineering and Medical Technology** *89*
Maya Rajan, Shahena S., Vinaya Chandran, and Linu Mathew

5 **Biopolymers and Composites in Tissue Engineering** *101*
Trinath Biswal

6 **Biopolymers, Composites, Nanocomposites, and Gels in Biotechnology** *139*
Ayush Bhandari, Karishma Shetty, Anubhav Wadhwa, and Khushwant S. Yadav

7 **Biopolymers, Blends, Composites, Gels, and Thin Films in Drug Delivery and Drug Design** *167*
Saman Noor, Nazim Hussain, and Areej Shahbaz

8 Biopolymers and Their Composites for Biotechnological Applications *189*
Md Enamul Hoque and Md Ushama Shafoyat

9 Biobased Polymers, Their Composites and Blends in Electronics *219*
Divyanshu Thakur and Maneesh Kumar Poddar

10 Polymers and Biopolymers in Sensing *243*
Greeshma U. Chandran, Hema S., Malavika Sajith, Rashid Sulthan, and Sreedha Sambhudevan

11 Applications of Biopolymers in Construction and Civil Engineering *265*
Eapen Philip and Reshmy R.

12 Biopolymers and Functional Biopolymers in Food Technology *291*
Sanem Argin and Y. Martin Lo

13 Biopolymers in Food Packaging *309*
Sneha Sara Varghese, Vaisakh P.H., Eapen Philip, Tressia Alias Princy Paulose, and Akhina H.

14 Nanofiber Composites for Packaging Applications *331*
Vandana A. Mooss, Nimitha K.C., Rinu Tresia P.X., Jiji Abraham, and Arunima Reghunadhan

15 Polymers, Their Composites, Blends, and Nanocomposites for the Fabrication of Prosthetics *361*
Md Enamul Hoque, Md Ushama Shafoyat, and Fatiha Tabassun

Index *391*

List of Contributors

Jiji Abraham
Department of Chemistry
Vimala College (Autonomous)
Thrissur
Kerala
India

Sanem Argin
Department of Food Engineering
Yeditepe University
Istanbul
Turkey

Mubeen Ashraf
Department of Microbiology
University of the Central Punjab
Lahore
Pakistan

Ayush Bhandari
Shobhaben Pratapbhai Patel
School of Pharmacy & Technology
Management
SVKM'S NMIMS (Deemed to be
University)
Mumbai
India

Trinath Biswal
Department of Chemistry
VSS University of Technology
Burla
India

Greeshma Chandran
Department of Chemistry
Amrita Vishwa Vidyapeetham
Amritapuri
Kerala
India

Vinaya Chandran
School of Biosciences
Mahatma Gandhi University
Kottayam
Kerala
India

Reeba Mary Cherian
Department of Chemistry
Newman College
Thodupuzha
Kerala
India
and

School of Chemical Sciences
Mahatma Gandhi University
Kottayam
Kerala
India

Akhina H.
International and Inter University
Centre for Nanoscience and
Nanotechnology
Kottayam
Kerala
India

Enamul Hoque
Department of Biomedical
Engineering
Military Institute of Science and
Technology
Dhaka
Bangladesh

Hafiz Muhammad Husnain
Institute of Biotechnology
Faculty of Environment and
Natural Sciences
Brandenburg University of
Technology Cottbus-Senftenberg
Universitätsplatz
Senftenberg
Germany

Nazim Hussain
Centre for Applied Molecular
Biology (CAMB)
University of the Punjab, Quaid-e-
Azam Campus
Lahore
Pakistan

Anwar Iftikhar
Centre for Applied Molecular
Biology (CAMB)
University of the Punjab, Quaid-e-
Azam Campus
Lahore
Pakistan

Cintil Jose
Department of Chemistry
Newman College
Thodupuzha
Kerala
India

Jithin Joy
Department of Chemistry
Newman College
Thodupuzha
Kerala
India

Nimitha K.C.
Department of Chemistry
Vimala College (Autonomous)
Thrissur
Kerala
India

Y. Martin Lo
Institute for Advanced Study
Shenzhen University
Shenzhen
China

Uttam Manna
Department of Chemistry and
Center for Nanotechnology
Indian Institute of Technology
Guwahati
Assam
India

Linu Mathew
School of Biosciences
Mahatma Gandhi University
Kottayam
Kerala
India

Vandana Mooss
Department of Chemistry
Dr. Viswanath Karad MIT-World
Peace University
Pune
Maharashtra
India

Mehvish Mumtaz
Centre for Applied Molecular
Biology (CAMB)
University of the Punjab, Quaid-e-
Azam Campus
Lahore
Pakistan

Saman Noor
Atta-ur-Rehman School of Applied
Biosciences (ASAB)
National University of Sciences
and Technology (NUST)
Islamabad
Pakistan

Vaisakh P.H.
Post Graduate and Research
Department of Chemistry
Bishop Moore College
Mavelikara
Kerala
India

Arunima Reghunadhan
Department of Chemistry
TKM College of Engineering
Karicode
Kollam
Kerala
India

Rinu Tressia P.X.
Department of Chemistry
Christ College
Irinjalakuda
Kerala
India

Tressia Alias Princy Paulose
Post Graduate and Research
Department of Chemistry
Bishop Moore College
Mavelikara
Kerala
India

Eapen Philip
Post Graduate and Research
Department of Chemistry
Bishop Moore College
Mavelikara
Kerala
India

Maneesh Kumar Poddar
Department of Chemical
Engineering
National Institute of Technology
Karnataka
Surathkal
Mangalore
India

Maya Rajan
School of Biosciences
Mahatma Gandhi University
Kottayam
Kerala
India

Reshmy R.
Department of Science and Humanities
Providence College of Engineering
Chengannur
Kerala
India

Malavika Sajith
Department of Chemistry
Amrita Vishwa Vidyapeetham
Amritapuri
Kerala
India

Ushama Shafoyat
Department of Biomedical Engineering
Military Institute of Science and Technology
Dhaka
Bangladesh

Areej Shahbaz
Centre for Applied Molecular Biology (CAMB)
University of the Punjab, Quaid-e-Azam Campus
Lahore
Pakistan

Hema S.
Department of Chemistry
Amrita Vishwa Vidyapeetham
Amritapuri
Kerala
India

Shahena S.
School of Biosciences
Mahatma Gandhi University
Kottayam
Kerala
India

Ushama Shafoyat
Department of Biomedical Engineering
Military Institute of Science and Technology
Dhaka
Bangladesh

Karishma Shetty
Shobhaben Pratapbhai Patel School of Pharmacy & Technology Management
SVKM'S NMIMS (Deemed to be University)
Mumbai
India

Sreedeep S
Department of Civil Engineering
Indian Institute of Technology
Guwahati
Assam
India

Abhisekh Saha
Department of Civil and Environmental Engineering
C.V. Raman Global University
Bhubaneswar
Odisha
India

Sreedha Sambhudevan
Department of Chemistry
Amrita Vishwa Vidyapeetham
Amritapuri
Kerala
India

Rashid Sulthan
Department of Chemistry
Amrita Vishwa Vidyapeetham
Amritapuri
Kerala
India

Fatiha Tabassun
Department of Biomedical Engineering
Military Institute of Science and Technology
Dhaka
Bangladesh

Divyanshu Thakur
Department of Chemical Engineering
National Institute of Technology
Karnataka
Surathkal
Mangalore
India

Sabu Thomas
School of Chemical Sciences
Mahatma Gandhi University
Priyadarshini Hills P.O.
Kottayam
Kerala
India
and
School of Energy Materials
Mahatma Gandhi University
Kottayam
Kerala
India

Rini Thresia Varghese
Department of Chemistry
Newman College
Thodupuzha
Kerala
India
and
School of Chemical Sciences
Mahatma Gandhi University
Kottayam
Kerala
India

Sneha Sara Varghese
Post Graduate and Research Department of Chemistry
Bishop Moore College
Mavelikara
Kerala
India

Anubhav Wadhwa
Shobhaben Pratapbhai Patel School of Pharmacy & Technology Management
SVKM'S NMIMS (Deemed to be University)
Mumbai
India

Kushwant S. Yadav
Shobhaben Pratapbhai Patel School of Pharmacy & Technology Management
SVKM'S NMIMS (Deemed to be University)
Mumbai
India

Preface

The term "biopolymers," which is well-known in today's research trends, has numerous applications. They are extensively chosen all over the world in an effort to investigate the possibilities of replacing traditional polymeric materials that have negative environmental effects. Because they are affordable, offer outstanding qualities, and are fully biodegradable, biopolymers originating from natural sources in particular draw interest. Since a few decades ago, materials like cellulose, starch, alginate, soy protein, etc. have been employed in the composite preparation. Nanofillers of biological origin like chitin and chitosan, are frequently used to improve the characteristics of virgin biopolymers.

The main focus of this book is on how biopolymers are used in engineering and biotechnology-related domains. Biopolymers are being used in biomedicine, drug delivery, and drug production since they are less toxic and extremely biocompatible. The field of engineering is broad, interwoven, and has permeated every aspect of our life. The writers of this book have contributed to the use of biopolymers in several fields, including electronics, civil engineering, bioengineering, etc. The book's 15 chapters cover a range of technical and technological investigations using biopolymers and composites. The introduction to biopolymers, their composites, and mixes is found in Chapter 1. The processing facets of synthetic biopolymers were covered in Chapter 2. Chapter 4 discusses the use of biopolymers in agriculture, and Chapter 5 discusses their use in bioengineering and medical technology. Chapters 6 to 8 go into detail into tissue engineering and other biotechnological applications. Biopolymers are finding use in the expanding field of electronics engineering. Applications for electronics and sensing are covered in chapters 9 and 10. Biodegradable, antimicrobial, and intelligent packaging are becoming more and more popular in the packaging sector. In this case, bioderived polymers are carefully chosen. Chapters 12 through 14 provide a review of these applications. Chapter 15 mentions applications in prosthetics.

The culmination of many people's tireless efforts from all across the world is the book. All of the book's contributors gave tremendous support and assistance, which is greatly appreciated. Sincere gratitude is due to everyone who provided timely contributions and recommendations. A particular thank you to the editorial team at Wiley for their perseverance, direction, and continued support during this project. The target audience for the book includes researchers at all career stages, research scientists, and business professionals. We hope that choosing biopolymers as a component of the research domains will be made easier with the aid of this book.

January, 2024

Arunima Reghunadhan
Akhina H.
Sabu Thomas

1

Introduction to Biopolymers, Their Blend, IPNs, Gel, Composites, and Nanocomposites

Mehvish Mumtaz[1], *Nazim Hussain*[1,*], *Mubeen Ashraf*[2], *Hafiz Muhammad Husnain Azam*[3], *and Anwar Iftikhar*[1]

[1] *Centre for Applied Molecular Biology (CAMB), University of the Punjab, Quaid-e-Azam Campus, Lahore, Pakistan*
[2] *Department of Microbiology, University of the Central Punjab Lahore, Pakistan*
[3] *Institute of Biotechnology, Faculty of Environment and Natural Sciences, Brandenburg University of Technology Cottbus-Senftenberg, Universitätsplatz Senftenberg, Germany*
* *Corresponding author*

1.1 Introduction

The search to employ biomaterials in preventing the use of non-renewable commodities and to lessen the waste created by composite polymers is increasing. In the environment, a variety of bacteria and plants make biopolymers. Microbes need appropriate nutrition and a regulated microenvironment to generate biodegradable polymers. Biopolymers are categorized in a variety of ways based on their size and can be classified according to their degradation rate and the type of the subunit from which they are formed, as well as their polymeric framework [1].

Thermoplastics, thermosetting polymers, and synthetic rubber are all terms for natural polymers that become pliable at a certain elevated temperature and solidify upon cooling. Biobased thermoplastic natural polymers now outnumber thermoplastic polyamides in terms of quantity. Mixtures, hybrids, and composite materials are all classifications for natural polymers based on their content. Biobased blends are made up of polymers from several sources; for instance, the Ecovio (BASF AG), which is made up of PLA and PBAT [2]. Biocomposites are biopolymers strengthened with

Applications of Biopolymers in Science, Biotechnology, and Engineering, First Edition.
Edited by Arunima Reghunadhan, Akhina H., and Sabu Thomas.

ordinary fibers and/or resins and preservatives, such as sisal, flax, hemp, jute, banana, wood, and other grasses [3]. A biodegradable matrix is supplemented with fibers in new composite materials. Polymeric biopolymers are polymers that have been synthesized or manipulated for a variety of uses.

Blending is a good technique for making advanced products with the right combination of qualities and may be conducted in a manufacturing environment utilizing standard machinery, thus there is no need for a large investment. Polymer mixes are being employed in a growing variety of industrial applications. The goal of blending might be to increase a material's efficiency, reduce its susceptibility to moisture, reduce costs, or strengthen the qualities for a particular function. Fragile biosurfactants have properties that are extremely similar to polystyrene (PS), a commonly used commercial thermoplastic. Inadequate mechanical properties led to the creation of several PS-based mixtures and polycaprolactone to address this issue, and PLA, as well as other natural polymers, are likely to become more common [4, 5].

Depending on the chemical ingredients, gels often comprise a material with a low molecular weight that is usually a simple solution in its perfect state (e.g., water). The existence of an additional considerably larger molecular mass constituent, which is frequently found in considerably smaller proportions than the "solvent" ingredient, is therefore attributed to the gel's flexible nature. This greater molecular weight component might be a polymeric, a colloid particle, or something else entirely; nonetheless, the gel's flexibility generally indicates that at least some of it has been formed into a full three-dimensional network across the whole hydrogel matrix. The properties are therefore substantially responsible for the processing of mechanical power throughout compression, notably the feature that when the gelatin is deformed the system filaments change in both energetic and entropic terms, contributing to a rise in the rate of energy. Loosening of this situation is generally blocked as long as the network's links stay intact, although such persistence across extended periods is far from assured. This second feature is crucial because differences in the intrinsic stability of the underlying infrastructure account for a large part of the difficulties in characterizing gels as a group of substances. In addition, when it comes to physiological gels – and those are the major topics for consideration here – great differences in cross-link durations and the ability of these connections to withstand massive deflection without collapsing are possibilities.

Nevertheless, it is widely accepted that all gels (physiological or other) have a system "component" (or "stage") and a soluble constituent, the latter of which is frequently present in significant amounts. Additional chemical constituents, such as polymers or particulates that have not yet gotten linked to the matrix, may be present in the solvent. The existence of all this low molecular weight substance, on the other hand, is one of the most (but not the most) crucial element of the gel state, because the gelatinization mechanism invariably looks to have solidified a huge volume of liquid. However, this isn't true, because the molecules are normally free to travel quickly via the channel's gaps [6].

Heterogeneous composites have already been designed to optimize the qualities of a variety of manufacturing components. Composite, blends, and IPNs all fall within the category of heterogeneous materials. Blends and IPNs are polymeric mixtures that contain two or more polymeric materials. IPNs, unlike blends, are made up of two distinct polymer systems that remain cross-linked to provide a characteristic shape. IPNs are created via interlacements among polymeric chains, while normal polymers are strengthened through inserting load bearing, strengthening fibers within their matrix.

Among the most inherent benefits of interpenetrating polymer network (IPN) composites is that they may integrate the benefits of their constituent polymerization in a single platform, resulting in improved characteristics. Because of their unique features, biopolymers have been extensively explored. IPN-based biomaterials with specialized features for a broad variety of products have been created using their chemical and structural characteristics. IPN constituents also associate the benefits of ordinary and synthetic polymers in a single organization, resulting in improved physical qualities. These types of mizes are known as blended IPNs. Hybrids IPNs have been investigated for improvement of artificial polymer characteristics. IPNs have been made from biopolymers such as polysaccharides, proteins, and polyhydroxyalkanoates (PHAs). To create hybrid IPNs or semi-IPNs, biopolymer channels have been interleaved through manufactured matrices [7].

Biobased materials are composed of compounds derived from biological material as their basic components. "Green components" are biopolymers bonded with synthetic materials and are biopolymer compounds that may be damaged by climate factors including air, light, heat, or bacteria. Natural fibers are more appealing than synthetic fibers, despite their poor ultimate tensile strength. Natural fibers provide several advantages, including simple availability, combustibility, biodegradability, and non-toxicity. Natural fibers suffer from an extensive array of reliability issues, limited computational temperatures, and excessive moisture penetration, all of which harm their use. Much research on fabricating synthetic fibers has been published showing increasing substances using organic fiber-reinforced biopolymer mixtures.

The biopolymer conditions influence the organization, conservational tolerance, and strength of a bio-based combination, whereas the reinforcing fiber affects the elasticity and properties of the composites. Biopolymer materials with commercial modern innovations have significant positive effects in global markets. The attempts to produce environmentally considerate composite materials with enhanced efficiency have yielded some significant worldwide results but are still ongoing. The aim of this analysis will be on composite materials made from polymers such as cellulose, starch, PLA, and PHA, in addition to others that are presently commercialized as well as commercially accessible, and those that are looking encouraging as mixtures for biocomposites in the coming decades [8].

Biopolymer-based nanocomposites are substances that are mostly made up of biopolymer frameworks with nanofillers scattered throughout. Natural polymers are biodegradable polymers derived from living creatures. They also feature a broad range of biological activities that enable the regulation of the boundary by nanofillers and multiresolution fabrication. They've been used in a range of ways due to their versatility in operating parameters and the low cost of their finished products. The majority of the research has concentrated on their electrical and biological applications. We will concentrate on biopolymer-based mixtures containing inert nanofillers in this study [9].

1.2 Biopolymers

Biopolymers are biomaterials that comprise monomeric components that are covalently bound and organized into custom larger compounds. The prefix "bio" indicates that the components are biodegradable and created by biological organisms. The word "biopolymer" refers to an inclusive assortment of polymers that are generally obtained from biological sources such as bacteria, plants, or forests. Biopolymers are substances created by artificial organic chemistry from biological entities like vegetable oils, carbohydrates, lipids, resins, proteins, amino acids, etc. [10]. Figure 1.1 shows the biological and artificial sources of biopolymers. Biopolymers are sophisticated chemical aggregates with exact and determined 3D forms and structures, as opposed to manufactured polymers, which have a lower and more unpredictable composition. This is one of the key characteristics that distinguish biopolymers as active molecules in vivo. Their purpose is dependent on their established composition and form. Haemoglobin, for instance, may not be able to transport oxygenated blood if it was not packed into a cubic symmetry.

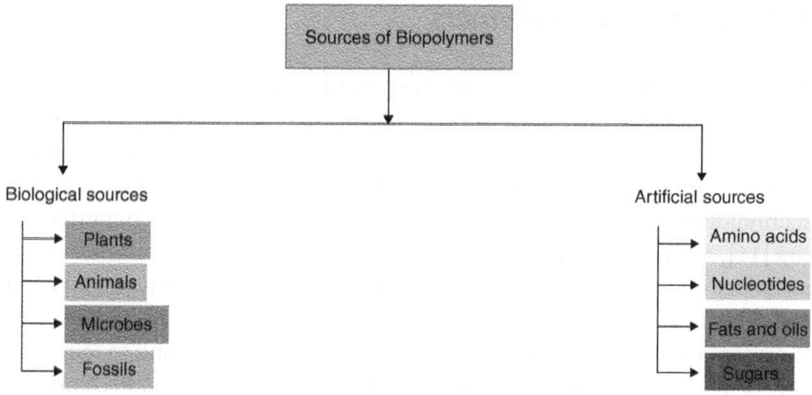

Figure 1.1 Biological and artificial sources of biopolymers.

The major feature that separates biopolymers from undegradable monomers is their long-term viability, which is enhanced by their degradability. Alternatives to fossil-fuel-based monomers were developed using disposable biopolymers made from renewable resources. They are usually made from flour, glucose, natural fibers, or other organic compostable ingredients in a variety of formulations. Interaction with microorganisms in the soil, compost, or sea debris degrades the biodegradable polymers. Additionally, compared to traditional burning, exposing biodegradable biopolymers to waste disposal by leveraging their property to be digestible by microorganisms in the subsurface greatly decreases CO_2 emissions. As a result, the usage of recyclable biopolymers is being highlighted as a way to combat climate change. With the catastrophic condition of the world environment increasing as a consequence of global warming, the development of renewable packaging has increased in recent years, to efficiently use restricted carbon reserves and preserve constrained energy supplies. Additionally, the expense of petroleum feedstocks has risen considerably, and the interest of society in adopting "ecofriendly" (or renewable) resources as the foundation for commercial products is increasing.

Polymer manufacture from renewable energies has been one of the fastest-growing materials fields in recent years. The potential for these polymers to substitute for hydrocarbon monomers is driving their advancement. The following are the primary motivations behind this drive: (1) insufficient nonrenewable supplies; (2) hydrocarbon economic volatility; (3) significance of conventional fuels as a precursor to global warming; (4) its use as a political tool on occasions; and (5) its link to the indiscriminate dumping caused by undegradable plastics [1].

1.3 Types of Biopolymers

Biopolymers come in a variety of shapes and sizes. Organic and inorganic natural polymers provide a simple approach to categorizing bioplastics based on their source. Organic biopolymers come from polysaccharides, peptides, and microorganisms, while artificial biopolymers come from biotechnological processes, microbiological, and petrochemical manufacturing [11]. As shown in Figure 1.2 there are various types of biopolymers such as biodegradable, non-biodegradable, biobased (polyester, polyamides) and non-biobased (vinyl polymer, polysaccharide). Elastomers, thermosets, and thermoplastics are the three types of biopolymers that respond to temperature changes [12]. In terms of manufacturing and use, PLA and PHA are the two most extensively utilized biopolymers [13].

Glycosyl connections connect polysaccharides mainly composed of monosaccharide cellulose, chitin, chitosan, starch, hyaluronic acid (HA), dextrin, alginate, carrageenan, and different gums; all are polysaccharide-based biopolymers. The most common carbohydrate, cellulose, is a potential biodegradable polymer that can be found in nearly all plant components [14]. It is excellent material for active chemical administration and synthetic biology because of its degradability, water permeability, specific strength, biocompatibility, cross-linking capabilities, and low immunogenicity [15]. Gelatin is a dehydrated protein complex that is made by incomplete or regulated hydrolysis of an impermeable protein called collagen [16]. Gelatin is a favored biomaterial having functional compounds like OH, NH_2, COOH, and others that help in the development of biomaterials. Due to its availability, noteworthy biodegradability, bioactivity, film-forming capacity, gelatinization, exceptional gas barrier

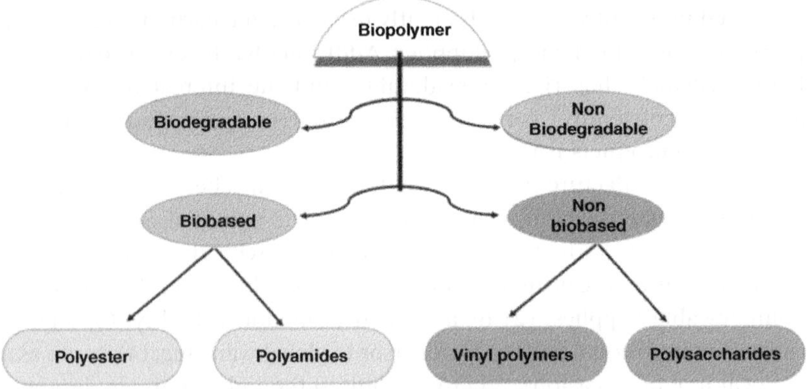

Figure 1.2 Graphical representation of types of biopolymers.

qualities, and lower costs, gelatin is a widely acknowledged peptide biomaterial [16]. Silk fibroin is a protein from the cocoon of silkworms. Alanine, glycine, and serine make up the majority of it [15].

Fibroid makes up the core, which is coated with sericin. Wound dressings and other medicinal items include sericin [17]. SPI (soy protein isolate) is a classified metabolic end of soybean oil with a protein content of above 90%. SPI is a biomaterial that has the potential to replace petroleum-based polymers. Since the hydrophilic and useful assemblies of amino acids aid in inter- and intra-molecular associations, soy-based biopolymers have excellent mechanical characteristics [14]. Wheat gluten is a biopolymeric substance made up of about 50% gliadin and 50% glutenin proteins that are formed as a by-product of food processing. Because of its inter- and intra-molecular disulfide connections, cysteine regulates the functional and mechanical characteristics of gluten [18]. Heat treatment aids in the improvement of intermolecular disulfide bridges, resulting in gluten propagation and thermoplastics [19]. To make gluten-based bioplastics, polymer methods of processing such as diffusion, contraction, and metal injection are employed. Disposable composite materials with specified tensile properties are made by combining natural fibers with gluten conditions, which are employed in the manufacturing, healthcare, and transportation segments [18].

PLA (polylactide) is a manufactured biomaterial with distinctive properties such as high stiffness and excellent transparency that can be used to replace traditional polymers [20]. It is a hydrophobic aliphatic polyester that is biodegradable, renewable, recyclable and has great processing properties [21]. PLA synthesis is a multi-step procedure that begins with the generation of lactic acid (LA) and ends with its polymerization [22]. LA, the fundamental component and primary antecedent in PLA manufacture, is primarily formed by fermentation by lactic acid bacteria (LAB) [15]. PLA's excellent mechanical and tribological qualities make it ideal for making a variety of materials for use in the fabric, packing, pharmaceutical, and auto manufacturing sectors. Renewal, reusability, excellent stability, robust degradability, elevated resistance, and recyclability have all been found in studies on natural fiber-reinforced PLA mixtures [20].

PHAs are the only polyesters generated from microbiological glucose or lipid digestion. They are made by subjecting bacteria to a carbon base while restricting nutrients [23]. Even with its hydrophobic nature, biocompatibility, degradability, and flexible characteristics, PHAs are a popular biomaterial in pharmacological, biomedical engineering, and conventional healthcare devices [15, 22]. PHAs include polyhydroxy valerate (PHV) or polyhydroxy butyrate (PHB), as well as their copolymers poly (hydroxybutyrate-co-hydroxy valerate) (PHBV) with various molar ratios, which are all recyclable and biocompatible [24]. PHBs have mechanical qualities that are comparable to petroleum-based polymers like polyethene (PE) and polypropylene (PP) [15]. PHAs are created without the use of farmland and can even be made from CO_2 in water ecosystems [25, 26].

1.4 Sources and Preparation of Biopolymers

Biopolymer materials are polymers that have been synthesized or manipulated for use in a variety of ways. The development of innovative biopolymer biosynthesis in plants offers a genuinely renewable natural route for its manufacture. Such polymers, like other thermoplastics, are made in quantity and afterwards moulded for a defined end application. Microorganisms are also elaborated in the production of an assortment of natural polymers, comprising carbohydrates, polyester blends, and polycaprolactone, which can vary from thick fluids through resins which are illustrated in Table 1.1. Their physiological qualities are determined by the polymer's constitution and molecular weight [27]. The genetic research of microbes may allow customization of the characteristics of diverse natural polymers generated with the help of organisms, making them ideal for high-value clinical uses such as biochemical regeneration and medication administration.

Biopolymers made by microbes require specialized nutrition and regulated climatic conditions to form. They are made via cultivation perhaps through mechanical copolymerization that is generated by fermenting. The majority of natural polymers are biodegradable, meaning they have no negative impact on living organisms. The creation of biopolymers from bacteria is thought to occur as a defense measure or as a barrier layer [28].

Table 1.1 Some biopolymers their sources and bacterial synthesis [1].

Polymer	Sources	Microbes for their production
Cellulose-based plastic	Glucose synthetic polymer	Estimated that 30% of the detritus generated following the separation of algal oil contains cellulose.
Polyester	Biomass	*Akaligenes eurrophus, Escherichia coli*
Polylactic acid	Lactic acid polymerization	Algal biomass fermentation through bacteria
Bio polyethene	Ethylene is formed from ethanol through cracking	Algal biomass fermentation through bacteria

Biological cycles, microbes, and proteins can break down biopolymers, allowing them to be reabsorbed into the surroundings. Biopolymers, also known as biological materials, are a type of plastic made from sustainable biological sources including vegetable oil, maize flour, and peas malt dextrin. The preservation of fossils fuels and reduction of CO_2 emissions can be helped by focusing more on natural polymers, hence encouraging sustainable development [29]. Because of their increased performance and capacity to thrive in a variety of settings, algae are an effective substrate for thermoplastic manufacture amongst microorganisms. The usage of algae allows for the utilization of carbon and the reduction of greenhouse gas emissions from companies and hydroelectric dams. Compared to conventional techniques of using maize and potato renewable sources as polymers, algae-based resins have been the latest phenomenon in the realm of bioplastics. Whereas algae-based polymers are still in their development, if commercialized they are expected to find use in a variety of sectors. Microbial plastics are currently regarded as a significant source of polymeric materials with high economic opportunities. They can change the way fluids stream, stabilize mixtures, flocculate granules, encapsulate things, and make dispersions [30].

1.5 Alteration of Biopolymers

Emulsification, integration of additives and stabilizers, mixing, and contact manipulation are a few of the methods utilized to adapt the characteristics of natural polymers to the desired use.

1.6 Blending

Blending is the process of combining composite components during a chemical process. Blending is a good technique to make advanced products with the right combination of qualities. This approach may be performed in an industrial context utilising standard machinery, thus there is no need for a massive investment. In a growing number of industrial applications polymeric mixes are being employed. This technology, when contrasted to certain other polymerization processes and the synthesis of novel monomers, may provide a broad variety of characteristics, and fulfil the criteria of the desired application at a cheap cost and in less time. The goal of blending might be to increase a material's efficiency, reduce its susceptibility to water, reduce costs, or enhance the qualities for a particular function. The primary motivation behind blending might be different from that of other approaches [4, 5]. The various methodologies used in the case of various polymers highlight the most important qualities that need to be addressed. Polyhydroxybutyrate (a PHA) is a flexible but pricey substance; therefore, it is no surprise that its alteration via mixing commonly incorporates starch to save money. The same technique may be seen in PLA, albeit the enhancement of toughness by the incorporation of elastomeric polymers appears to be more essential. The cost of flour is low, but its mechanical qualities are poor, and it is susceptible to water; therefore, it is critical to enhancing these features. Effective mixtures, in these circumstances, should deliver equal performance to the beginning material at a lower cost or provide substantial added value [31, 32]. Fragile biopolymers have properties that are remarkably similar to polystyrene (PS), a broadly utilized thermoplastic [33].

1.7 Plasticization

Plasticization is frequently used to increase the functional properties and various characteristics of biopolymers to meet the needs of certain uses. Thermoplastics convert interfacial tension within polymer matrix into bonds between macromolecules and tiny molecular mass compounds, allowing for more morphological shifts and enhanced compressibility. The phase separation and operating temperature of the substance are also reduced, allowing heat-sensitive polymers like poly(3-hydroxybutyrate) to be melt-processed at lower temperatures [34]. Gelatinization, or the destruction of indigenous starch crystallization by the application of a softener, is required for melt processing of starch [35]. The product created in this modus is identified as thermoplastic starch (TPS). TPS's rheological and mechanical qualities are comprehensively affected through the plasticizer's content and configuration. Because of the restricted structural flexibility of its inflexible chains, neat starch has a high glass conversion temperature, and its comparatively great elasticity and toughness are associated with poor deformability and dimensional stability [36]. Elastomeric behavior would be beneficial for a variety of purposes, and it may be obtained by lowering the glass transition temperature (Tg) below room temperature by deformation [37].

1.8 Biopolymer Gels and Their Establishment

So far, the topic of gels has been considered in a fairly broad sense. The network-forming substance, the solution, and the types of interaction between components have not been detailed in great detail. The structure of the gels of concern is described in the following sections. In actuality, such devices would be limited to mechanical gels generated in aquatic media from biopolymeric materials. In practice, this implies that we will be primarily interested in networks made up of amino acids or carbohydrates and that the systems will evolve in an aqueous solution with only simple ions as the other members. In principle, the connections amongst polymers that form networks will range in strength from simple chain interconnections on one end of the union strength measure to chemical bonding on another. This entails a broad range of permeability reactions, as described in the introduction [6]. The production of biopolymer gels is the first challenge to address. There are a variety of techniques available; therefore this is a big topic. If we start with a decently persistent biopolymer suspension in H_2O, one apparent strategy is to increase or drop the temperature of the sample significantly. A shift in the polymeric morphological configuration may result from such a temperature change, which is frequently accompanied by, or consequently includes, an interaction event. The hydrogel may occur in the end (if concentrations are sufficient).

When temperatures are dropped, the most likely transformation process is reordering, and hydrogel may be thought of as a frustrated polymer crystallization occurrence in this scenario. As technological management separates from solutions, the network emerges kinetically as a metastable intermediary stage. In some circumstances, such as the thermal properties of polypeptide chains, the polymerization could be disordered, and structure creation is the result of a complicated combination of molecular forces that includes hydrophobic and electrostatic connections. Various techniques to produce biocomposite gels exist in addition to thermodynamic gelatinization processes. An example is the addition of a component to the initial solution. A biopolymer solution may frequently be dissolved by introducing an adequate salt solution or adjusting the pH with acid or alkali.

Certain tiny neutrality compounds, such as urea, may cause aggregation, and in some specific circumstances, the presence of a greater molecular weight component, such as an enzyme, may cause combination. In certain circumstances, the additional component merely accelerates the conformational modification and attachment processes in the same way as a temperature shift would through determining the ability of interconnections inside and between molecules. In some cases, the adjuvant is reported to become actively engaged in the association process. It is conceivable that a mixture of the aforementioned mechanisms is at work. Conventional gels can also be used to make new gels. A gel created using one of the methods above can be subsequently processed by expanding it in a suitable flush. Furthermore, the salve might be created at one degree (or through using a somewhat complicated heating or cooling pattern) and then employed in various situations in practice. Because the gelatinization method is normally kinetically regulated, gels with distinct physical qualities (rigidity, transparency, etc.) might exist under seemingly comparable circumstances if the thermal or chemical processes used to generate them were different. This highlights the challenging subject of gel thermodynamic state (volatile, thermodynamically stable, or unstable), as concerns of thermally, physical, or biochemical evolution would affect whether such substances truly approximated to a global free potential optimum. When the idea of a movement towards stability is used in this study, it should be considered that, in most circumstances, a thermodynamic steady state is meant [6].

1.9 IPNs

IPNs are made up of a combination of two or more polymers. Two distinct polymer systems are cross-linked to generate IPNs, which have a characteristic shape. Composites constructed on an IPN have specialized characteristics for a variety of applications. There exist also IPNs that associate the benefits of ordinary and synthetic polymers in unified coordination, resulting in improved physical qualities. These are known as hybrid IPNs. Hybrid IPNs have been investigated for research into the role of natural polymers in the improvement of thermoplastic characteristics.

1.10 IPNs Consequential from Biopolymers

IPNs have been made by way of biobased polymers such as polysaccharides, proteins, and poly hydroxyalkanoates (PHAs) for a range of uses. To create hybrid IPNs or semi-IPNs, biopolymer systems have been intersected with synthetic channels.

1.11 Alginate

Brown algae create alginate, which is a naturally occurring carbohydrate. Alginate has remained connected with fibrin to customize the capacity of the matrix for in vitro ovarian follicle advancement [38]. As an IPN biomaterial, this biopolymer was used as an organic scaffold for tissue expansion in which alginate has been connected by fibrin to plasticize interpenetrating backgrounds on behalf of in vitro ovarian follicle advancement. IPN hydrogels with electrically sensitive behavior were created using sodium alginate (SA) and poly (diallyl dimethylammonium chloride) (PDADMAC). Whenever an electrostatic attraction was applied to this inflated IPN hydrogel amongst a couple of electrodes, the IPN displayed folding behavior [39]. With calcium alginate (Ca(II)-Alg) hydrogel and a dextran methacrylate (Dex-MA) copied, new biological and pharmaceutical IPNs have been studied. The UV cross-linking of the methacrylate moieties in this semi-IPN resulted in an IPN robust hydrogel that might be exploited for regulated delivery of bioactive compounds [40]. Wang et al. created additional alginate-based IPNs. They generated a pH-sensitive semi-IPN superabsorbent hydrogel by a linkage of sodium alginate-g-poly (NaAlg-g-PNaA) and direct polyvinylpyrrolidone (PVP) [41]. These polymers are sensitive to external pH stimuli and exhibit a revocable on–off distension feature. To make semi-IPNs for the discharge of protein and encapsulation of cell, Pescosolido et al. employed calcium alginate and hydroxyethyl methacrylate-derivatized dextran (Dex-HEMA). They were able to encapsulate chondrocytes and release bovine serum albumin using these semi-IPNs [42].

1.12 Agarose

Agarose is a polymer derived from red algae [43]. This polymer is utilized to create moisturizers consuming increased automated features that are ideal for tissue regeneration as an IPN biopolymer. DeKosky et al. employed agarose to create improved mechanically stable hydrogels for cell entrapment. The shear modulus of agarose–poly diacrylate IPN hydrogels was 4 times higher than that of an unadulterated PEG diacrylate linkage (39.9 kPa against 9.9 kPa) and 4.9 times higher than that of a clean agarose system (8.2 kPa) [44]. In agarose, Lomakin et al. developed an IPN of PEG diacrylate. This substance resembles the cuticle of insects, which is made up of chitin fibers intertwined with a protein base. PEG diacrylate remained utilized to replicate character of cuticular proteins; however, agarose was employed to simulate the part of chitin. These agarose–PEG diacrylate IPNs were reported to have 100 intervals of the resilience of agarose and five intervals the durability of a cross-linked PEG diacrylate system [45].

1.13 Chitosan

Chitosan is a deacetylated imitative of chitin, which is a long-chain polymer of N-acetylglucosamine derived from the exoskeleton of arthropods. Biopolymer j215 is a nautical material derived from crabs and shrimps.

Because of its possible uses as a biopolymer in tissue built-up, parting membranes, and surface reactive materials, it has attracted a lot of attention. Peng et al. designed a process for making semi-IPN chitosan and polyether copolymer. In different pH conditions, they looked at pH sensitivity, inflammation and proclamation kinetics, and organizational variations in the gel [46]. Chitosan and poly(vinyl alcohol) (PVA) IPNs have been used to make super porous hydrogels [47]. Poly (vinyl alcohol) adds to the molecular mass of the reinforcer in this situation. It was discovered that adding a tiny quantity of high molecular weight PVA to the mix increased tensile stability while somewhat reducing compressibility. Poly (ethylene glycol) [48], polyallylamine [49], polyvinylpyrrolidone [50], acrylate polylactide (PLA) [51], and glycine–glutamic acid [52] have all been described as chitosan-based IPNs.

1.14 Starch

Starch is a polymer made up of anhydroglucose components that are found in abundance in vegetation. Amylose and amylopectin are the two primary polymers found in native starch granules. Amylopectin is diverged into (a (1! 4)-linked glucose components interconnected by a(1! 6)-D-glucosidic connection), while amylose is mainly straight (a(1! 4)-linked carbohydrate components accessible in boiling water). Because of its gelatin properties, starch is utilized for a variety of industrial purposes, such as a thickening and scaling agent in pulp and leather companies. Murthy et al. [53] made semi-IPN hydrogels out of starch and a variable polymerization of polyethene glycol. The expanding behavior of those same IPN hydrogels was investigated in a variety of functional, pH, and ionic/salty conditions, and its ionic nature resulted in a high level of adaptability. Furthermore, starch has been observed to have certain drawbacks, including poor degradability and fragility [54]. Furthermore, starch-based hydrogels have several drawbacks, such as low structural rigidity, which restricts their use. Li et al. employed a novel cationic starch-g-acrylic acid amphoteric hydrogel that demonstrated improved enlargement measurements and reversibility. Significant compliance, swellability, and mechanical properties were discovered in these semi-IPN hydrogels [55, 56]. Bajpai and Saxena created semi-IPN gels with drug conveyance solicitations using absorbent starch and poly(acrylic acid) (PAA) [57]. They utilized riboflavin to figure out how much medication they could release. The amount of medication released was shown to rise when the gel's carbohydrate content and original water concentration increased.

Benzyl starch (BS) is another chemically modified starch. The use of benzyl starch as an acrylic resin additive, cleaner element, bonding agent,

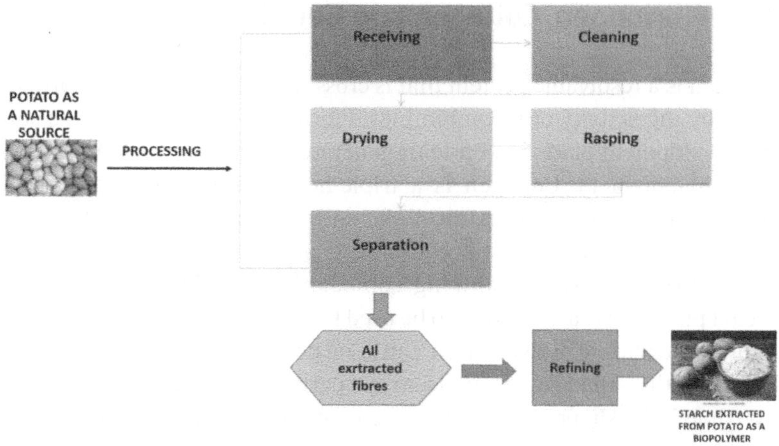

Figure 1.3 Biodegradable biopolymers obtained from a natural source (potato). *Source:* Left photo, DLeonis/Adobe Stock; right photo, Natallia/Adobe Stock.

and 216j 8 IPN has been suggested. It is manufactured from biopolymers surfactant [58], which is used in textile completion, paper coatings, soil amendments, and other uses. Semi-IPNs were created by Cao and Zhang using castor oil-based polyurethane (PU) and BS flicks. Starch is also produced from potatoes through a number of processes like receiving, cleaning, drying, rasping, separation and refining as shown in Figure 1.3. With BS contents of 5–70 wt%, their research indicates that these semi-IPN films exhibited virtuous or definite miscibility [59]. Through a rise in the concentration of such semi-IPNs, their structural rigidity and Young's modulus rose.

1.15 Dextran and Gum Arabic

Dextran is an ordinary hydrophilic compostable polysaccharide that has been used in the creation of hydrogels as an alternative to PEG [60]. With good biocompatibility with vascular cells, dextran-based IPNs have been investigated as potential constituents for 3D scaffolds for vascular tissue work as well as renewal [61]. Gum arabic (GA) is a polysaccharide that derives from Acacia (*Senegalia senegal* and *Vachellia seyal*) plants [62]. Its excellent solubility, pH constancy, non-toxicity, and antioxidant property are only a few of its notable characteristics [62, 63]. GA has been employed in an assortment of uses, including food, textiles, ceramics, lithography, face paint, and drugs [64].

1.16 Fibrinogen, Collagen and Gelatin

Fibrinogen is a resolvable protein that is crosslinked to fibrin in the form of calcium by the activity of thrombin [38]. Fibrin has been utilized as a soft tissue development platform. Proteins generate collagen and gelatin, which exist as biopolymers. Collagen is a triple-helical protein of amino acid chains that are found in the extracellular surrounding substance of animal tissues. It has been employed because of its ability to generate thermoreversible physical gels. Crosslinking compounds like glutaraldehyde and diphenyl phosphoryl azide can also be used to create collagen chemical gels. Hydrolyzed collagen derived from cattle or porcine skin is used to make gelatin. Gelatin is made up of single-strand particles, unlike collagen, which has a triple-helix shape. Gelatin has been employed as an IPN biomaterial in the usage of IPN microgels to be evaluated as a possible medication distribution mechanism [65].

1.17 Cellulose and its Derivatives

The most common polymer on the planet is cellulose. This biopolymer is found in large amounts in wool and woods, which are used to make newspaper, fabrics, and cellulose byproducts such as cellophane and rayon. Williamson et al. used a 9% LiCl/N, N-dimethylacetamide solvent solution to make semi-IPNs of cellulose and N, N-dimethyl acrylamide (DMA) [66]. The IPNs that resulted were highly transparent and had a modulus that was six times greater than the DMA reference. Close molecular communications and complexation among cellulose and DMA were shown to increase hydraulic flexibility. The utilization of cellulose derived by long indicators for the manufacture of IPN biomaterials has been described by Chauhan et al. [67]. They employed hydroxypropyl cellulose, cyanoethyl cellulose, hydroxyethyl cellulose (HEC), hydrazinodeoxy cellulose, and cellulose phosphate, among other cellulosic compounds. The cellulose acetate butyrate (CAB) is another cotton product that is employed. With the aim to make translucent constituents with greater absorption capabilities than the particular polycarbonate system, Laskar et al. utilized CAB to construct polycarbonate/CAB IPNs. They discovered that when the CAB concentration increases, the storage moduli of IPNs do not change much, but the damping qualities, and UV and natural conflict, improve [68]. CAB connections have also been exploited by Fichet et al. [69]. Scientists used a one-pot shot method to make polydimethylsiloxane–CAB IPNs, in which all of the ingredients are mixed combined initially and then the connections are created separately and quasi-simultaneously.

1.18 Polyhydroxyalkanoates

Polyhydroxyalkanoates are a class of biodegradable and biocompatible polymers found in nature that are known as aliphatic polyesters [70]. Bacterial thermoplastics and microbial polyesters are other names for them. They were accumulated as cytoplasmic grains into some species of bacteria as a consequence of physiologic problems generated by an uneven growth owing to a lack of a necessary nutrient and the existence of an overabundance of a carbon foundation [71]. Stitches, cardiovascular coverings, wound bandages, steered skin maintenance technologies, and tissue engineering are only a few of the biomedical applications [70]. PHAs are created by bacteria such as *Ralstonia eutropha* as well as *Pseudomonas oleovorans*. Certain PHAs are overly stiff and inelastic, while others are elastomeric but have poor automatic properties. PHAs' physical and mechanical qualities must be varied and enhanced for packing ingredients, biomedical uses, and other unambiguous presentations [72–74]. By using UV light, Hao and Deng created semi-IPNs of PEG and PHB. When compared to pure PEG hydrogels, the semi-IPNs had better mechanical characteristics [75]. Gursel et al. used photopolymerization of HEMA in the existence of PHBV to create PHBV–HEMA IPN barriers [76]. On the same surface, the membranes showed unique features including partial hydrophilicity and hydrophobicity, as well as dramatically changed material performance. Martellini et al. used X-rays from a 60 °C basis with an overall dosage of 10–100 kGy to create semi-IPNs based on PHB with PEG diacrylate at various proportions via radiation-induced polymerization. The results revealed that the hydrophilic ingredient increased water intake by 25% [77].

1.19 Lactide-Derived Polymers

Polylactides are translational aromatic polyamide synthetic materials made from lactic acid, which is commonly generated from maize processing utilising appropriate organisms such as lactobacilli [78]. Polylactides are among the most significant alternative energy natural polymers, and they have been studied for a spectrum of uses such as biological, pharmacological, and manufacturing. The L- and D-enantiomeric forms of lactate are both biologically active. Isotactic semicrystalline polylactides are produced by polymerizing L- or D-lactic acid. A combination of both kinds of L- or D-lactic acid or meso-lactide, on the other hand, can produce atactic polylactide, which is amorphous and has variable characteristics. In terms of developing biomaterials, Rohman et al. attained semi-IPNs established on poly (DL-lactide)/poly(methyl methacrylate) (PMMA) [79, 80]. Poly (lactic-co-glycolic acid) (PLGA) is a polyglycolic acid (PGA) and polylactic acid (PLA) copolymer. Hasircia et al. employed PLGA to create IPNs that enhanced the surface strength of PLGA being used in bone plate production [81, 82]. They employed PLGA that was reinforced with a 3D poly (propylene fumarate) network (PPF). This treatment increased the biomechanical qualities of the skeletal plates as well as their specific stiffness.

1.20 Biopolymer Composites

Composites are diverse specific structural mixes made up of two or more constituent materials with a wide range of characteristics [13]. Biopolymer combinations are disposable compounds that are manufactured by connecting numerous environmental fibers from animal and plant bases to usual and/or artificial biopolymers [23]. Normal fiber (conditioning ingredient) is introduced to the biopolymer matrix in a dispersion medium to increase the flexibility and impact resistance of the manufactured aggregate in biopolymer compounds. The goal of making this type of combination remains to create products with strong mechanical characteristics, which are provided by natural fiber and biodegradable polymers. Nanocomposite typically has optimum rigidity and impact resistance of 1 to 4 GPa and 20 to 200 MPa [13]. Thermophysical conductivity, ionic properties, structural traits, crystallinity, degradability, and invention price of biopolymer mixtures are all affected by natural fiber reinforcements [83].

Sustainable development, affordability, ultra-light qualities, considerable tensile stiffness, biocompatibility, renewability of biofuels, and producer and user safety are among the significant advantages of biocomposite 56 K. Biocomposites are composite materials formed by a matrix and a reinforcement of natural fibers. They have several advantages, including being renewable, cheap, recyclable, and biodegradable [84]. One advantage of natural fibers is their low density, which results in a higher specific tensile strength and stiffness than glass fibers, as well as its lower manufacturing costs. Biocomposites can be used alone or as a complement to standard materials, such as carbon fiber. Advocates of bio composites state that the use of these materials improves health and safety in their production, are lighter in weight, have a visual appeal similar to that of wood, and are environmentally superior [84]. Crude and laminated mixtures are two ways of producing fibers and matrices. In bulk mixtures, the fibers are randomly aligned in a three-dimensional framework, virtually displaying isotropic behavior. Laminated compounds, on either side, are higher scores, with fibers aligned in many sheets and linked together in the substrate. Each fiber layer in this combination has a two-dimensional direction [13].

Lightly, desorption, complex formation, reactive connectedness, electrostatic forces, and biomechanical bond formation can all be used to adhere natural fiber to the polymeric substrate. Chemical structure and molecular configuration, geometrical quality of synthetic fibers, the friction coefficient of component substances, and microscopic organization of fiber and matrix are all aspects that influence sticking processes [23].

The type of fiber, the proportion of fiber content, humidity permeation of fiber, surface treatment approach of fiber, arrangement and scheme of the compound, interfacial bonding among fiber and conditions, the existence of vacuums, the inclusion of preservatives such as compatibilizers, nanofiller, and binder, all influence the value of biocomposites [20, 85]. The densities, moisture susceptibility, permeability, process ability, and lifespan of biobased hybrids are all affected by the reinforcing materials and plasticizers used. As a consequence of the hydrophilic nature of fiber and the superficial chemistry of biocomposite, researchers decided to chemically alter the fibers to increase their adherence to a compound. Biochemical alteration enhances the effectiveness of biobased composites based on the treatment type, proper technology, and ambient variables [86]. Bioavailability and stability of biobased composites are key concerns, as there has been no effective method to fully manage those two characteristics to date.

1.21 Processing Techniques of Biopolymer Composites

With biocomposites, a variety of manufacturing processes have been developed and used. Compression moulding, investment casting, and extraction are often used processes for biopolymer compound formulations. Synthetic fiber-reinforced thermoplastics are also made via thermoforming, mixing, and long fiber thermoplastic-direct (LFT-D) processes. Flat semi-finished goods or hybrid fleeces, which are cut precisely to the required dimensions are most commonly utilized in compression moulding [87]. Compression moulding with in situ polymerization produced Ramie fiber included PLA/PCL matrix-based polymeric sustainable composite with silane coupling agent for increased bond strength [88]. The raw ingredients are integrated into the injection moulding machine as fiber-reinforced grains and warmed to a liquid condition for the injection moulding process. The plasticized polymeric substance is then injected into the shape at constant pressure.

1.22 Mechanical Assets of Biopolymer Mixtures

Usual fiber-reinforced biopolymer complexes have dynamic qualities that are suitable for a variety of purposes. Synthetic constituents characteristics of the material fibers and biomaterial, exterior adjustment of fiber, polymeric post processing (temperature and pressure applied), processing environment, fiber loading ability to focus, fiber orientation in the mixture, polycondensation and depolymerization, are all factors that influence the material characteristics of biopolymer blends [20, 86]. The inclusion of waxy compounds in fiber impacts the adhesive and water sorption qualities of composites [89], which influences the material characteristics. Tensile, flexural strengths, contact, static mechanical thermal conductivity, hardness, rigidity, ductility, fragility, creep, and endurance are fundamental characteristics of biopolymer mixtures.

1.23 Biopolymer Nanocomposites

Various biomaterials, which are thermoplastics made after regular foundations or chemically after a biomaterial, are also alternative options owing to their excellent characteristics, which include good barrier effectiveness, biodegradability, or reduced density. They do, nevertheless, have weak mechanical qualities, a small exhaustion lifespan, small biological conflict, deprived longstanding strength, as well as the ability to be molded, shaped, or otherwise manipulated during manufacturing. Nanomaterials, for instance, can have larger specific surface regions, thermal properties, and concentration than standard microfillers, resulting in substances with novel and enhanced characteristics due to synergistic effects that outperform the basic rule of mixtures. Biopolymers combined with nanoparticles and nanocomposites are formed as shown in Figure 1.4. As a result, nanocomposites are useful in a variety of fields, including healthcare, pharmacy, aesthetics, packaged food, farming, forest management, communications, transportation, and architecture. This Special Issue, which includes 17 scientific papers, highlights current breakthroughs in the manufacture, characterization, and uses of biopolymer composites and nanocomposites that are environmentally benign and sustainable.

PLA, cellulose esters, PHAs, and starch-based plastics are the most extensively utilized biomaterials for the present development of other biomaterials [2, 90]. PLA is a recyclable monomer that may be absorbed by humans, but which is degraded by recycling plants. It provides excellent options for petroleum-based composites in the wrapping, agrarian, particular services,

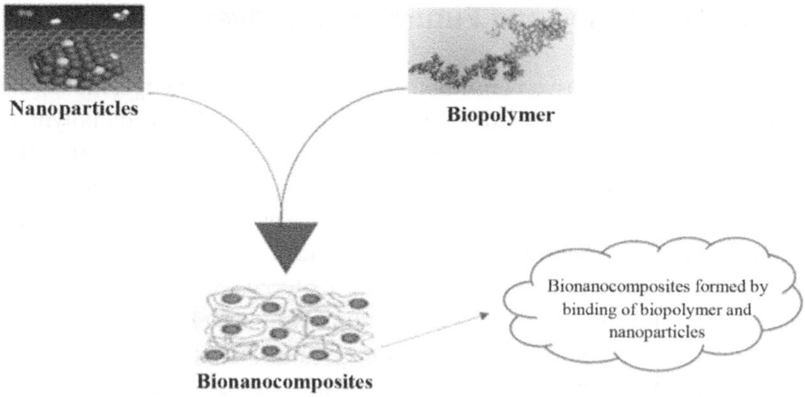

Figure 1.4 Formation of nanocomposites through the combination of biopolymer and nanoparticles.

skincare products, biomaterials, and tissue-engineered sectors [2, 3, 91]. It has bioactivity due to its propensity to solubilize on the exterior, generating lactic acid, which is one of the greatest choices for petroleum-based polymeric materials in the labelling, agrarian, hygiene products, facelifts, health sciences, and tissue engineering industries. By strengthening it with diverse nanofillers, such as cellulose nanocrystals, chitin nanofibers, metal oxide nanoparticles, or clays, a huge proportion of research has been committed to extending its machinability and variety of uses [90, 91].

Due to the combined interactions, the coupling of PLA with chitin nanofibers offers promising potential for the creation of bioplastic materials with better organizational and more efficient features. At the nano scale, the homogeneous distribution of these nanofibers inside the PLA matrix is challenging to achieve. Coltelli et al. employed poly (ethylene glycol) (PEG), a biocompatible polymer, to make precomposites, which were then combined with PLA in the extruder to make translucent nanocomposites [92]. Although the nanocomposites retained great standards of elongation at break (>150%), the tensile characteristics did not demonstrate a beneficial influence of up to 12 wt% chitin addition. This process has the benefit of being industrially applicable and not affecting the thermomechanical characteristics of polymeric PLA. This is in distinction to the consequences obtained when several kinds of cellulose microfibers with varied dimensions were added, whereby elasticity decreased as filler loading increased [93]. Potato pulp dust, which is a byproduct of starch manufacturing and phase separation, is an intriguing nanofiller for PLA. It is mostly made up of lignocellulosic fibers, starch, and proteins, and is inexpensive, making it ideal for commercial uses.

1.24 Conclusion and Future Perspectives

In the field of innovative continuous release composite compositions, the usage of naturally generated biopolymers has risen. These biopolymers have a one-of-a-kind ability to combine materials to generate composites. Because of environmental concerns and the needed reduction in use of petroleum-based materials, green mixtures have gathered a lot of attention. Cellulose, chitin and chitosan, amino acids, and other biomaterials found in nature are frequently employed to create materials and technologies for industry. As a result of this new information, scientists have made several attempts to functionalize biopolymers for biomedical applications, wastewater treatment, pharmaceutical implementations, and other uses. Versatility, biocompatibility, and adjustable degradability are all benefits of biopolymer composites that make them valuable for several commercial goods. Through the introduction of novel nanofiller and the production of sophisticated hybrid substances that may be used in a larger variety of disciplines, it is apparent that the domain of biopolymer composites will flourish. While research on biopolymer composites is yet in its early stages, it has already demonstrated that they can substitute conventional composites made of petroleum-derived plastics. Some issues, such as precise restriction of decomposition period, extended processing life, and ensuring products are equivalent in function to contemporary complexes, should be addressed in the future.

References

1 Mohan, S., Oluwafemi O., Kalarikkal, N. et al. (2016). Biopolymers–application in nanoscience and nanotechnology. *Recent Advances in Biopolymers* 1 (1): 47–66.

2 Dean, K., Li, L., and Yu, L. (2006). Polymer blends and composites from renewable sources. *Progress in Polymer Science* 6 (31): 576–602.

3 Saini, P., Arora, M., and Kumar, M.R. (2016). Poly (lactic acid) blends in biomedical applications. *Advanced Drug Delivery Reviews* 107: 47–59.

4 Clark, A.H. (1991). Structural and mechanical properties of biopolymer gels. In: *Food Polymers, Gels and Colloids* (ed. E. Dickinson), 322–338. Amsterdam: Elsevier.

5 Torres, F.G., Troncoso, O.P., and Torres, C. (2013). IPNs derived from biopolymers. In: *Handbook of Biopolymer-Based Materials: From Blends and Composites to Gels and Complex Networks* (ed. S. Thomas, D. Durand, C. Chassenieux et al.), 211–233.

6 Basheer, A., Sunooj, K., and Lackner, M. (2021). Biopolymer composites: a review. *International Journal of Biobased Plastics* 3: 40–84.

7 Díez-Pascual, A.M. (2019). Synthesis and Applications of Biopolymer Composites. *International Journal of Molecular Sciences* 20 (9): 2321.

8 Hernández, N., Williams, R.C., and Cochran, E.W. (2014). The battle for the "green" polymer. Different approaches for biopolymer synthesis: bioadvantaged vs. bioreplacement. *Organic & Biomolecular Chemistry* 12 (18): 2834–2849.

9 Calori, I.R., Braga, G., da Costa Carvalho de Jesus, P. et al. (2020). Polymer scaffolds as drug delivery systems. *European Polymer Journal* 129: 109621.

10 George, A., Sanjay, M.R., Srisuk, R. et al. (2020). A comprehensive review on chemical properties and applications of biopolymers and their composites. *International Journal of Biological Macromolecules* 154: 329–338.

11 Christian, S.J. (2020). Natural fibre-reinforced noncementitious composites (biocomposites). In: *Nonconventional and Vernacular Construction Materials* (ed. K.A. Harries and B. Sharma), 169–187. Amsterdam: Elsevier.

12 Montinaro, N., Fustaino, M., Bellisario, D. et al. (2022). Testing the dispersion of nanoparticles in a nanocomposite with an ultra-low fill content using a novel non-destructive evaluation technique. *Materials* 15 (3): 1208.

13 Park, S.-B., Lih, E., Park, S.-K. et al. (2017). Biopolymer-based functional composites for medical applications. *Progress in Polymer Science* 68: 77–105.

14 Shankar, S. and Rhim, J.-W. (2018). Bionanocomposite films for food packaging applications. *Reference Module in Food Science* 1: 1–10.

15 Wankhade, V. (2020). Animal-derived biopolymers in food and biomedical technology. In: *Biopolymer-Based Formulations* (ed. K. Pal, I. Banerjee, P. Sarkar et al.), 139–152. Elsevier.

16 Kim, N.K., Bruna, F.G., Das, O. et al. (2020). Fire-retardancy and mechanical performance of protein-based natural fibre-biopolymer composites. *Composites Part C: Open Access* 1: 100011.

17 Wu, Q., Rabu, J., Goulin, K. et al. (2017). Flexible strength-improved and crack-resistant biocomposites based on plasticised wheat gluten reinforced with a flax-fibre-weave. *Composites Part A: Applied Science and Manufacturing* 94: 61–69.

18 Getme, A.S. and Patel, B. (2020). A review: bio-fibers as reinforcement in composites of polylactic acid (PLA). *Materials Today: Proceedings* 26: 2116–2122.

19 Siakeng, R., Jawaid, M., Ariffin, H. et al. (2019). Natural fiber reinforced polylactic acid composites: a review. *Polymer Composites* 40 (2): 446–463.

20 Sarasini, F. (2017). Thermoplastic biopolymer matrices for biocomposites. In: *Biocomposites for High-Performance Applications* (ed. D. Ray), 81–123. Amsterdam: Elsevier.

21 Azammi, A.N., Ilyas, R.A., Sapuan, S.M. et al. (2020). Characterization studies of biopolymeric matrix and cellulose fibres-based composites related to functionalized fibre-matrix interface. In: *Interfaces in Particle and Fibre Reinforced Composites* (ed. K.L. Goh, M.K. Aswathi, R.T. De Silva et al.), 29–93. Amsterdam: Elsevier.

22 Terzopoulou, Z.N., Papageorgiou, G.Z., Papadopoulou, E. et al. (2015). Green composites prepared from aliphatic polyesters and bast fibers. *Industrial Crops and Products* 68: 60–79.

23 Lackner, M., Kamravamanesh, D., Krampl, M. et al. (2019). Characterization of photosynthetically synthesized poly (3-hydroxybutyrate) using a randomly mutated strain of *Synechocystis* sp. PCC 6714. *International Journal of Biobased Plastics* 1 (1): 48–59.

24 Kamravamanesh, D., Kiesenhofer, D., Fluch, S. et al. (2019). Scale-up challenges and requirement of technology-transfer for cyanobacterial poly (3-hydroxybutyrate) production in industrial scale. *International Journal of Biobased Plastics* 1 (1): 60–71.

25 Bozell, J.J. and Petersen, G.R. (2010). Technology development for the production of biobased products from biorefinery carbohydrates—the US Department of Energy's "Top 10" revisited. *Green Chemistry* 12 (4): 539–554.

26 Sukan, A., Roy, I., and Keshavarz, T. (2015). Dual production of biopolymers from bacteria. *Carbohydrate Polymers* 126: 47–51.

27 Koyama, N. (1995). Continuous production of poly (3-hydroxybutyrate-co-3-hyhroxyvalerate) by *Alcaligenes eutrophus*. *Biotechnology Letters* 17 (3): 281–284.

28 Tirrell, J.G. and Tirrell, D.A. (1996). Synthesis of biopolymers: proteins, polyesters, polysaccharides, and polynucleotides. *Current Opinion in Solid State and Materials Science* 1 (3): 407–411.

29 Zhang, M. and Thomas, N.L. (2010). Preparation and properties of polyhydroxybutyrate blended with different types of starch. *Journal of Applied Polymer Science* 116 (2): 688–694.

30 Zeng, J.-B., Jiao, L., Li, Y.-D. et al. (2011). Bio-based blends of starch and poly (butylene succinate) with improved miscibility, mechanical properties, and reduced water absorption. *Carbohydrate Polymers* 83 (2): 762–768.

31 Liu, H. and Zhang, J. (2011). Research progress in toughening modification of poly (lactic acid). *Journal of Polymer Science Part B: Polymer Physics* 49 (15): 1051–1083.

32 Bibers, I., Tupureina, V., Dzene, A. et al. (1999). Improvement of the deformative characteristics of poly-β-hydroxybutyrate by plasticization. *Mechanics of Composite Materials* 35 (4): 357–364.

33 Li, H. and Huneault, M.A. (2011). Comparison of sorbitol and glycerol as plasticizers for thermoplastic starch in TPS/PLA blends. *Journal of Applied Polymer Science* 119 (4): 2439–2448.

34 Bajpai, P. (2019). *Biobased Polymers: Properties and Applications in Packaging*. Elsevier.

35 Lourdin, D., Coignard, L., Bizot, H. et al. (1997). Influence of equilibrium relative humidity and plasticizer concentration on the water content and glass transition of starch materials. *Polymer* 38 (21): 5401–5406.

36 Shikanov, A., Xu, M., Woodruff, K.T. et al. (2009). Interpenetrating fibrin–alginate matrices for in vitro ovarian follicle development. *Biomaterials* 30 (29): 5476–5485.

37 Kim, S.J., Yoon, S.G., Lee, S.M. et al. (2003). Characteristics of electrical responsive alginate/poly (diallyldimethylammonium chloride) IPN hydrogel in HCl solutions. *Sensors and Actuators B: Chemical* 96 (1–2): 1–5.

38 Matricardi, P., Pontoriero, M., Coviello, T. et al. (2008). In situ cross-linkable novel alginate-dextran methacrylate IPN hydrogels for biomedical applications: mechanical characterization and drug delivery properties. *Biomacromolecules* 9 (7): 2014–2020.

39 Wang, W. and Wang, A. (2010). Synthesis and swelling properties of pH-sensitive semi-IPN superabsorbent hydrogels based on sodium alginate-g-poly (sodium acrylate) and polyvinylpyrrolidone. *Carbohydrate Polymers* 80 (4): 1028–1036.

40 Pescosolido, L., Vermonden, T., Malda, J. et al. (2011). In situ forming IPN hydrogels of calcium alginate and dextran-HEMA for biomedical applications. *Acta Biomaterialia* 7 (4): 1627–1633.

41 Bao, X., Hayashi, K., Li, Y. et al. (2010). Novel agarose and agar fibers: fabrication and characterization. *Materials Letters* 64 (22): 2435–2437.

42 DeKosky, B.J., Dormer, N.H., Ingavle, G.C. et al. (2010). Hierarchically designed agarose and poly (ethylene glycol) interpenetrating network hydrogels for cartilage tissue engineering. *Tissue Engineering Part C: Methods* 16 (6): 1533–1542.

43 Lomakin, J. (2008). Mechanical performance and microstructure of biomimetic PEG-agarose interpenetrating networks as determined by dynamic mechanical analysis and AFM. *AIChE Annual Meeting, Philadelphia*.

44 Peng, T., Yao, K.D., Yuan, C. et al. (1994). Structural changes of pH-sensitive chitosan/polyether hydrogels in different pH solution. *Journal of Polymer Science Part A: Polymer Chemistry* 32 (3): 591–596.

45 Park, H. and Kim, D. (2006). Swelling and mechanical properties of glycol chitosan/poly (vinyl alcohol) IPN-type superporous hydrogels. *Journal of Biomedical Materials Research Part A* 78 (4): 662–667.

46 Lee, S.J., Kim, S.S., and Lee, Y.M. (2000). Interpenetrating polymer network hydrogels based on poly (ethylene glycol) macromer and chitosan. *Carbohydrate Polymers* 41 (2): 197–205.

47 Shin, M.S., Kim, S.J., Park, S.J. et al. (2002). Synthesis and characteristics of the interpenetrating polymer network hydrogel composed of chitosan and polyallylamine. *Journal of Applied Polymer Science* 86 (2): 498–503.

48 Risbud, M.V., Hardikar, A.A., Bhat, S.V. et al. (2000). pH-sensitive freeze-dried chitosan–polyvinyl pyrrolidone hydrogels as controlled release system for antibiotic delivery. *Journal of Controlled Release* 68 (1): 23–30.

49 Yang, H. et al. (2007). Synthesis and drug release behavior of polylactide/chitosan IPN film. *Journal of Jinan University Natural Science and Medicine Edition* 28 (5): 494.

50 Rani, M., Agarwal, A., Maharana, T. et al. (2010). A comparative study for interpenetrating polymeric network (IPN) of chitosan-amino acid beads for controlled drug release. *African Journal of Pharmacy and Pharmacology* 4 (2): 035–054.

51 Murthy, P.K., Mohan, Y.M., Sreeramulu, J. et al. (2006). Semi-IPNs of starch and poly (acrylamide-co-sodium methacrylate): preparation, swelling and diffusion characteristics evaluation. *Reactive and Functional Polymers* 66 (12): 1482–1493.

52 Dufresne, A. and Vignon, M.R. (1998). Improvement of starch films performances using cellulose microfibrils. *Macromolecules* 31 (8): 2693–2696.

53 Li, B.-Z., Wang, L.-J., Li, D. et al. (2009). Fabrication of starch-based microparticles by an emulsification-crosslinking method. *Journal of Food Engineering* 92 (3): 250–254.

54 Li, X., Wang, J., Chen, X. et al. (2009). Structure and characterization of amphoteric semi-IPN hydrogel based on cationic starch. *Carbohydrate Polymers* 75 (4): 688–693.

55 Bajpai, S. and Saxena, S. (2004). Dynamic release of riboflavin from a starch-based semi IPN via partial enzymatic degradation: part II. *Reactive and Functional Polymers* 61 (1): 115–129.

56 Lee, M.R., Swanson, B.G., and Baik, B.K. (2001). Influence of amylose content on properties of wheat starch and breadmaking quality of starch and gluten blends. *Cereal Chemistry* 78 (6): 701–706.

57 Cao, X. and Zhang, L. (2005). Effects of molecular weight on the miscibility and properties of polyurethane/benzyl starch semi-interpenetrating polymer networks. *Biomacromolecules* 6 (2): 671–677.

58 Cadée, J., van Luyn, M.J.A., Brouwer, L.A. et al. (2000). In vivo biocompatibility of dextran-based hydrogels. *Journal of Biomedical Materials Research* 50 (3): 397–404.

59 Liu, Y. and Chan-Park, M.B. (2009). Hydrogel based on interpenetrating polymer networks of dextran and gelatin for vascular tissue engineering. *Biomaterials* 30 (2): 196–207.

60 Dror, Y., Cohen, Y., and Yerushalmi-Rozen, R. (2006). Structure of gum arabic in aqueous solution. *Journal of Polymer Science Part B: Polymer Physics* 44 (22): 3265–3271.

61 Ali, B.H., Ziada, A., and Blunden, G. (2009). Biological effects of gum arabic: a review of some recent research. *Food and Chemical Toxicology* 47 (1): 1–8.

62 Abd-Allah, A.R., Al-Majed, A.A., Mostafa, A.M. et al. (2002). Protective effect of arabic gum against cardiotoxicity induced by doxorubicin in mice: a possible mechanism of protection. *Journal of Biochemical and Molecular Toxicology* 16 (5): 254–259.

63 Kumar, P. and Singh, I. (2010). Formulation and characterization of tramadol-loaded IPN microgels of alginate and gelatin: optimization using response surface methodology. *Acta Pharmaceutica* 60 (3): 295–310.

64 Williamson, S.L., Armentrout, R.S., Porter, R.S. et al. (1998). Microstructural examination of semi-interpenetrating networks of poly (N, N-dimethylacrylamide) with cellulose or chitin synthesized in lithium chloride/N, N-dimethylacetamide. *Macromolecules* 31 (23): 8134–8141.

65 Chauhan, G.S., Guleria, L.K., and Mahajan, S. (2001). A study in sorption of some metal ions on novel hydrogels based on modified cellulosics and 2-acrylamido-2-methyl propane sulphonic acid. *Desalination* 141 (3): 325–329.

66 Laskar, J., Vidal, F., Fichet, O. et al. (2004). Synthesis and characterization of interpenetrating networks from polycarbonate and cellulose acetate butyrate. *Polymer* 45 (15): 5047–5055.

67 Vidal, F., Fichet, O., Laskar, J. et al. (2006). Polysiloxane–Cellulose acetate butyrate cellulose interpenetrating polymers networks close to true IPNs on a large composition range. Part II. *Polymer* 47 (11): 3747–3753.

68 Misra, S.K., Valappil, S.P., Roy, I. et al. (2006). Polyhydroxyalkanoate (PHA)/inorganic phase composites for tissue engineering applications. *Biomacromolecules* 7 (8): 2249–2258.

69 Lenz, R.W. and Marchessault, R.H. (2005). Bacterial polyesters: biosynthesis, biodegradable plastics, and biotechnology. *Biomacromolecules* 6 (1): 1–8.

70 Balaji, S., Gopi, K., Lavanya, B. et al. (2012). Isolation and optimization of poly β hydroxybutyrate producing cyanobacterial strains. *Int J Appl Biol Pharm Technol* 3 (1): 137–145.

71 Hazer, B. (2002). Chemical modification of bacterial polyesters. *Current Trends in Polymer Science* 7: 131–138.

72 Hazer, B. (2005). Chemical modification of synthetic and biosynthetic polyesters. *Biopolymers Online: Biology• Chemistry• Biotechnology• Applications* 10.

73 Hao, J. and Deng, X. (2001). Semi-interpenetrating networks of bacterial poly (3-hydroxybutyrate) with net-poly (ethylene glycol). *Polymer* 42 (9): 4091–4097.

74 Gursel, I., Balcik, C., Arica, Y. et al. (1998). Synthesis and mechanical properties of interpenetrating networks of polyhydroxybutyrate-co-hydroxyvalerate and polyhydroxyethyl methacrylate. *Biomaterials* 19 (13): 1137–1143.

75 Martellini, F., Mei, LHI, Lora, S. et al. (2004). Semi-interpenetrating polymer networks of poly (3-hydroxybutyrate) prepared by radiation-induced polymerization. *Radiation Physics and Chemistry* 71 (1–2): 257–262.

76 Gupta, A. and Kumar, V. (2007). New emerging trends in synthetic biodegradable polymers–Polylactide: a critique. *European Polymer Journal* 43 (10): 4053–4074.

77 Rohman, G., Grande, D., Laupêtre, F. et al. (2005). Design of porous polymeric materials from interpenetrating polymer networks (IPNs): poly (DL-lactide)/poly (methyl methacrylate)-based semi-IPN systems. *Macromolecules* 38 (17): 7274–7285.

78 Rohman, G., Laupêtre, F., Boileau, S. et al. (2007). Poly (D, L-lactide)/poly (methyl methacrylate) interpenetrating polymer networks: synthesis, characterization, and use as precursors to porous polymeric materials. *Polymer* 48 (24): 7017–7028.

79 Hasirci, V., Lewandrowski, K.U., Bondre, S.P. et al. (2000). High strength bioresorbable bone plates: preparation, mechanical properties, and in vitro analysis. *Bio-Medical Materials and Engineering* 10 (1): 19–29.

80 Hasırcı, V., Lewandrowski, K.U., Gresser, J.D. et al. (2001). Versatility of biodegradable biopolymers: degradability and an in vivo application. *Journal of Biotechnology* 86 (2): 135–150.

81 Hosseini, S.B. (2020). Natural fiber polymer nanocomposites. In: *Fiber-Reinforced Nanocomposites: Fundamentals and Applications* (ed. B. Han, T.A. Nguyen, and K.S. Bhat), 279–299. Amsterdam: Elsevier.

82 Mohamed, S., Zainudin, E.S., Spauan, S.M. et al. (2018). Introduction to natural fiber reinforced vinyl ester and vinyl polymer composites. In: *Natural Fibre Reinforced Vinyl Ester and Vinyl Polymer Composites* (ed. S.M. Sapuan, H. Ismail, and E.S. Zainudin), 1–25. Amsterdam: Elsevier.

83 Kabir, M., Wang, H., Lau, K.T. et al. (2012). Chemical treatments on plant-based natural fibre reinforced polymer composites: an overview. *Composites Part B: Engineering* 43 (7): 2883–2892.

84 Thyavihalli Girijappa, Y.G., Mavinkere Rangappa, S., Parameswaranpillai, J., and Siengchin, S. (2019). Natural fibers as sustainable and renewable resource for development of eco-friendly composites: a comprehensive review. *Frontiers in Materials* 6: 481024.

85 Faruk, O. and Bledzki, A.K. (2011). Processing of biofiber-reinforced composites. In: *Wiley Encyclopedia of Composites* (ed. L. Nicholais, A. Borzacchellio, S.M. Lee), 1–15. Chichester: Wiley.

86 Xu, H., Wang, L., Teng, C. et al. (2008). Biodegradable composites: ramie fibre reinforced PLLA-PCL composite prepared by in situ polymerization process. *Polymer Bulletin* 61 (5): 663–670.

87 Ramamoorthy, S.K., Åkesson, D., Rajan, R. et al. (2019). Mechanical performance of biofibers and their corresponding composites. In: *Mechanical and Physical Testing of Biocomposites, Fibre-reinforced Composites and Hybrid Composites* (ed. M. Jawaid, M. Thariq, and N. Saba), 259–292. Amsterdam: Elsevier.

88 Díez-Pascual, A.M. and Díez-Vicente, A.L. (2014). High-performance aminated poly (phenylene sulfide)/ZnO nanocomposites for medical applications. *ACS Applied Materials & Interfaces* 6 (13): 10132–10145.

89 Naffakh, M., Díez-Pascual, A.M., and Marco, C. (2016). Polymer blend nanocomposites based on poly (l-lactic acid), polypropylene and WS 2 inorganic nanotubes. *RSC Advances* 6 (46): 40033–40044.

90 Jonoobi, M., Harun, J., Harun, J., Mathew, A.P., and Oksman, K. (2010). Mechanical properties of cellulose nanofiber (CNF) reinforced polylactic acid (PLA) prepared by twin screw extrusion. *Composites Science and Technology* 70 (12): 1742–1747.

91 Nakagaito, A.N., Kanzawa, S., and Takagi, H. (2018). Polylactic acid reinforced with mixed cellulose and chitin nanofibers—effect of mixture ratio on the mechanical properties of composites. *Journal of Composites Science* 2 (2): 36.

92 Coltelli, M.-B., Cinelli, P., Gigante, V. et al. (2019). Chitin nanofibrils in poly (lactic acid)(PLA) nanocomposites: dispersion and thermo-mechanical properties. *International Journal of Molecular Sciences* 20 (3): 504.

93 Aliotta, L., Gigante, V., Coltelli, M.B., et al. (2019). Evaluation of mechanical and interfacial properties of bio-composites based on poly (lactic acid) with natural cellulose fibers. *International Journal of Molecular Sciences* 20 (4): 960.

2

Synthetic Biopolymers

Properties, Fabrication, and Applications

Reeba Mary Cherian[1,2], *Rini Thresia Varghese*[1,2], *Jithin Joy*[1], *Cintil Jose*[1,*], *and Sabu Thomas*[2,3]

[1] *Department of Chemistry, Newman College, Hopbush, Kerala, India*
[2] *School of Chemical Sciences, Mahatma Gandhi University, Kottayam, Kerala, India*
[3] *School of Energy Materials, Mahatma Gandhi University, Kottayam, Kerala, India*
[*] *Corresponding author*

2.1 Introduction

The commercial world, requiring various energy sources and various derived multiple materials has been driven by the exorbitant dependency on petroleum-based products [1, 2]. The unending demand in routine lives tangled with its non-biodegradable nature, adverse impacts on environment, the noxious life cycle from initial phase to disposal, the monetary domains and the finite dependency, motivates the scientific communities to shift towards the philosophies of sustainability, green chemistry and ecofriendly engineered products [3, 4]. The scarcity of these petroleum products and the environmental issues have shown a need for the design of new materials that are acclimatized with the environment and free from the reliance on fossil fuels [5]. One such way to enable this is the advancement of biobased materials.

Applications of Biopolymers in Science, Biotechnology, and Engineering, First Edition.
Edited by Arunima Reghunadhan, Akhina H., and Sabu Thomas.

2.2 Biobased Materials, Biomaterials, and Biopolymers Definition

Biobased materials are those products which are derived from living matter or as a synthetic form [6, 7]. The terms "biobased" also and "biomaterials" are often using synonymously. In fact, biomaterials are those which are biocompatible in nature and are especially utilized in biomedical applications like genetic and tissue engineering [8]. These biobased materials are now being investigated by researchers to fabricate a greener substitute for petroleum-based products. Biodegradability is the ability to be degraded or broken down hydrolytically or enzymatically by different microorganisms. The biobased materials are not essentially biodegradable, but efforts are being made to tailor ecofriendly greener products. These materials are derived either from biomass or fabricated with biotechnological procedures [6].

Biopolymers refers to those polymers that are sourced from living matter. They are chain structured polymers, either linear or cross-linked, composed of natural monomers like nucleic acid, amino acids or saccharides [9, 10]. Cellulose, chitin, starch, peptides and proteins, DNA and RNA are examples of biopolymers [10]. Even though the cost of manufacture and functional properties of biopolymers are quite appealing, they have to be improved to compete with petroleum-based products. Recently, biopolymers have been found from blends or composites to edible, personal hygiene, and medicinal products.

2.3 Classification of Biopolymers

Biopolymers can be classified on the basis of their degradability, that is, biodegradable and nonbiodegradable being the factor of degradability, which is imperative when designing greener products. Considering the usability, the nondegradable ones exceed the degradable biopolymers [11].

The source of the biopolymer is also of parallel importance based on which biopolymers come from which source: biopolymer from natural sources (like plants, animals, or micro-organisms); micro-organic type; and artificially processed type. Natural biopolymers can be classified based on their chemical structure such as proteins-based biopolymers, polyester-based biopolymers, polynucleotide-based biopolymers, polyisoprenes, and polysaccharide-based biopolymers [10].

They can be further divided by responsiveness to thermal condition as thermosets, elastomers, and thermoplastics. Based on composition, the biopolymers have been classified as blends, laminated or composites [6, 12].

2.4 Synthetic Biopolymers

The artificially processed biopolymers or micro-organic type biopolymers are grouped under the category of synthetic biopolymers. The artificially processed biopolymers include poly-lactic acid (PLA), poly-butylene succinate, poly-glycolide acid, poly-caprolactone etc., whereas the micro-organic type include copolymers of 3-hydroxyhexanoate, poly-3-hydroxybutyrate etc. [6, 10]. They can be degradable or nonbiodegradable as shown in Figure 2.1 [6]. They can also be classified based on the mode of fabrication like addition and condensation polymerization [13] and are discussed in the following sections.

These can be chemically synthesized polymers obtained from renewable sources whose structural behavior determines its degradation: photochemically, thermally or erosion driven [14]. Researchers are focusing on the manufacture of synthetic yet biodegradable polymers with enhanced properties that can be readily synthesized and share properties that are competitive with those of petroleum-derived products.

Figure 2.2 shows the chemical hydrolysable backbone structures of a few commercial biodegradable synthetic biopolymers.

2.4.1 Polyesters

Polyesters encompass the most comprehensively researched group of synthetic biopolymers. The molecular structure and the functional groups

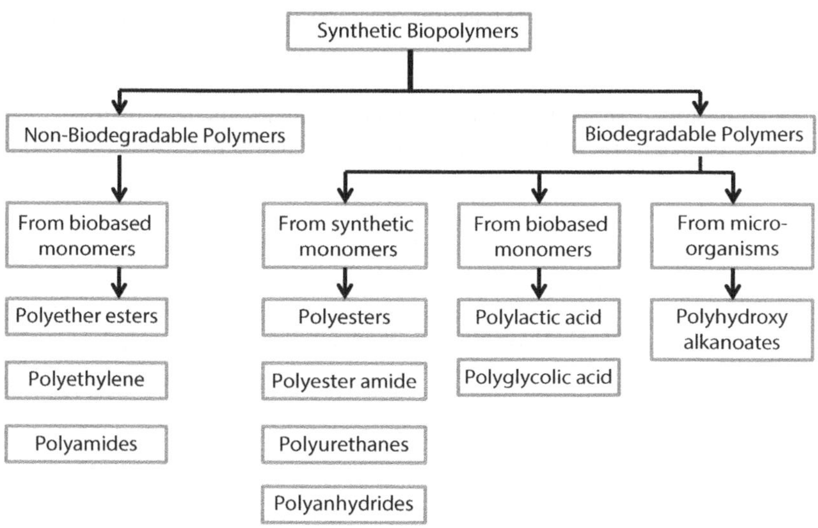

Figure 2.1 Classification of synthetic biopolymers.

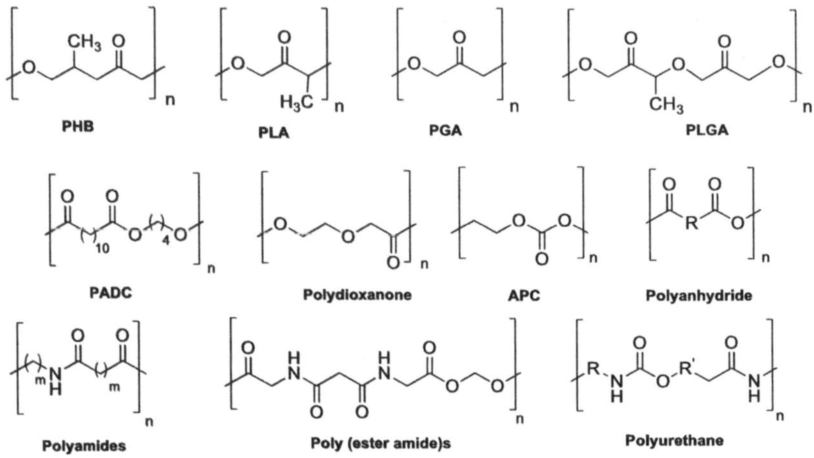

Figure 2.2 Chemical structures of a few synthetic biopolymers.

categorize the polyester group in to two categories: poly-(hydroxy acid)s and poly-(ether ester)s.

2.4.1.1 Poly-(hydroxy acid)s

Poly-(hydroxy acid)s contain the hydroxyl acid repeating functional group forming hydroxyalkanoic acids. They can be differentiated based on the position of hydroxyl group α, β, γ, δ, ε or ω with respect to the carboxyl acid group. The α positioned hydroxyl group constitute the group of poly-(α-hydroxyalkanoic acid)s whereas the other positioned OH groups in poly-(hydroxy acid)s, (except ω), constitutes the group of polyhydroxyalkanoates (PHAs) [6].

Commonly used poly-(α-hydroxyalkanoic acid)s include hydroxyalkanoic acids like lactic acids, glycolic acid, tartaric acids, malic acids, and benzylic acids and also their derivatives.

Polyhydroxyalkanoates (PHAs) are classified into two categories based on carbon chain length. The PHAs with fewer carbon atoms (3–5) and those with medium number of carbon atoms (6–14 atoms). The source of the PHAs is naturally occurring from bacteria as granules in their cells [15], and it has competency with the conventional polymers. Poly-(3-hydroxybutyrate), poly(4-hydroxybutyrate), poly-(β-malic acid) and poly-(3-hydroxybutyrate-co-4-hydroxybutyrate) are examples of some commercially used PHAs.

2.4.1.1.1 Poly-lactic Acid

Poly-lactic acids or PLAs the most promising degradable ecofriendly biopolymer materials. They are characterized by crystalline nature, glass transition temperature of about 60–65°C and a melting temperature of 175°C. L-lactide (LPLA) has a semi-crystalline structure with high tensile strength, high modulus and low elongation [10, 12, 16].

The monomer lactic acid is synthesized by the carbohydrate fermentation which makes them a biopolymer. The polymerization of lactic acid can be carried out via three different synthetic routes: (1) condensation polymerization followed by the aid of a chain coupling agent; (2) azeotropic dehydrative condensation; and (3) ring opening polymerization [12, 17].

2.4.1.1.2 Polyglycolic Acid

Polyglycolic Acid (PGA) biopolymer is the simplest fossil fuel derived synthetic linear aliphatic polyester compound and is crystalline (45–55%) [18, 19] with high tensile modulus and low solubility in organic solvents. PGA is more hydrophilic than PLA due to the lack of the asymmetrical methyl groups [20]. PGA has good gas barrier properties to CO_2 and O_2, glass transition temperature between 35°C and 40°C, melting point greater than 200°C, stiffness (7 GPa) and high mechanical strength (115 MPa) [21, 22]. The 100% degradable nature boosts its environmental friendliness, and it is known to be degraded within approximately 12 months [18].

The monomeric unit of PGA that is glycolic acid can be synthesized by different techniques which require extreme reaction conditions, processing cost and toxicity of the reacting species. The synthetic routes for glycolic acid includes: (1) hydrolysis of chloroacetic acid; (2) carbonylation of formaldehyde; (3) enzymatic synthesis from glycolonitrile [23]; (4) electrocatalytic reduction of CO_2 [24]; and (5) hydrogenation or hydrolysis of dimethyl oxalate (DMO)-DMO synthesized by the esterification and carbonylation of methanol, CO and O_2 [25]. However, the production cost of PGA is very high [18]. There are different synthetic routes for the fabrication of PGA, which may include:

1) Polymerization by direct poly-condensation of glycolic acid via an esterification reaction. This method requires high temperature, prolonged time, and high vacuum for the removal of water, the byproduct of the reaction. But these reaction conditions cause depolymerization, which causes the formation of low molecular weight PGA. The transesterification and increase in the viscosity of the reaction medium by the polymerization also limits the formation of high molecular weight PGA. However, azeotropic polycondensation and metal/solid state polycondensation mechanisms have been reported to synthesize fairly high molecular weight PGA [22].

   Melt condensation followed by solid state polymerization at 190°C in vacuum achieved PGA of molecular weight of 91,000 g mol^{-1} M_W using zinc dehydrate as catalyst has been reported [26]. Solution-based polycondensation with methane sulfonic acid as catalyst was done by Ayoob et al. resulted in high molecular weight PGA. Azeotropic mixture of anisole and water was taken for the solution-based polycondensation which yielded PGA of M_n of 27,000 g mol^{-1} with triflic acid as catalyst [27]. All these polycondensation reaction bears a disadvantage of the presence of catalytic impurities due to the essentiality of catalyst in every reaction [22].

2) Ring opening polymerization synthesizes PGA with high molecular weight of about $M_n > 100{,}000$ g mol^{-1} and high conversion. Initially, the glycolic acid is converted to glycolide via the formation of PGA oligomer under reduced pressure, which demands high energy and processing cost. The formed glycolide is removed by distillation and is removed traces of glycolic acid and water as they cause depolymerization. The ROP is done at 150–230°C for 2–6 hrs with tin(II) catalysts under inert atmosphere. The initiator concentration determines the molecular weight of the PGA in ROP [28].
3) Other synthetic routes include the synthesis of PGA from trioxane using triflic acid catalyst which yields low molecular weight PGA, which then subjected to postpolymerization with zinc acetate dehydrate as catalyst yielding high molecular weight PGA of $M_n > 45{,}000$ g mol^{-1} [23].

2.4.1.1.3 Poly(lactic-co-glycolic acid)

The copolymerization of PLA and PGA yields poly (lactic-co-glycolic acid) or PGLA which can show properties of both the monomers. PGLAs are commonly synthesized by the ring opening copolymerization of PLA and PGA. The asymmetric α-carbon in the lactic acid results in the formation of enantiomers of PLA that is poly(D-lactic acid) and poly(L-lactic acid). The commonest PLGA comprises equal mixtures of both the enantiomers with PGA. Different types of PLGA are developed with different monomeric ratios of PLA and PGA. For instance, PLGA 75:25 means 75% PLA and 25% PGA [29].

The direct condensation reaction between the PLA and PLGA yields only low molecular weight PLGA because of the reversible condensation reaction and high extent of the reaction. Hence, ring opening polymerization stands as a door for the synthesis of high molecular weight PLGA via forming cyclic dimers. The method uses tin compounds as catalyst such as tin (II) alkoxides, tin (II) 2-ethylhexanoate or aluminum isopropoxide [30]. However, the complete elimination of toxic catalyst seems to be difficult. Hence many other biofriendly catalysts like zinc lactate [31] and calcium [32] and zirconium acetylacetonates [33] have been used, of which only zirconium acetylacetonates gave good yield of high molecular weight PLGA [34].

The PLGA has dissolution in a wide range of common polar and non-polar solvents such as tetrahydrofuran, acetone, ethyl acetate or chlorinated solvents. PLGA is less stable when compared to PLA and PGA as it degrades easily by hydrolysis in aqueous environments and those monomeric polymers will be the degraded products. The PLGA has tailored degradation properties because the degradation rate is affected by the monomeric ratios, the enantiomeric ratios of lactic acid, molecular weight of the polymer and the end group functionalization (carboxyl end group promotes good degradation than ester end groups). The copolymer has a glassy behavior as it has a glass transition temperature of 35°C and it decreases with reduction in the PLA ratio in PLGA [35].

These properties of PLA help them to be used in biomedical applications as drug delivery systems, scaffolds for tissue engineering and sutures for wound closure [34].

2.4.1.1.4 Poly (hydroxybutyrates)

Polyhydroxybutyrate (PHB) is an important class of thermoplastic biopolymer among polyhydroxyalkanoates, which are closely analogous to plastics. They are biosynthesized by microorganism fermentation which includes natural isolates, recombinant bacteria etc. *Bacillus* spp., *Pseudomonas* spp., *Aeromonas* spp., *Cupriavidus* spp. are all bacterial species used for the industrial production of PHB. They act as a carbon reserve in different bacteria strains. The advantage comes in the way that the anaerobic microbes completely degrade to water, CO_2, and CH_4 [36].

Poly(3-hydroxybutyrate) (P3(HB)) was the first isolated PHA [37]. It is found as secondary metabolite in the cells of microbes under unfavorable conditions like nutrient stress with carbon excess [38]. This is possible with both gram positive and negative bacteria. Per work done so far, they accumulate as intercellular granules in the bacterial cells of about 75 different genera [37, 39]. Depending on the microorganism, different forms of PHB can be synthesized. Isotactic PHB was synthesized by bacterial processes with PHA synthase polymerizing enzyme whereas the syndiotatic PHB with stereo regularity has been achieved through chemical synthesis [40, 41].

There are different approaches for the synthesis of PHB. The first approach is the ring opening polymerization of β- butyrolactone [41]. The second approach utilizes transgenic plant cells exploiting the availability of acetyl-CoA, the primary substrate for PHA biosynthesis for example with flax, *Linum usitatissimum* [42]. The third approach is by bacterial fermentation where it is assumed that 90% of dry cell weight is comprised of PHB [43].

PHBs are linear chain structures with a high degree of crystallinity. They can be used as both neat polymer or as copolymer, and industrially they are generated through bacterial fermentation. PHB has its major application in the packaging field as it shares close analogous properties with petroleum-based products. They have superior barrier permeability compared with polyethylene (PE) and polypropylene (PP), rigidity, less flexibility than PP, better barrier properties than polyvinylchloride (PVC) and polyethylene terephthalate (PET) [44], and biodegradability making them an ecofriendly alternative for synthetic polymers. The glass transition temperature is approximately 5°C to 6°C and melting temperature of about 160°C to 180°C [37].

2.4.1.2 Poly-(alkene dicarboxylate)s

Poly-(alkene dicarboxylate)s (PADCs) are biodegradable polyesters formed by the reaction of dihydroxy compounds and dicarboxylic compounds. They can be divided in to three different groups: (1) aliphatic polyesters (2) aromatic polyesters (3) aliphatic–aromatic polyesters [6]. Aliphatic polyesters or poly (alkylene alkanoates or PAAs) are prepared by the thermal polycondensation of aliphatic dicarboxylic acids or either their diesters or dichlorides. The disadvantages lie in the factors like high vacuum and temperature conditions, high cost of dicarboxylic acid chlorides and side reactions. An ecofriendly polycondensation can be carried out in the presence of non-specific lipases but the high cost of the enzyme is a disadvantage [45]. Other efficient catalysts have also been reported which include titanium compounds, bismuth compounds, zirconium, and germanium compounds, triflates etc. The ring opening polymerization method is more advantageous than polycondensation reactions, but the low availability and high cost of cyclic ester oligomers (CEOs) are disadvantages. PACDs obtained from biobased monomers, specifically those obtained from succinic acid (SA), adipic acid, sebacic acid, fatty acids and carbohydrates and the diols like 1,3-propane diol, 1,4- butanediol (BDO), 1,10- decanediol, are the main significant biopolymers of this group. SA can be obtained by bacterial fermentation and BDO as the reduced product of SA [46].

Aliphatic-aromatic polyesters are synthesized by the polycondensation reaction of aliphatic-diols, aliphatic dicarboxylic acids/esters, and aromatic carboxylic acids/esters. Aromatic polyesters are biobased polyesters which can be a proper substitute for fuel-based commodities, the most important of which is poly (ethylene terephthalate) or PET [6].

2.4.2 Poly-(ether ester)s

Poly-(ether ester)s are those thermoplastic elastomers prepared from alkane diol, alkylene terephthalate and poly-(alkylene glycol ether). The polymeric chain is comprised of alternate hard and soft segments contributed by alkylene terephthalate and poly-(alkylene glycol ether) respectively [47]. These biopolymers can be prepared via a two-step trans esterification process. The biocopolymer has many desirable properties such as high breaking stress, impact resistance, high yield stress, outstanding toughness and elasticity, and broad spectrum of working temperature allowing them to be used for a wide array of dynamic applications in biomedical applications [6].

Polydioxanone comprises an important class among poly-(ether ester) biopolymer compounds with colorless, crystalline nature and ether–ester repeating units. They are prepared by the ring opening polymerization of *p*-dioxanone under heat and organometallic catalyst like zirconium acetylacetonate or zinc L-lactate [48]. They degrade to low toxic monomers and therefore can be used in biomedical applications like orthopedics, drug delivery, tissue engineering and plastic surgery. Also, they have been found to have lower modulus than PLA and PGA, and hence were, for the first time, used to make degradable monofilament sutures [49].

2.4.3 Aliphatic Polycarbonates

Polycarbonates are those biodegradable polymers with carbonate (-O-C(O)-O-) backbone linkages. Polycarbonates have received much attention in regard of the incorporation of CO_2 in to the materials thus mitigating the amount of greenhouse gas in the atmosphere [50, 51].

Aliphatic polycarbonates (APC) are biopolymers without any aromatic groups throughout the polymeric chain. The properties of APC depend on the average molecular weight (M_n) and the polydispersity index (PDI) [52, 53]. APCs with different properties are made by three different techniques: (1) polycondensation. The polycondensation reaction involved a two-step melt state process of condensation between dialkyl carbonates and aliphatic diols followed by the chain growth polymerization via transesterification process in presence of catalysts. The catalyst, reaction temperature and the reactant ratio influence the polymerization outcome [51]. TiO_2/ SiO_2- poly(vinyl pyrrolidone) based catalysts [54], and enzyme catalysts are all reported [54, 55]. (2) By alternating copolymerization of carbon dioxide and epoxy. The copolymerization was initiated by the epoxide ring opening in presence of metal catalyst followed by the CO_2 insertion [50, 56]. (3) Ring opening polymerization of cyclic carbonate monomers yield high quality polycarbonates with high molecular weight and low PDI. The catalyst involve in ROP includes transition metal catalysts, organo-catalysts, alkyl halides and enzyme catalyst and they can be done in melt or solution state [51].

Low melting point, low glass transition temperature, low elasticity, and high susceptibility to hydrolysis (degradation) add up to the advantages of this biopolymer [51]. They find opportunities in biomedical applications like drug delivery, scaffold engineering, and as coating resins, binders, and surfactants and also in the packaging field. Poly-(ethylene carbonate) and poly- (propylene carbonate) are the two main developed APCs for high end product processing [6].

2.4.4 Polyamides

The amide linkages (-CO-NH-) constitute the formation of polyamides with enhanced durability, flexibility, and strength. Polyamides are conventionally prepared by the direct condensation reaction between amines and carboxylic acids or their derivatives like acid chlorides or esters. The condensation reaction can be between (1) diamines and dicarboxylic acids, (2) ω-amino carboxylic acids, (3) α-amino carboxylic acids. The high temperature melt condensation is the commonly employed method for polyamide formation, but the disadvantage is the generation of toxic waste and harsh conditions. The other employed method is the ring opening polymerization by lactams or N-carboxy-amino acid anhydrides. A dehydrogenative amidation of alcohols and amines using ruthenium catalyst [57], rhodium catalyst [58], silver catalyst [59] and NHC ligands(LU [60]) has also been reported [6, 61].

Nylon forms an important example of polyamides with competent properties like great strength, resistance to abrasion and great resilience enabling them to be used in the fabrication of clothing, engineered plastic, and other consumer products. The α-amino acids form poly-(α-amino acids) which can be categorized in to three kinds like poly-(γ-glutamic acid), ε-poly-(L-lysine) and poly-(α-aspartic acid). Polyamides, despite their water absorbent property, act as a plasticizer and hence regulate plastic properties. Also, they find applications in the biomedical field such as in the fabrication of dilation catheters because of its compatibility and inertness to the body tissues and fluids [6].

2.4.5 Poly-(ester amide)s

Poly-(ester amide)s have ester and amide groups as their basic backbone linkages sharing the properties of both the polyester and polyamides. The degrading properties of the ester group under physiological conditions and the enhanced mechanical, thermal properties due to the hydrogen bonding of the amide group are combined in the poly-(ester amides) for the fabrication of beneficial products and mitigating the disadvantages of both [6, 62].

α-amino acids, α-hydroxyacids, α, ω-aminoalcohols, cyclic depsipeptides, fatty diacids, diols, carbohydrates are the different monomers employed for the synthesis of poly-(ester amides). The different methods employed for the synthesis of poly-(ester amides) are: (1) melt polycondensation of diol with diamide-diester, which in turn is synthesized from α-amino acid methyl ester and diacyl chloride. (2) Solution polycondensation to prepare high molecular weight polymer, removing the disadvantages of the melt condensation like high temperature and side reactions. This method either utilizes condensing agents or the activation of the carboxylic group. (3) Interfacial polycondensation which occurs at the interface of two immiscible liquids containing two di-functional reactants. (4) Ring opening polymerization of depsipeptides [62].

The hyper-branched polymers from cyclic anhydride and di-isopropanol amine find applications as additives in demulsification and as chemicals in the oil field etc. They serve as sustainable biomaterials for applications in the biomedical field [6].

2.4.6 Poly-(ether amide)s

Poly-(ether amide)s or polyether block amides (PEBAX) are thermoplastic elastomers made of hard segments of polyether and soft segments of polyamide. These are manufactured via extrusion techniques or injection molding [6]. They are characterized by light weight and great flexibility yet possess high strength and resilience with good impact resistance [63]. They have the gas separation property where it separates polar/non-polar gas mixtures [64]. This allows the permeation of water vapor molecules and hinders the liquid water transportation [65]. They find application in food packaging, sportswear, medical packaging for gas sterilization [65].

2.4.7 Polyurethanes

Polyurethanes (PUs) form a major division of synthetic biopolymers which comprise urethane (-NH-CO-O-) repeating units either linear or branched or crosslinked. They are synthesized by the polyaddition reaction between poly-isocyanate and polyol. They have a close similarity with nylon except for lower melting point and water solubility [66]. Based on the composition there are four types of polyurethane: (1) poly-(ester urethane)s; (2) poly-(carbonate urethane)s; (3) poly-(siloxane urethane)s; and (4) poly- (ether urethane)s. Poly-(ester urethane)s give a comparatively good degradation as they are more susceptible to microorganisms because of the natural polyols used in the fabrication of glucose, fructose and sucrose, and glycerol. Polyester polyols include di acids or di- or tri- functional polyols like succinic acid, adipic acid, azelaic acid, glutaric acid, 1,2-propane diol, 1,4-butane diol, 1,6-hexane diol and 1,10- dodecanediol. Biobased polyols are derived from castor oil, soy oil, sunflower oil, linseed, and rapeseed oil. Chain extenders or cross linkers are also used to induce crosslinking and branching. Poly-(ester urethane)s include polycaprolactone, polyglycolide and polylactide. The hard segments were formed by the polydiisocyanate, and the soft segments are formed by the polyols. The hard segments give high mechanical strength through hydrogen bonding whereas the soft segments give PU with elastic properties and the degrading property [67].

They find applications as shock absorbers in shoe soles, adhesives and insulators, automobile components and in biomedical engineering as implants and grafts owing to its toughness, flexibility, and surface- functional nature [6].

2.4.8 Vinyl Polymers

Attempts have made to fabricate biobased polyolefines to replace petroleum derived ones. The most important biobased derived olefin may include propylene and optionally butylene and ethylene from agricultural feed stocks. The polypropylene can be obtained by the gasification reaction of lignocellulosic materials containing starch, hemicellulose, lactose, and many others. The by-products, which may include ethylene and butylene and various olefinic compounds sourced from 100% renewable materials, can be the monomers for the production of polypropylene, polybutylene and polyethene. Another synthetic method for the production of olefins involves the anodic decarboxylation of carboxylic acids obtained by the fermentation of renewable raw materials like starch, cellulose, lignin, glycerol, sorbitol, sugars and lactates. The carboxylic acids like propionic acid, butyric acid, valeric acid and caproic acid yield ethylene, propene, butane, and pentene respectively upon anodic decarboxylation. The fermentation is done in the presence of bacteria of some genera like *Propionibacterium*, *Clostridium*, *Butyribacterium* and *Butyrivibrio*. Biobased polyvinyl chloride (PVC) can also be fabricated from the ethylene produced from renewable sources. Also, biobased water-soluble polyvinyl alcohol (PVOH) can also be synthesized from bioethylene by the hydrolysis of biobased polyvinylacetate. PVOH is thermoplastic material having application as adhesives and laminates for food containers, having excellent oxygen barrier properties and film forming properties. Biobased polyethene (PE) shows close resemblance with the properties of fossil-based PEs and bio-PE is not degradable and shows thermoplastic behavior hence could undergo mechanical recycling or otherwise called physical recycling. Biobased polyethene and polypropylene find applications in cosmetics and the automobile industry [6, 67].

2.4.9 Polyanhydrides

Polyanhydrides constitutes anhydride linkages as the repeating unit throughout the polymeric backbone chain. They can be fabricated by the melt condensation reaction between the monomeric carboxylic acid and superfluous portion of acetic anhydride under vacuum or high temperature conditions. The aliphatic polyanhydrides have a faster rate of degradation and the rate can be controlled by tuning the hydrophilic and hydrophobic copolymer monomer units. They are capable of enhancing the mechanical properties of the fabricating product and also find application in pharmaceuticals as they hydrolyze to non-mutagenic and non-toxic monomeric diacids that permits safe metabolism and its removal from the body [6, 67, 68].

2.5 Properties of Synthetic Polymers

The different properties of a polymer can be categorized into three groups: (1) intrinsic properties, which refer to the inherent property of the polymer itself; (2) processing properties are those properties which are related and affected by the process of the fabrication; and (3) product property which refers to the properties of the fabricated synthetic polymer itself.

2.5.1 Intrinsic Properties

Intrinsic properties depend on the polymer structure and chemical composition and are divided into two classes: intrinsic bulk properties and intrinsic molecular properties.

Density is one of the most important intrinsic properties, and which ranges from less than 1 g/cm^3 to 1.7 g/cm^3 and it differs from polymer to polymer. The density of synthetic biopolymers is greater than conventional fossil-based polymers, which causes higher energy consumption. The density depends on thermal history and the purity of the components. Even though crystallinity is not a self-declared property, it could be affected by the thermal nature and the orientation of the chain. Crystallinity affects other factors like modulus, melt viscosity, strength, heat resistance and transparency. The morphology in the sense of crystallinity and polarity, and the chemical composition, play role in determining the solubility of the polymer. The increase in crystallinity decreases the solubility of the polymer. For example, the amorphous PLA is soluble in organic solvents like THF, benzene, chlorinated hydrocarbons, and dioxin whereas crystalline polymer can dissolve only at elevated temperatures. The mechanical properties of a polymer are dependent on the molecular weight and optical purity of the polymer. The high molecular weight elevates the tensile strength and modulus, and lower elongation at break. The barrier properties are one of the most intensely studied properties of a fabricated polymer. The barrier properties are dependent on the molecular alignment, crystallinity, and density. Barrier properties provide a barrier to water vapor and oxygen and can be divided in to three types: (1) films which have low permeability to water vapor and oxygen; (2) films having low permeability with vapor and high permeability to oxygen; and (3) films having low permeability to oxygen and high permeability to gaseous vapor. Transparency is affected by the incorporation of transparent nucleating additive in the polymer matrix, which crystallizes to form miniscule sized crystallites with size smaller than the wavelength of the visible light and thereby permits the light to propagate through it giving it transparency.

2.5.2 Processing Properties

Processing properties are dependent on the polymerizing conditions and include melt flow index, melt strength and viscosity. The properties are greatly determined by the processing conditions at each stage of production. These processing properties greatly determine the ease of fabrication of that polymer and the cost effectiveness. PLA and PHA are very difficult to fabricate as they are very sensitive to moisture and temperature.

PLA is characterized with low melt viscosity, temperature instability and long moulding cycle and slow crystallization rate. The T_m of PLA ranges from between 155°C to 185°C and the processing temperature between 185°C to 190°C, but this temperature range causes the scission of the polymeric chain, reducing the molecular weight by thermal degradation. The degradation produces lactide, which reduces the melt viscosity and elasticity. The bridging can increase the molecular weight distribution and thereby increases the melt elasticity.

The narrow difference in the crystallization temperature and the melting temperature causes a rapid crystallization on cooling causing a decrease in the transparency (whitening), meaning there is difficulty in designing the shape and appearance of the fabricated product. Also, PGA has insufficient melt stability leading to the deterioration of the quality of the product. Also, the melting temperature of PHA is higher than the degradation temperature; thermal degradation occurs leading to the lowering of the molecular weight.

2.5.3 Product Properties

Product properties, or the properties of the resultant final polymer product, are derived from the cumulative contribution of both the intrinsic properties and the processing properties. The properties include the mechanical strength, heat resistance, water resistance, antistatic properties, flammability, or flame retardancy, aesthetic properties and environmental behavior. The wearing stress of a polymer depends on the coefficient of friction, stiffness, and brittleness. PLA fibers have low wearing resistance, lack of impact resistance which causes crack formation upon cutting, inferior heat resistance, hydrophobicity, water resistance, flammability with low flame retardancy, high surface tension, less color stability leading to the development of yellow color on heating and good gloss and clarity.

The major flaw of the biopolymer is the lack of adequate heat resistance and is known as the heat deformation temperature (HDT). The aesthetic properties of the biopolymer, the color, odor, and the tactile response, also demand attention. For a fully functional biopolymer, all these properties have to be adequately tuned to meet all the criteria at different dimensions [67].

2.6 Applications of Synthetic Biopolymers

Synthetic biopolymers are an important class of materials used in many fields for many years. They have been utilized for the production of many innovative products/materials, and hence they have made a significant mark. However, over the coming years, by making improvements by proposing novel ideas in their blending options and properties, they can reach the efficacy of their fossil fuel counterparts. Also, they possess many advantages over biopolymers in terms of properties like stability and flexibility to suit different applications. The properties strongly depend on the polymer structure, so fabricating the intended polymer structure can broaden the scope of application. They find application mainly in the medical field, packaging, agricultural, and environmental fields.

2.6.1 Biomedical Applications

Synthetic biopolymers have been reported to have a plethora of applications in the biomedical field due to their properties such as stability, non-toxicity, controlled release, and clearance from the body, which are the required properties for commercial application in biomedical field. Also, they are capable of undergoing degradation without releasing harmful substances in vivo. One of the most important synthetic biopolymers, PLA, is used in biomedical applications for production of surgical sutures, drug delivery systems and implants [69]. It is also widely used for preparation of scaffolds, tissue engineering applications due to its biocompatibility and the ease of fabrication of porous materials. Also, PLA blended with lignin and chitin are reported to be used in implant and healing application.

2.6.1.1 Scaffold-Tissue Engineering

Poly esters like, poly(lactic acid) (PLA), poly(glycolic acid) (PGA), and their copolymer poly(lactic acid-co-glycolic acid) (PLGA), and poly(ε-caprolactone) (PCL) are among the most used synthetic biopolymers for the preparation of three-dimensional scaffolds for use in tissue engineering applications due to their biocompatibility and good biodegradation kinetics [70]. Excellent polymeric scaffolds should possess properties like high porosity, proper pore size range, surface area, enhanced mechanical properties, biocompatibility, and proper degradation rate. These features are required for the use of scaffolds in cell seeding, tissue formation, growth etc. [71].

Conde et al. developed a poly(L-lactic acid) scaffold using solvent-casting/particulate leaching method. They have investigated the influence of pore size on the proliferation and differentiation of dental pulp stem cells (DPSC). The scaffolds were prepared in pulp chambers of 1 mm thick tooth slices and porosity was developed using salt crystals. Results suggested that the scaffolds allowed proliferation and differentiation of DPSC onto odontoblast-like cells [72].

Wang and Li proposed the preparation of a scaffold from biodegradable porous PLA polymer for utilization in potential tissue engineering applications. Using a solvent free method, they have ruled out the possibility of the use of organic solvents that reduces the ability of biological cells to form new tissues [73]. PBAT scaffolds have been reported to possess characteristic properties for bone tissue engineering applications due to supported cell attachment and proliferation [74]. Also, crosslinked biopolymer-based scaffolds have better properties like higher tensile strength, and also resist degradation by collagenase but possess the disadvantage of decreased porosity [75].

Poly-(lactic acid) and poly-(glycolic acid) and their co-polymers possess properties like exceptional flexibility and knot robustness; hence they are employed as sutures for biomedical applications [6].

2.6.1.2 Drug Delivery

Kim and co-workers developed a porous poly(lactide-co-glycolide)(PLGA)/HA-DEAP microparticle system for the release of an anti-cholinergic drug, tiotropium. The release of the drug was favored by destabilizing electrostatic interactions between carboxyl and amine groups of the system under acidic environments. The system is effective against treating chronic pulmonary and respiratory diseases [76]. Poly(caprolactone), a biodegradable polyester can be hydrolysed in physiological conditions, hence it is widely used as a long-term implantable biomaterial. It has been approved by the Food and Drug Administration to be used as an absorbable suture, which is sold under the brand name Monocryl, also it serves as an excellent drug delivery device [77].

In a study conducted by Farajikhah et al., the electrospun PCL fibers were shown to be effective in the propagation of MCF-7 breast cancer and also effective in supporting cell growth [74].

A block copolymer comprising of poly(N-isopropylacrylamide)-b-poly(L-lactide) (PNLA), and poly(L-lactide) (PLLA) was fabricated into a nanofibrous films for drug delivery applications. In-vitro studies showed that the drug release rate can be adjusted by the content and molecular weight of PNLA and environmental temperature. All the results indicated that the PNLA/PLLA electrospun nanofibrous films were a suitable candidate for controlled drug delivery systems [78].

2.6.2 Packaging Applications

Materials developed for packaging applications are mainly studied for permeability and tensile properties. PLAs obtained by condensation or ring opening polymerization are being utilized to prepare packaging material like cups, bottles, films, waste bags etc., due to its advantages of easy availability and lower price compared to other biodegradable polyesters. Also, PLA packaging materials are ideal source for fresh items whose quality can be maintained. PLA fibers are UV resistant and are less allergenic. PLA is increasingly known as an active packaging material due to its excellent combination of barrier properties like flavor and aroma barrier characteristics and also excellent permeability performance [69].

Polyhydroxy alkanoates (PHA) also function as a good biodegradable packaging material due to their properties like temperature stability, good tensile strength, flavor and odor barrier properties, oil, and grease resistance, and also ease to dye make it a suitable candidate in food industry. They have been used in coatings, bottles, containers, and films [79]. PHA blended with PHB and poly (3-hydroxy octanoate) has been used as food additives and packages under the name "metabolix PHA" [80]. Romani et al. prepared a blend from rice starch thermoplastic starch (TPS), fish protein and oregano essential oil for food packaging applications. A ratio of 50:50 starch/protein was suitable for water permeability, solubility parameters. Results showed low solubility (8.0%), water vapor permeability (0.18 g mm kPa^{-1} h^{-1} m^{-2}), tensile strength (5.69 MPa), elongation at break

(85.5%) values. Antioxidant studies confirmed that it prevents fruits and vegetables from browning when it is used as a packaging material. More studies on TPS blending with polyesters are studied to improve food applications [81].

2.6.3 Automotive Sector

Synthetic biopolymers and their composites have been used extensively in the automotive sector for various applications. Biopolymers have emerged as a source material for the development of innovative products for use in modern vehicles. Plastic components using bioderived materials are prepared from polyesters, polyamides, polypropylene etc., which possess the advantage over conventional plastics of natural and sustainable degradation [6]. They have been used for the production of seat belts and airbags, and also components on automobile dashboards. Natural fiber reinforced polymer composites are known to reduce the fuel consumption in automobiles. They possess high tensile strength and flexural strength and also have high strength to weight ratio. They are used in interior or exterior parts of automobiles like back and side door panel, spare tire lining, engine insulation, wheel box, bumper, boot-liners, etc. [82].

2.6.4 Agricultural Sector

Synthetic biopolymers have been extensively used in the agricultural field due to their properties of natural degradation and soil-enhancement. They have been employed for the development of polymer films required for mulch production, which enhances plant growth because the polymer films preserve moisture and maintain optimal soil temperature. They also possess the advantage of undergoing photodegradation or degradation aided by microorganisms. Polymer films are usually prepared from polyethylene, polybutylene. They have also been used for preparation of plant containers [6]. Polyacrylamide (PAM) is used as a soil additive to eliminate sediment run off and control soil erosion during irrigation. It has been applied to half of the irrigated land in the U.S. to control erosion and protect from sediment run-off since it is 90% effective in preventing erosion induced sediment runoff thus having a positive environmental effect. However, studies are ongoing to produce renewable, more rapidly degrading PAM alternatives that can be utilized for organic farming options [83] as shown in Table 2.1.

Table 2.1 An overview of applications of important synthetic biopolymers.

Category of biopolymer	Biopolymer	Application
Polyesters	Poly(lactic acid)	Implants, scaffolds, drug delivery, food packaging
	Poly(glycolic acid)	3-D scaffolds, surgical sutures, pharmaceutical supplies
	Poly(caprolactone)	Tissue engineering, implants, drug delivery device
	Polyhydroxybutyrate	High potential gas barrier properties, biocompatible controlled drug release
	Polyhydroxyvalerate	Biomedical applications, internal sutures
Polyamides	Nylon	Textiles, carpets, automotive industry
	Poly(α-amino acid)	Catheters, materials for transfusion equipment
Polyurethanes	Poly ester urethane	Porous scaffolds, grafting
	Poly ether urethanes	Biocompatible medical devices, medical implants
Polyhydroxyalkonoate	Poly(3-hydroxybutyrate)	Drug carriers, tissue engineering, anticancer agents
Poly-(ether ester)	Polydioxanone	Absorbable monofilament wound suture, healing
Poly-(ester amide)	α-amino acid derivatives	Drug delivery systems, tissue engineering
Vinyl polymers	Poly(vinyl alcohol)	Water soluble films, food packaging, coating, adhesives
	Poly(vinyl chloride)	Plastic pipes, bottles, medical devices, insulation materials

References

1 Chassenieux, C., Durand, D., Jyotishkumar, P., and Thomas, S. (2013). Biopolymers: state of the art, new challenges, and opportunities. In: *Handbook of Biopolymer-Based Materials: From Blends and Composites to Gels and Complex Networks* (ed. S. Thomas, D. Durand, C. Chassenieux, et al.), 1–6. doi: 10.1002/9783527652457.CH1.

2 Nagalakshmaiah, M., Afrin, S., Malladi, R.P. et al. (2019). Biocomposites: present trends and challenges for the future. In: *Green Composites for*

Automotive Applications (ed. G. Koronis and A. Silva), 197–215. doi: 10.1016/B978-0-08-102177-4.00009-4.

3 Koncar, V. (2018). *Smart Textiles for In Situ Monitoring of Composites*. Elsevier.

4 Mohanty, A.K., Misra, M., and Drzal, L.T. (2018). Sustainable bio-composites from renewable resources: opportunities and challenges in the green materials world. In: *Renewable Energy* (ed. B. Sorenson), 396–409. doi: 10.4324/9781315793245-107.

5 Netravali, A.N. and Chabba, S. (2003). Composites get greener. *Materials Today* 6 (4): 22–29. doi: 10.1016/S1369-7021(03)00427-9.

6 George, A., Sanjay, M.R., Srisuk, R. et al. (2020). A comprehensive review on chemical properties and applications of biopolymers and their composites. *International Journal of Biological Macromolecules* 154: 329–338. doi: 10.1016/j.ijbiomac.2020.03.120.

7 Mohanty, A.K., Misra, M., Drzal, L.T. (2005). *Natural Fibers, Biopolymers, and Biocomposites*. 15–50. doi: 10.1201/9780203508206.

8 Yin, J. and Luan, S. (2016). Opportunities and challenges for the development of polymer-based biomaterials and medical devices. *Regenerative Biomaterials* 3 (2): 129–135. doi: 10.1093/RB/RBW008.

9 Ashter, S.A. (2016). *Introduction to Bioplastics Engineering*. Elsevier.

10 Hassan, M.E., Bai, J., and Dou, D.Q. (2019). Biopolymers; definition, classification, and applications. *Egyptian Journal of Chemistry* 62 (9): 1725–1737. doi: 10.21608/EJCHEM.2019.6967.1580.

11 Ravenstijn, J. (2010). Bioplastics in consumer electronics. *Industrial Biotechnology* 6 (5): 252–263. doi: 10.1089/IND.2010.6.252.

12 Sampath, U.G.T.M., Ching, Y.C., Chuah, C.H. et al. (2016). Fabrication of porous materials from natural/synthetic biopolymers and their composites. *Materials* 9 (12): 1–32. doi: 10.3390/ma9120991.

13 Othman, S.H. (2014). Bio-nanocomposite materials for food packaging applications: types of biopolymer and nano-sized filler. *Agriculture and Agricultural Science Procedia* 2: 296–303. doi: 10.1016/J.AASPRO.2014.11.042.

14 Siracusa, V. (2019). Microbial degradation of synthetic biopolymers waste. *Polymers* 11 (6): 1066. doi: 10.3390/POLYM11061066.

15 Ross, G., Ross, S., and Tighe, B.J. (2016). Bioplastics: new routes, new products. In: *Brydson's Plastics Materials*, 8e (ed. M. Gilbert), 631–652. doi: 10.1016/B978-0-323-35824-8.00023-2.

16 Middleton, J.C. and Tipton, A.J. (2000). Synthetic biodegradable polymers as orthopedic devices. *Biomaterials* 21 (23): 2335–2346. doi: 10.1016/S0142-9612(00)00101-0.

17 Chee, C.Y., Rosiyah, Y.Y., and Li, G. (2013). Preparation and characterization of nano particle reinforced polyactides composite. *Journal of Nano Research* 25: 128–136. doi: 10.4028/WWW.SCIENTIFIC.NET/JNANOR.25.128.

18 Niaounakis, M. (2015). Building and construction applications. In: *Biopolymers: Applications and Trends*, Elsevier, 445–505. doi: 10.1016/B978-0-323-35399-1.00010-7.

19 Prieto, E.M. and Guelcher, S.A. (2014). Tailoring properties of polymeric biomedical foams. In: *Biomedical Foams for Tissue Engineering Applications*, 129–162. doi: 10.1533/9780857097033.1.129.

20 Gorth, D. and Webster, T.J. (2011). Matrices for tissue engineering and regenerative medicine. In: *Biomaterials for Artificial Organs* (ed. M. Lysaght and T.J. Webster), 270–286. doi: 10.1533/9780857090843.2.270.

21 Pascault, J.P., Höfer, R., and Fuertes, P. (2012). Mono-, di-, and oligosaccharides as precursors for polymer synthesis. *Polymer Science: A Comprehensive Reference* 10: 59–82. doi: 10.1016/B978-0-444-53349-4.00254-5.

22 Samantaray, P.K., Little, A., Haddleton, D.M. et al. (2020). Poly(glycolic acid) (PGA): a versatile building block expanding high performance and sustainable bioplastic applications. *Green Chemistry* 22 (13): 4055–4081. doi: 10.1039/D0Gc02394C.

23 Göktürk, E., Pemba, A.G., and Miller, S.A. (2015). Polyglycolic acid from the direct polymerization of renewable C1 feedstocks. *Polymer Chemistry* 6 (21): 3918–3925. doi: 10.1039/C5PY00230C.

24 Valderrama, M., van Putten, R., and Gruter, G.M. (2019). The potential of oxalic–and glycolic acid-based polyesters (review). Towards CO2 as a feedstock (Carbon Capture and Utilization–CCU). *European Polymer Journal*. doi: 10.1016/j.eurpolymj.2019.07.036.

25 Jem, K.J. and Tan, B. (2020). The development and challenges of poly (lactic acid) and poly (glycolic acid). *Advanced Industrial and Engineering Polymer Research* 3 (2): 60–70. doi: 10.1016/J.AIEPR.2020.01.002.

26 Takahashi, K., Taniguchi, I., Miyamoto, M., and Kimura, Y. (2000). Melt/solid polycondensation of glycolic acid to obtain high-molecular-weight poly(glycolic acid). *Polymer* 41 (24): 8725–8728. doi: 10.1016/S0032-3861(00)00282-2.

27 Ayyoob, M., Lee, D.H., Kim, J.H. et al. (2017). Synthesis of poly(glycolic acids) via solution polycondensation and investigation of their thermal degradation behaviors. *Fibers and Polymers* 18 (3): 407–415. doi: 10.1007/S12221-017-6889-1.

28 Gilding, D.K. and Reed, A.M. (1979). Biodegradable polymers for use in surgery—polyglycolic/poly(actic acid) homo- and copolymers: 1. *Polymer* 20 (12): 1459–1464. doi: 10.1016/0032-3861(79)90009-0.

29 Makadia, H.K. and Siegel, S.J. (2011). Poly lactic-co-glycolic acid (PLGA) as biodegradable controlled drug delivery carrier. *Polymers* 3 (3): 1377–1397. doi: 10.3390/POLYM3031377.

30 Astete, C.E. and Sabliov, C.M. (2012). Synthesis and characterization of PLGA nanoparticles. *J Biomater Sci Polym Ed.* 17 (3): 247–289. doi: 10.1163/156856206775997322.

31 Kreiser- Saunders, I. and Kricheldorf, H.R. (1998). Polylactones, 39. Zn lactate-catalyzed copolymerization of L-lactide with glycolide or ε-caprolactone. *Macromolecular Chemistry and Physics* 199 (6): 1081–1087.

32 Dobrzyński, P., Kasperczyk, J., and Bero, M. (1999). Application of calcium acetylacetonate to the polymerization of glycolide and copolymerization of glycolide with ε-caprolactone and L-lactide. *Macromolecules* 32 (14): 4735–4737. doi: 10.1021/MA981969Z.

33 Bero, M., Dobrzyński, P., and Kasperczyk, J. (1999). Application of zirconium (IV) acetylacetonate to the copolymerization of glycolide with ε-caprolactone and lactide. *Polymer Bulletin* 42 (2): 131–139. doi: 10.1007/S002890050444.

34 Azimi, B., Nourpanah, P., Rabiee, M., and Arbab, S. (2014). Poly (lactide - co- glycolide) fiber: an overview, *Journal of Engineered Fibers and Fabrics,* 9 (1): 47–66. doi: 10.1177/155892501-400900107.

35 Gentile, P., Chiono, V., Carmagnola, I., and Hatton, P.V. (2014). An overview of poly(lactic-co-glycolic) acid (PLGA)-based biomaterials for bone tissue engineering. *International Journal of Molecular Sciences* 15 (3): 3640–3659. doi: 10.3390/IJMS15033640.

36 Shivalkar Yadav, K. and Prabha, R. (2017). Polyhydroxybutyrate as bio-degradable plastic – a review. *IOSR Journal of Environmental Science, Toxicology and Food Technology* 11 (05): 10–12. doi: 10.9790/2402-1105011012.

37 McAdam, B., Fournet, M.B., McDonald, P., and Mojicevic, M. (2020). Production of polyhydroxybutyrate (PHB) and factors impacting its chemical and mechanical characteristics. *Polymers* 12 (12): 2908. doi: 10.3390/POLYM12122908.

38 Bharti, S.N. and Swetha, G. (2016). Need for bioplastics and role of biopolymer PHB: a short review. *Journal of Petroleum & Environmental Biotechnology* 7 (2): 7. doi: 10.4172/2157-7463.1000272.

39 Sindhu, R., Ammu, B., Binod, P. et al. (2011). Production and characterization of poly-3-hydroxybutyrate from crude glycerol by *Bacillus sphaericus* NII 0838 and improving its thermal properties by blending with other polymers. *Brazilian Archives of Biology and Technology* 54 (4): 783–794. doi: 10.1590/S1516-89132011000400019.

40 Sudesh, K., Abe, H., and Doi, Y. (2000). Synthesis, structure, and properties of polyhydroxyalkanoates: biological polyesters. *Progress in Polymer Science* 25 (10): 1503–1555. doi: 10.1016/S0079-6700(00)00035-6.

41 Vroman, I. and Tighzert, L. (2009). Biodegradable polymers. *Materials* 2 (2): 307–344. doi: 10.3390/MA2020307.

42 Dobrogojski, J., Spychalski, M., Luciński, R., and Borek, S. (2018). Transgenic plants as a source of polyhydroxyalkanoates. *Acta Physiologiae Plantarum* 40 (9): 1–17. doi: 10.1007/S11738-018-2742-4.

43 Batista, M.B., Teixeira, C.S., Sfeir, M.Z.T. et al. (2018). PHB biosynthesis counteracts redox stress in *Herbaspirillum seropedicae*. *Frontiers in Microbiology* 9999: 472. doi: 10.3389/FMICB.2018.00472.

44 Markl, E. (2018). PHB - bio based and biodegradable replacement for PP: a review. *Novel Techniques in Nutrition & Food Science* 2 (5): 206–209. doi: 10.31031/NTNF.2018.02.000546.

45 Kobayashi, S., Uyama, H., and Kimura, S. (2001). Enzymatic polymerization. *Chemical Reviews* 101 (12): 3793–3818. doi: 10.1021/CR990121L.

46 Díaz, A., Katsarava, R., and Puiggalí, J. (2014). Synthesis, properties and applications of biodegradable polymers derived from diols and dicarboxylic acids: from polyesters to poly(ester amide)s. *International Journal of Molecular Sciences* 15 (5): 7064–7123. doi: 10.3390/ijms15057064.

47 Zhan, J., Ma, L., Liu, X. et al. (2017). Mechanical, thermal, and flame-retardant behaviors of thermoplastic polyether–ester elastomer composites with polyphenylene oxide and aluminum hypophosphite. *Journal of Polymer-Plastics Technology and Engineering*, 56 (10): 1096–1107. doi: 10.1080/03602559.2016.1253737.

48 Kariduraganavar, M.Y., Kittur, A.A., and Kamble, R.R. (2014). Polymer synthesis and processing. In: *Natural and Synthetic Biomedical Polymers* (ed. S.G. Kumbar, C.T. Laurencin and M. Deng), 1–31. doi: 10.1016/B978-0-12-396983-5.00001-6.

49 Pachence, J.M. and Kohn, J. (2000). Biodegradable polymers. In: *Principles of Tissue Engineering* (ed. R.P. Lanza, R. Langer and J. Vacanti), 263–277. doi: 10.1016/B978-012436630-5/50026-X.

50 Sugimoto, H. and Inoue, S. (2006). Recent progress in the synthesis of polymers based on carbon dioxide. *Pure and Applied Chemistry* 78 (10): 1823–1834. doi: 10.1351/PAC200678101823.

51 Xu, J., Feng, E., and Song, J. (2014). Renaissance of aliphatic polycarbonates: new techniques and biomedical applications. *Journal of Applied Polymer Science* 131 (5). doi: 10.1002/APP.39822.

52 Feng, J., Zhuo, R.X., and Zhang, X.Z. (2012). Construction of functional aliphatic polycarbonates for biomedical applications. *Progress in Polymer Science* 37 (2): 211–236. doi: 10.1016/J.PROGPOLYMSCI.2011.07.008.

53 Tempelaar, S., Mespouille, L., Coulembier, O. et al. (2013). Synthesis and post-polymerisation modifications of aliphatic poly(carbonate)s prepared by ring-opening polymerisation. *Chemical Society Reviews* 42 (3): 1312–1336. doi: 10.1039/C2CS35268K.

54 Zhu, W., Huang, X., Li, C. et al. (2011). High-molecular-weight aliphatic polycarbonates by melt polycondensation of dimethyl carbonate and aliphatic diols: synthesis and characterization. *Polymer International* 60 (7): 1060–1067. doi: 10.1002/PI.3043.

55 Jiang, Z., Liu, C., Xie, W., and Gross, R.A. (2007). Controlled lipase-catalyzed synthesis of poly(hexamethylene carbonate). *Macromolecules* 40 (22): 7934–7943. doi: 10.1021/MA070665M.

56 Fukuoka, S., Kawamura, M., Komiya, K. et al. (2003). A novel non-phosgene polycarbonate production process using by-product CO2 as starting material. *Green Chemistry* 5 (5): 497–507. doi: 10.1039/B304963A.

57 Gunanathan, C., Ben-David, Y., and Milstein, D. (2007). Direct synthesis of amides from alcohols and amines with liberation of H2. *Science* 317 (5839): 790–792. doi: 10.1126/SCIENCE.1145295.

58 Zweifel, T., Naubron, J.-V., and Grützmacher, H. (2009). Catalyzed dehydrogenative coupling of primary alcohols with water, methanol, or amines. *Angewandte Chemie* 121 (3): 567–571. doi: 10.1002/ANGE.200804757.

59 Shimizu, M., Saito, T., and Isogai, A. (2016). Water-resistant and high oxygen-barrier nanocellulose films with interfibrillar cross-linkages formed through multivalent metal ions. *Journal of Membrane Science* 500: 1–7. doi: 10.1016/j.memsci.2015.11.002.

60 Nordstrom, L.U., Vogt, H., and Madsen, R. (2008). Amide Synthesis from Alcohols and Amines by the Extrusion of Dihydrogen. *Journal of the American Chemical Society* 130 (52): 17672–17673.

61 Zeng, H. and Guan, Z. (2011). Direct synthesis of polyamides via catalytic dehydrogenation of diols and diamines. *Journal of the American Chemical Society* 133 (5): 1161. doi: 10.1021/JA106958S.

62 Fonseca, A.C., Gil, M.H., and Simões, P.N. (2014). Biodegradable poly(ester amide)s – a remarkable opportunity for the biomedical area: review on the synthesis, characterization, and applications. *Progress in Polymer Science* 39 (7): 1291–1311. doi: 10.1016/j.progpolymsci.2013.11.007.

63 Nguyen, Q.T., Germain, Y., Clément, R., and Hirata, Y. (2001). Pervaporation, a novel technique for the measurement of vapor transmission rate of highly permeable films. *Polymer Testing* 20 (8): 901–911. doi: 10.1016/S0142-9418(01)00019-8.

64 Kim, J.H., Ha, S.Y., and Lee, Y.M. (2001). Gas permeation of poly(amide-6-b-ethylene oxide) copolymer. *Journal of Membrane Science* 190 (2): 179–193. doi: 10.1016/S0376-7388(01)00444-6.

65 Thanakkasaranee, S., Kim, D., and Seo, J. (2018). Preparation and characterization of poly(ether-block-amide)/polyethylene glycol composite films with temperature- dependent permeation. *Polymers* 10 (2). doi: 10.3390/polym10020225.

66 Heath, D.E., Guelcher, S.A., and Cooper, S.L. (2013). Polyurethanes. In: *Biomaterials Science: An Introduction to Materials*, 3e (ed. W.R. Wagner,

S.E. Sakiyama-Elbert, G. Zhnag et al.), 79–82. doi: 10.1016/B978-0-08-087780-8.00009-7.

67 Niaounakis, M. (2015). *Biopolymers: Applications and Trends.* Elsevier.

68 Karandikar, S., Mirani, A., Waybhase, V. et al. (2017). Nanovaccines for oral delivery-formulation strategies and challenges. In: *Nanostructures for Oral Medicine* (ed. E. Andronescu and A. Mihai), 263–293. doi: 10.1016/B978-0-323-47720-8.00011-0.

69 Jamshidian, M., Tehrany, E.A., Imran, M. et al. (2010). Poly-lactic acid: production, applications, nanocomposites, and release studies. *Comprehensive Reviews in Food Science and Food Safety* 9 (5): 552–571. doi:10.1111/j.1541-4337.2010.00126.x.

70 Okamoto, M. and John, B. (2013). Synthetic biopolymer nanocomposites for tissue engineering scaffolds. *Progress in Polymer Science* 38 (10–11): 1487–1503. doi: 10.1016/j.progpolymsci.2013.06.001.

71 Ma, P.X. (2004). Scaffolds for tissue fabrication. *Materials Today* 7 (5): 30–40. doi: 10.1016/S1369-7021(04)00233-0.

72 Conde, C.M., Demarco, F.F., Casagrande, L. et al. (2015). Influence of poly-l-lactic acid scaffold's pore size on the proliferation and differentiation of dental pulp stem cells. *Brazilian Dental Journal* 26 (2): 93–98. doi: 10.1590/0103-6440201300032.

73 Wang, X., Li, W., and Kumar, V. (2006). A method for solvent-free fabrication of porous polymer using solid-state foaming and ultrasound for tissue engineering applications. *Biomaterials* 27 (9): 1924–1929. doi: 10.1016/j.biomaterials.2005.09.029.

74 Mtibe, A., Motloung, M.P., Bandyopadhyay, J., and Ray, S.S. (2021). Synthetic biopolymers and their composites: advantages and limitations—an overview. *Macromolecular Rapid Communications* 42 (15): 1–28. doi: 10.1002/marc.202100130.

75 Reddy, N., Reddy, R., and Jiang, Q. (2015). Crosslinking biopolymers for biomedical applications. *Trends in Biotechnology* 33 (6): 362–369. doi: 10.1016/j.tibtech.2015.03.008.

76 Kim, S., Kwag, D.S., Lee, D.J., and Lee, E.S. (2016). Acidic pH-stimulated tiotropium release from porous poly(lactic-co-glycolic acid) microparticles containing 3-diethylaminopropyl-conjugated hyaluronate. *Macromolecular Research* 24 (2): 176–181. doi: 10.1007/s13233-016-4022-8.

77 Adikwu, M.U. and Esimone, C.O. (2010). *Biopolymers in Drug Delivery.* Bentham Books.

78 Li, J., Peng, C., Wang, Z., and Ren, J. (2018). Preparation of thermo-responsive drug-loaded nanofibrous films created by electrospinning. *RSC Advances* 8 (31): 17551–17557. doi: 10.1039/c8ra02442a.

79 Yadav, A., Mangaraj, S., Singh, R. et al. (2018). Biopolymers as packaging material in food and allied industry. *International Journal of Chemical Studies* 6 (2): 2411–2418.

80 Fakirov, S. (2015). Biodegradable polyesters. In: *Biodegradable Polyesters* (ed. S. Fakirov). Wiley. doi: 10.1002/9783527656950.

81 Romani, V.P., Prentice-Hernández, C., and Martins, V.G. (2017). Active and sustainable materials from rice starch, fish protein and oregano essential oil for food packaging. *Industrial Crops and Products* 97: 268–274. doi: 10.1016/j.indcrop.2016.12.026.

82 Rashad, A.M. (2019). The effect of polypropylene, polyvinyl-alcohol, carbon, and glass fibers on geopolymers properties. *Materials Science and Technology (United Kingdom)* 35 (2): 127–146. doi: 10.1080/02670836.2018.1514096.

83 Orts, W.J., Roa-Espinosa, A., Sojka, R.E. et al. (2007). Use of synthetic polymers and biopolymers for soil stabilization in agricultural, construction, and military applications. *Journal of Materials in Civil Engineering* 19 (1): 58–66. doi: 10.1061/(asce)0899-1561(2007)19:1(58).

3

Role of Biopolymers and Their Composites in Sustainable Agriculture

Recent Developments and Future Perspectives

Abhisekh Saha[1], *Sreedeep S.*[2,*], *and Uttam Manna*[3]

[1] *Department of Civil and Environmental Engineering, C.V. Raman Global University Bhubaneswar, Odisha, India*
[2] *Department of Civil Engineering, Indian Institute of Technology Guwahati, Assam, India*
[3] *Department of Chemistry and Center for Nanotechnology, Indian Institute of Technology Guwahati, Assam, India*
[*] *Corresponding author*

3.1 Introduction

A major part of the world's agricultural production is facing a huge challenge to ensure food security due to the increase in global population, and the unavailability of arable land and freshwater. This situation is further exacerbated by the current scenarios of extreme climate events, unpredictable rainfall patterns, and an increase in global temperature, which result in climate-induced drought and other abiotic stress to plants. Apart from these, the low water and nutrient retention capacity of soil, excess leaching of nutrients to deeper layers, soil erosion, and land degradation are also major challenges to meet the growing food demand [1]. In 2015, the United Nations (UN) adopted the 17 sustainable development goals (SDGs) to ensure food security, peace, and prosperity for all, out of which goal 2 (zero hunger), goal 3 (good health and well-being), goal 6 (clean water and sanitation), goal 13 (climate action), goal 14 (life below water), and goal 15 (life on land) are directly related to soil ecology, soil biochemical properties, and soil quality. Therefore, it is essential to develop sustainable soil management strategies

Applications of Biopolymers in Science, Biotechnology, and Engineering, First Edition.
Edited by Arunima Reghunadhan, Akhina H., and Sabu Thomas.

to improve the water retention properties of soil, enhance nutrient use efficiency, minimize desertification and degradation, and ecological restoration of environmentally polluted land.

Past studies have highlighted that application of soil conditioners could improve soil pore structure, soil-water retention capacity, reduce surface runoff and downward leaching of nutrients, and thereby conserve adequate moisture and nutrient in the plant-root zone [2]. Material science, along with polymer chemistry and nanotechnology, are emerging as innovative solutions for the development of smart and green materials through valorization of renewable resources that can be effectively used as soil conditioners to improve the physicochemical properties of soil. The usage of waste, as well as natural materials for the preparation of novel soil conditioners, supports the interdependent components of sustainable agriculture, as presented in Figure 3.1. Moreover, utilization of waste materials for soil and water management would reduce the burden of waste disposal facilities and contributes towards a circular economy by improving crop production at low per capita consumption of water and fertilizers. The low consumption of chemical fertilizer and usage of biodegradable green materials would result in conservation of soil ecosystem and self-sustaining agricultural systems without changing environmental conditions.

In this context, the application of biopolymers has been recognized as a viable soil management technique for the improvement of soil-pore structure and soil retention properties. The properties of biopolymers can be fine tuned by varying monomer units, structure (linear chain or three-dimensional interpenetrating network) depending on their application. Superabsorbent biopolymer composite (SBC), a three-dimensional (3D)

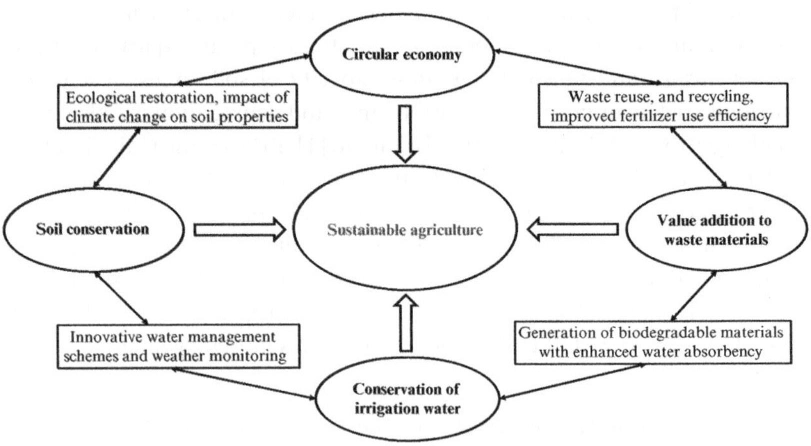

Figure 3.1 A basic framework to achieve sustainability in agriculture.

crosslinked polymer network, is gaining a lot of interest in the field of agriculture and horticulture due to its ability to absorb and retain higher amounts of water and solute molecules [3]. The presence of different hydrophilic groups, such as amine, hydroxyl, and carboxyl, attached to the polymer network of SBC, effectively contributes toward its higher water absorbency. Figure 3.2 illustrates a conceptual diagram to show the effectiveness of SBC based agriculture over chemical fertilizer-based agriculture, considering both the aspects of soil environment and plant health. The utilization of chemical fertilizers can improve crop growth and productivity by providing essential plant nutrients as compared to the conventional agricultural system. However, the intensive use of chemical fertilizer for a prolonged duration has serious environmental consequences such as soil degradation, water eutrophication, and contamination of surface water. Moreover, the over-application of chemical fertilizers increases the cost of agricultural production and reduces the profit margin for crop growers. On the other hand, the application of SBC into agriculture fields can increase the water retention properties of soil by absorbing and storing the rainfall and irrigation water within their polymer structure. With the evaporation of water during drying, the stored water within SBC is released to the soil, which is available for plant roots uptake. This will reduce the downward migration of water and nutrient into deeper layers and reduce the requirement of irrigation and chemical fertilizers. The application of SBC can also increase the population of soil bacteria and soil microbial activity under water stress conditions, which is beneficial for nutrient cycling [4].

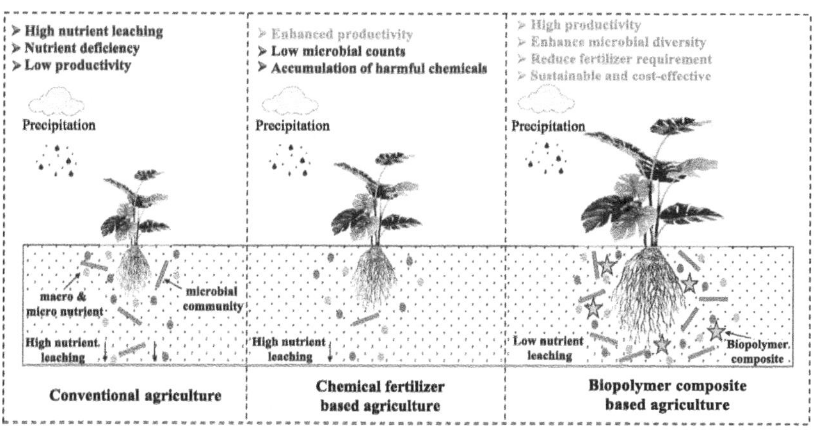

Figure 3.2 Comparison of possible advantage of biopolymer composite based agriculture with respect to conventional agriculture and chemical fertilizer-based agriculture.

The application of SBC for the improvement of soil and plant properties under water stress conditions is investigated in the previous literature [5–7]. Past studies indicate that the SBC addition in soil improves the aggregate stability, bulk density, soil porosity, and aeration for coarse-textured soils. However, there are some contrasting observations reported in the literature on the influence of SBC on fine-textured soils with high clay content [8]. Moreover, some studies pointed out that the total amount of stored water inside the hydrophilic polymer is not easily available to plant roots [9, 10], regardless of the improvement in the soil-water retention properties. Therefore, it is necessary to understand the complex interaction between the soil, SBC, plant, and atmosphere to evaluate the effectiveness of SBC for sustainable agriculture. This chapter aims to critically review the possibility of SBC application for agricultural applications considering the soil–SBC–plant–atmosphere interaction, as shown in Figure 3.3. The study tries to understand the influence of SBC amendment (with different SBC compositions, swelling characteristics) on different soil properties (such as soil

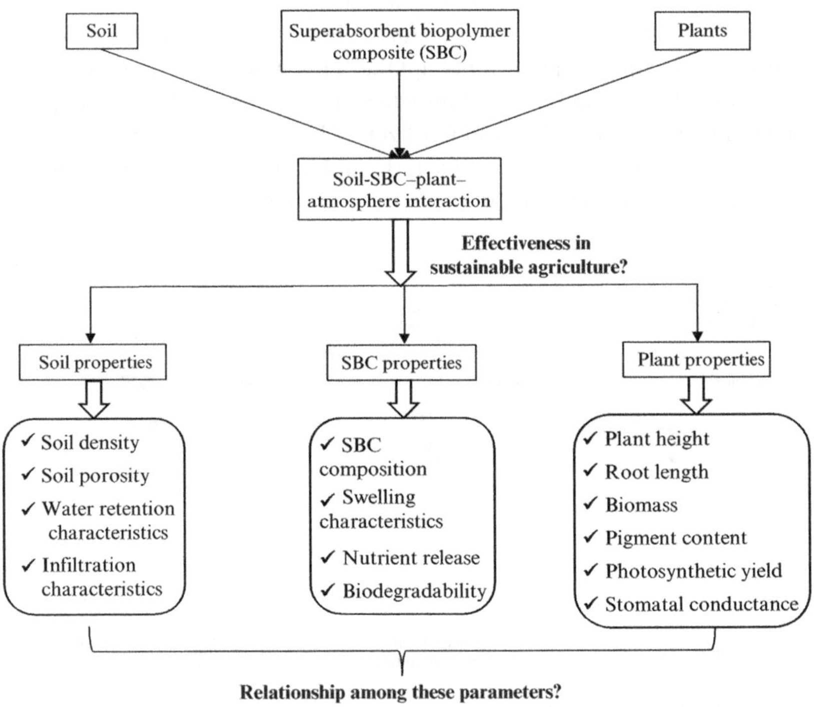

Figure 3.3 Important aspects of soil-SBC-plant-atmosphere interaction for sustainable agriculture.

density, soil porosity, water retention characteristics, field capacity, and infiltration characteristics) and plant properties (plant height, root length, biomass, stomatal conductance, and photosynthetic yield). The last section of this chapter highlights the limitations of the current understanding on soil–SBC–plant–atmosphere and future perspectives of SBC for sustainable agriculture.

3.2 Biopolymers

Biopolymers are a versatile group of materials that are either derived from biological sources, such as plants, animals, and microorganisms, or produced chemically from natural sources, such as sugars, amino acids, natural fats, and oils. These biological polymers or biopolymers are made of repetitive monomer units, and there can exist varieties of biopolymer depending on their folding characteristic and network structure. The monomer units of biopolymers generally have re-occurring molecules of amino acid proteins, nucleotides, and saccharides from sugar [11].

The advantages of biopolymer over synthetic polymer are demonstrated in Figure 3.4. Biopolymers are biorenewable as they can be derived from natural materials (i.e., plants, microbes, animals), which are easily available and cost-effective. In contrast, synthetic polymers are mostly produced from petrochemicals (i.e., non-renewable resources). Most of the

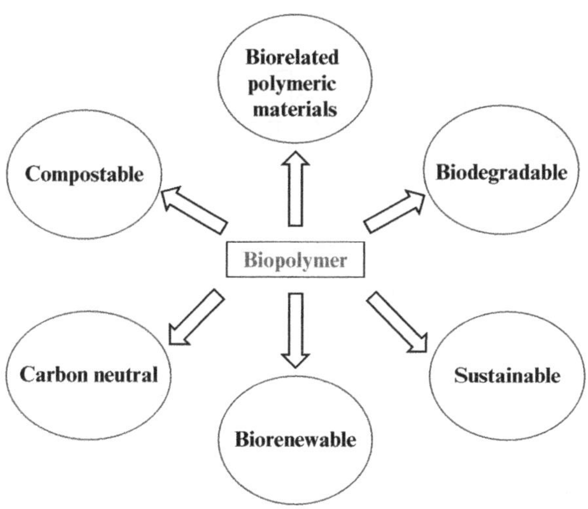

Figure 3.4 Advantages of biopolymers over synthetic polymers.

biopolymers are biodegradable in soil due to the presence of biobased materials and decomposed to CO_2 and water by soil microorganisms. Biopolymers are also compostable through the industrial composting process and can be degraded by 90% within six months [12]. In addition, the application of biopolymer can also reduce the carbon emission into the atmosphere as the CO_2 produced during biopolymer degradation can be reabsorbed by the plants, making them carbon neutral. Therefore, the utilization of biopolymers in the agricultural sector is one of the viable methods for achieving sustainability.

3.2.1 Classification of Biopolymers

Biopolymers can be classified based on different criteria, including their origin, composition, preparation methods, degradability, and physical properties. Based on their origin, biopolymers are broadly classified into natural biopolymers and hybrid biopolymers. Natural biopolymers have natural sources, such as plants, animals, and microorganisms, whereas synthetic biopolymers are synthesized from biobased materials through a chemical or biological process. Natural biopolymers can be roughly categorized as polypeptide and protein-based (collagen, fibrin, gelatin, silk, protein, etc.), polysaccharide-based (chitin, chitosan, alginate, carrageenan, hyaluronic acid, cellulose, starch, etc.), and polyphenol-based (lignin) biopolymers (Figure 3.5). Polypeptide and

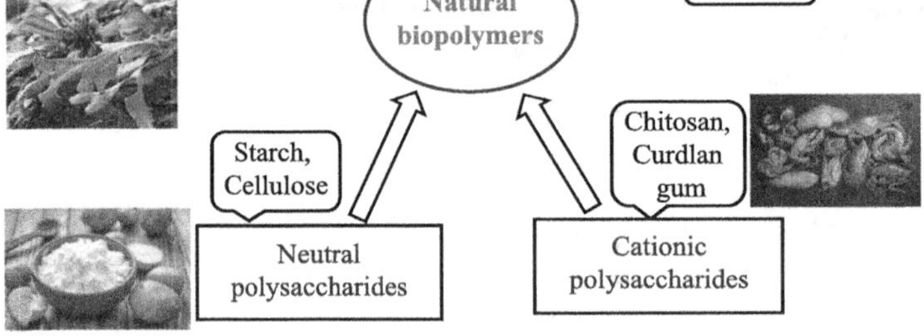

Figure 3.5 Overview of natural biopolymers and their classification. *Source:* Algae image: arenacreations/Adobe Stock. Starch image: Sea Wave/Adobe Stock. Woodchip image: Nikul/Adobe Stock Photos. Crustacean image: New Africa/Adobe Stock. Jelly image: 2ragon/Adobe Stock.

protein-based biopolymers are derived from naturally occurring α-L-amino acids. Polypeptides consist of shorter chains of amino acids (≤100 amino acids), whereas protein contains larger amino acid (>100 amino acids) chains [13]. On the other hand, polysaccharides are made of different monomer units of monosaccharides as well as disaccharides, and their branching and molecular weight sequences are not consistent. Based on their ionic nature, polysaccharides can be categorized as neutral polysaccharides (example: starch, cellulose), anionic polysaccharides (example: alginate, carrageenan), and cationic polysaccharides (example: chitin, chitosan, curdlan gum). Polyphenol biopolymers, which are most abundant in plants and have diverse polymer structures, consist of naturally occurring organic compounds characterized by multiple units of phenol. The synthetic biopolymers can also be classified into two categories, including biopolymers derived by biosynthesis and biopolymers derived by chemical synthesis. Some of the examples for biopolymers derived by biosynthesis (i.e., derived from living cells) are polylactides, polyglycolide, polyhydroxyalkanoates (PHA), xanthan gum, and gellan gum. Similarly, biopolymers, such as polyurethanes, polyesteramides, polyanhydrides, vinyl polymers, acrylic resins, are chemically synthesized from biobased materials or a combination of synthetic chemical and biobased materials. It may be noted that synthetic biopolymers are also ecofriendly and biodegradable in nature.

Based on the response to thermal conditions, biopolymers can be classified into thermoplastics, thermosets, and elastomers [11]. Thermoplastic biopolymers are formed when they are molten, and their shape and internal structure are determined after cooling. These polymers can be reheated and reshaped depending on their applications. In contrast, thermoset biopolymers are three-dimensional polymer networks of highly crosslinked polymer chains and cannot be remolded after heating. Elastomers have the characteristics of both thermoset and thermoplastic biopolymers. Elastomers act as thermoplastic biopolymers with distinct chains initially and later develop covalent crosslink polymer networks like thermosets. In the last few decades, thermoset biopolymers have gained more interest among researchers as compared to thermoplastic biopolymers, whereas elastomers have shown their potential to develop sustainable materials with specific properties.

3.2.2 Biopolymer Composite

Biopolymers have gained remarkable attention in material science for the fabrication of novel materials that offer great market value owing to their ecofriendliness and biorenewability as compared to the petrochemical-derived synthetic polymers. However, the thermal stability and other important mechanical properties of biopolymers are lower than the synthetic polymers. To overcome this problem, a new approach has been introduced to enhance the stability and performance of the biopolymer by integrating synthetic polymers within the micro regime of the polymer structure. This results in a new generation of materials termed as biopolymer composites, which consists of petrochemical-based material within the polymeric network of environmental-friendly biopolymer. For agricultural and horticultural applications, where soil water and nutrient retention are important parameters for plant growth and yield, superabsorbent biopolymer composites (SBC) are emerging as sustainable materials with their improved water absorbency, lower salt sensitivity, mechanical and thermal stability. These SBCs are synthesized using polysaccharide-based biopolymers, such as chitosan, starch, cellulose, and alginate along with the synthetic vinyl monomers, such as acrylic acid, methacrylate, vinyl alcohol, and sulfonic acid, in the presence of crosslinker using graft polymerization technique. Graft polymerization is a process in which one or more polymer chain is covalently attached with another polymer chain (i.e., backbone polymer). It is an effective method to modify the properties of the backbone polymer, such as water absorbency, water repellency, acid–base resistance, or salt sensitivity, depending on the specific applications. The properties of the synthesized SBC can be maintained by controlling the synthesis parameters, known as grafting percentage (GP) and grafting efficiency (GE), which are calculated using the following expression [14],

$$GP(\%) = \frac{weight\ of\ grafted\ polymer - weight\ of\ biopolymer}{weight\ of\ biopolymer} \times 100 \quad (3.1)$$

$$GE(\%) = \frac{weight\ of\ grafted\ polymer - weight\ of\ biopolymer}{weight\ of\ synthetic\ polymer} \times 100 \quad (3.2)$$

The graft polymerization process is usually initiated with free radical reaction to create active sites on the biopolymer, where the synthetic monomers are attached. A detailed description of the graft polymerization process is presented in Figure 3.6. The polymerization reaction starts with the decomposition of initiator to free radicals upon heating, which creates free radical sites on the biopolymer (polysaccharides) chain. The initiator also creates active sites on the synthetic monomers and initiates the grafting of synthetic monomers on the polymeric structure of biopolymer, which results in the growth of grafted polymer chain. During the polymerization process, the synthetic polymer chain reacts with the end vinyl group of the crosslinker to produce a three-dimensional (3D), interpenetrating polymer network. The purpose of crosslinker is to link two or more polymer chains

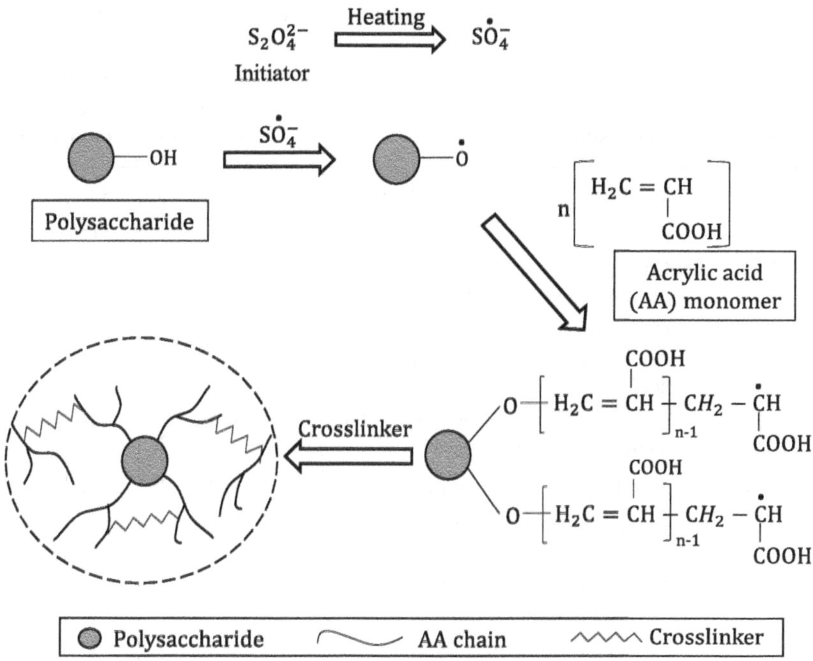

Figure 3.6 Details of the graft polymerization technique for development of superabsorbent biopolymer composite (SBC). *Source:* Adapted from [1].

through covalent bonds and produce a branched polymer network with higher molecular weight. The process of crosslinking makes the synthesized polymer water-insoluble and forms a gel-like consistency when in contact with water. After the completion of polymerization process, the homopolymer by-product (which form during homo-polymerization of synthetic monomer) need to be separated by washing the resultant product with ethanol and water to obtain the pure graft copolymer (i.e., superabsorbent biopolymer composite). It can be noted that the free radical site generation (initiation process) on the monomer and biopolymer chain can be performed without the free-radical initiators. In such cases, the creation of active sites is achieved through UV irradiation, microwave irradiation, high energy radiation method (gamma rays and electron beams), or enzymes. Table 3.1 reports the details of various biopolymers and synthetic monomers used in the past studies to produce SBC by graft polymerization method along with their water absorbency (or swelling ratio) in distilled water. As it can be observed from Table 3.1, the swelling ratios of SBC are in the range of 30–1000 g/g depending on the composition (both biopolymer and monomer).

Table 3.1 Details of various superabsorbent biopolymer composites (SBC) and swelling capacity.

Reference	Composition	Used crosslinker	Swelling ratio[a] (g/g)
[15]	Yeast, sodium alginate, polyvinyl alcohol	Citric acid	32
[16]	Corn starch, polyacrylamide, urea	MBA[b]	215
[17]	Corn starch, sulfanilic acid, polyacrylonitrile	–	250
[18]	Corn starch, acrylic acid, acrylamide, 2-acrylamido-2-methylpropane sulfonic acid	PEGDGE[c]	25
[19]	Carboxymethyl tamarind kernel gum, sodium acrylate	MBA	648
[20]	Starch, acrylic acid, amino sulfonic acid	MBA	1026
[21]	Cassava starch, ZnO, SiO_2, sodium acrylate	MBA	201
[22]	Polyvinylpyrrolidone, starch, polyacrylamide	–	26
[23]	Cassava starch, polyacrylamide	MBA	82
[24]	Starch, acrylic acid, 2-Acrylamido-2-methylpropanesulfonic acid, ferric chloride, ferrous sulfate	MBA	162
[25]	Corn starch, natural char, acrylamide, acrylic acid,	MBA	390
[26]	Potato starch, acrylic acid, acrylamide, polyvinyl alcohol	MBA	921
[27]	Cassava starch, rice husk, Sodium acrylate	MBA	420
[28]	Wheat starch, polyacrylic acid, montmorillonite nanoclay	MBA	400
[29]	Sodium carboxymethyl cellulose, oil shale semi-coke, acrylic acid	MBA	421
[30]	Microcellulose, acrylic acid, sodium periodate, ethylene glycol, urea	MBA	942
[31]	Acrylic acid, urea, sodium dihydrogen phosphate, dibasic sodium phosphate, ammonium polyphosphate	MBA	681
[32]	Polyaryl polymethylene isocyanate, castor oil, urea	–	160

[a] swelling ratio reported in grams per gram of SBC.
[b] N,N′-methylene-bis-acrylamide.
[c] poly(ethylene glycol) diglycidyl ether.

Figure 3.7 shows a coconut fiber-based SBC at dry state (granular form) and swollen state (gel-like consistency), which has a maximum swelling capacity of 342 g/g in distilled water [3]. The surface morphology (as obtained with field emission scanning electron microscope [FESEM]) of the SBC at two different magnifications are also presented in Figure 3.7. It can be observed that the SBC particles consist of several micropores and cavities, which act as water-absorbing sites and support water molecules diffuse into the polymer network. Hence, these SBC are mostly hydrophilic in nature and able to absorb and retain a large amount of water inside their polymer structure, which can improve the water retention properties of soil and reduce the abiotic stresses (such as drought stress) on plants.

3.3 Effect of SBC Addition on Soil Parameters

The effect of SBC application on basic physical properties of soil, such as bulk density, porosity, and pore size distribution, have been studied by previous researchers [6–8]. A detailed summary of relevant literature highlighting the influence of SBC application on the soil properties is presented in Table 3.2.

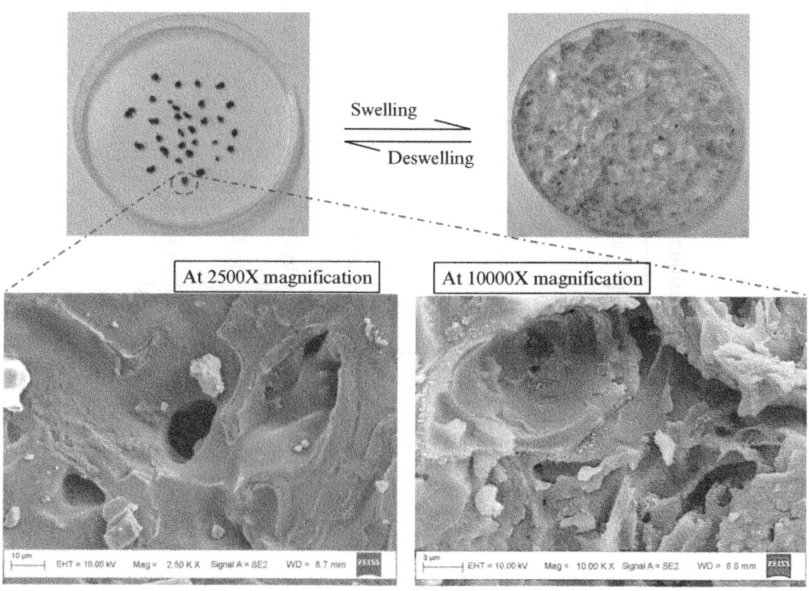

Figure 3.7 Pictorial representation of dry and water swollen SBC along with its surface morphology (obtained with FESEM analysis) at 2500X and 10000X magnification. *Source:* [3] / Springer Nature.

Table 3.2 Summary of relevant literature highlighting the influence of SBC application on the soil properties.

Reference	Soil texture [USDA[a] classification]	SBC composition	SBC concentration(w/w)	Key observations
[8]	Sand Sandy loam Clay loam	Cellulose, polyacylate	0.5%, and 0.7%	• Significant improvement in plant available water content (PAWC) with SBC amendment. • SBC is unsuitable for clay loam as the aeration pores drastically reduced with increasing SBC concentrations.
[34]	Loamy sand	Lignin, corn starch, acrylamide, and acrylic acid	200[b]	• Aggregate stability of soil increased with the application of SBC. • Water holding capacity was increased by 30% with SBC amendment.
[35]	USDA classification not reported	Carboxymethylcellulose sodium salt, hydroxyethylcellulose	0.2%, 0.5%, 1% and 1.5%	• Water retention capacity of both the soils increased with the addition of SBC. • Application of SBC in soil delayed water evaporation rate and more water available to plant for longer duration.
[36]	Sand	Carboxymethylcellulose sodium salt, hydroxyethylcellulose	0.5%, 1%, and 2%	• Soil-moisture content at field capacity (FC) was increased by 60% with 0.5% SBC amendment. • Moisture content at permanent wilting point remains same as bare soil up to an SBC concentration of 1%.
[37]	Sand	Gum Arabic, agar, acrylic acid	1%	• Water holding capacity of sandy soil improved with the addition of hydrophilic SBC. • Water evaporation loss ratio showed that SBC amended soil took more time for complete desaturation.

[7]	Sandy clay loam	Guar gum, Ethylene glycol dimethacrylic acid, acrylic acid	0.1%, 0.2%, and 0.3%	• Saturated water content (at zero matric suction) was increased by 52% in 0.3% SBC amended soils. • Improvement in moisture retention curve was reduced at higher matric suction.
[6]	USDA classification not reported	Starch, urea, acrylic acid	0.25%, and 0.1%	• Bulk density of the used soil was significantly decreased with SBC addition. • Water holding capacity of the soil was increased due to increase in porosity with SBC amendment.
[3]	Silt loam	Coconut fiber, acrylic acid	0.1%, 0.2%, and 0.4%	• Addition of SBC in soil showed an increase in water availability of 125% with 0.4% SBC amendment. • Irrigation water requirement was reduced by 29% as compared to bare soil at an optimum SBC application rate of 0.2%.
[38]	Sandy loam	Rice husk, gelatin, acrylic acid, acrylamide	0.25%, 0.5%, 0.75%, and 1%	• Application of SBC has doubled the time required for complete evaporation with 1% SBC concentrations.

[a] USDA-United States Department of Agriculture.
[b] Concentrations reported in kg/ha.

The bulk density of soil was found to be decreasing with the increasing SBC concentrations due to the higher swelling ability of SBC inside the soil matrix [6]. The swelling of SBC exerts pressure inside the soil matrix, which tends to induce volume changes in the soil samples and an increase in total pore volume of the soil. The decrease in bulk density and increase in the total volume of soil sample (for the same weight of soil) translates to an increase in soil pore space and soil aeration for microbial count [7]. Therefore, porosity of the soil, which is a representation of total pores available inside the soil matrix, also increases due to the SBC amendment [19]. An increase in porosity of soil enhances the water storage capacity followed by nutrient retention. However, the increase in soil porosity with the addition of hydrophilic polymer may depend on the soil texture and particle size distribution [8, 33]. The soil with higher amount of clay particles (negatively charged) can interact with the polymer particles and restrict their swelling potential [9].

Previous studies have revealed that the addition of SBC in soil can improve the water retention properties, which is attributed to the improvement in the soil pore structure and increase in soil porosity [8, 35–37]. The water retention characteristics of soil is quantitatively defined by the soil-water retention curve (SWRC), which a constitutive relationship between the soil suction (negative pore water pressure) and the amount of water stored inside the soil matrix (defined in terms of gravimetric water content, volumetric water content, or degree of saturation) [39, 40]. Soil suction can be defined as the complex energy state due to the interaction of air, water, solid, and/or any other phase present in the soil. It indicates how tightly the soil matrix retains the pore water or how easily the stored water can be extracted from the soil matrix [41]. SWRC of soil can be obtained by simultaneously measuring soil suction and water content of the soil sample during drying (designated as drying SWRC) or wetting (wetting SWRC). The drying SWRC is instrumental for estimation of irrigation water requirement and irrigation scheduling. Drying SWRC can be classified into three different zones: (1) saturation zone, (2) desaturation zone, and (3) residual zone. In the saturation zone, the soil remains saturated with the pore water held under tension as a result of capillary forces and extends up to air entry value (AEV). The AEV is the suction at which air enters the largest pore present in the soil sample during drying [42]. The desaturation zone is in between the AEV and residual point, wherein pore air replaces pore water within the pores of the soil [43]. Soil pore-water is retained in this zone in the form of thin films on the particle surfaces under the influence of the short-range solid–liquid interaction mechanism. The residual zone starts from the residual point and is terminated at oven-dry conditions (i.e., water content equal to zero). An increase in suction in this zone does not result in significant changes in water content. The saturation and desaturation zone are the most important phase in the SWRC for agricultural application. The saturated water content (θ_s), which denotes the water content at near zero suction (or saturated state), is an important parameter to obtain the soil-water storage. On the other hand, field capacity (denotes the water content held inside the soil matrix after drainage of excess water under gravitational force) and permanent wilting point (denotes the minimum water content below which the plant species wilt permanently and fail to recover after irrigation) are important for irrigation scheduling and estimation of irrigation water requirements [9, 44].

The addition of crosslinked SBC can significantly influence the field capacity (FC) of soil by accommodating the deep draining water inside the polymer network. Montesano et al. [36], have reported that the application of a cellulose-based SBC increased the field capacity of coarse-grained sandy soil up to 400% with SBC concentration of 2%. Narjary et al. [8] reported that the huge increase in FC could be a result of reduction in pore diameters and the number of large pore channels as the SBC amendment act as pore blocking agent inside the soil matrix. As coarse-grained soils have large pore diameters as compared to fine-grained soil, the addition of hydrophilic polymers is more effective for coarse-grained soils [45]. For fine-grained soils, SBC particles cannot achieve their swelling equilibrium inside the soil matrix due to the smaller pore diameters and ionic nature of the clay particles, which resulted in a lesser improvement in FC. The difference between FC and permanent wilting point (PWP) is known as plant available water content (PAWC), which indicates the amount of water available for plant roots.

Past studies have shown that the SBC addition improves the PAWC for coarse-textured soils due to the improvement in FC [3, 8]. However, the addition of SBC also increases the PWP of soil, which suggests that plant roots cannot extract all the stored water in SBC amended soil [3, 36]. The water inside the soil and polymer matrix are so tightly bonded that the plant roots are not able to extract it. Therefore, there exists a range of SBC amendment rates beyond which the increase in water retention of soil does not result in higher water availability to plants. Saha et al. [3] have reported an optimum application rate of 0.2% (w/w) for a natural fiber-based SBC in silt loam soil based on the irrigation water requirement. There are not many studies reported in the literature that explore the optimum SBC application rate for different texture soils under field conditions. Efforts should be made to develop proper guidelines for SBC applicate rate based on the soil texture to optimize the usage of SBC in the agricultural field.

3.4 Effect of SBC Addition on Plant Properties

The effect of SBC application on plant properties has been extensively studied in the literature [27, 46–49]. A brief overview of the literature highlighting the influence of SBC applications on plant parameters are presented in Table 3.3. The addition of SBC in soil improves the plant growth considerably under normal irrigation and deficit irrigation conditions due to the improvement in water and nutrient holding capacity of soil and reduction in leaching of nutrients. Moreover, the incorporation of SBC significantly improved the seed germination rate, root length, and germination index [34, 50]. Rashad et al. [47] studied the effect of starch-gelatin-rice husk-acrylic acid-based SBC application on the photosynthetic pigments (chlorophyll a, chlorophyll b, and carotenoids) of sweet pepper (*Capsicum annuum*) plant. According to their results, the total chlorophyll content of the plant was improved by five times at the same irrigation water level due to the addition of SBC. The addition of the SBC also improved the leaf area index (LAI), number of leaves per plant, and plant height with 1% SBC concentrations. This was due to the low nutrient leaching in SBC amended soil and slow (optimal) release of water and nutrient from SBC for a prolonged duration.

Another important advantage of SBC for agricultural application is its effectiveness in minimizing the effect of water stress conditions on plant health and crop yield during drought. It is a well-known fact that long-term water stress conditions have serious consequences on plant health. Mazloom et al. [5] reported that the accumulation of proline, a water stress biomarker in plants, could reach a level of 70 times higher than the normal conditions. The biomarker prevents dehydration of plant cell membranes and reduces the relative water content of plant leaves. The reduction of relative water content, which is the balance between water absorption and transpiration from leaves, causes serious damage to plant leaves and restricts plant growth. Moreover, the lower water content in soil restricts the plant uptake of nutrients such as phosphorous (P), nitrogen (N), potassium (K), which are essential for plant growth. Past studies have highlighted that the addition of SBC in soil can alleviate the negative influence of drought and prolong the survival of plant species under water stress conditions [5, 36, 47]. It can also reduce the evaporation loss from the soil, which is an added advantage for increasing water availability to plants under drought conditions. The absorbed water inside the SBC structure is tightly bonded, which results in a controlled release of water to dry surface soil. However, it is not clear from the previous studies whether all the stored water inside the SBC network is available for plant root uptake. There are not many studies that

Table 3.3 Brief overview of the literature highlighting the influence of SBC applications on plant parameters.

References	Plant species	SBC composition	SBC concentrations	Key conclusions
[46]	*Raphanus sativus* L.	Starch, aluminum sulfate octadecahydrate	0.12%, 0.24%	• Plant height and biomass was increased with increasing SBC concentrations. • Plants in bare soil were showed signs of dehydration with moisture evaporation.
[36]	*Raphanus sativus* L. *Cucumis sativus* L.	Carboxymethylcellulose sodium salt, hydroxyethylcellulose	0.5%, 1%, and 2%	• Germination index of seeds did not show any symptoms of phytotoxicity in presence of SBC. • Results confirmed the positive effect of SBC on plant growth and reduction in detrimental effects of water stress on plant health.
[48]	*Eucalyptus urophylla* *Eucalyptus Grandis*	Cellulose acetate, ethylenediaminetetraacetic dianhydride	Not reported	• NPK concentrations in leaves were increased with the addition of SBC as it can release the NPK for plant root uptake. • Lower seedling mortality under drought conditions in SBC amended soil as compared to commercial polymer amended soil.
[5]	*Zea mays* L.	Lignin, poly ethylene glycol diglycidyl ether	0.3%, and 0.6%	• Plant height and biomass was improved with SBC addition in deficit irrigation conditions (90% FC, 65% FC, 45% FC). • Under extreme drought conditions, nutrient uptake of phosphorus (P) was increased with SBC amendment.
[27]	*Cucumis melo* L.	Rice husk, starch, sodium acrylate	0.25%, 0.5%, 0.75%, and 1%	• Plant height, stem diameter was increased up to an SBC concentration of 0.75% and thereafter decreased with increasing SBC concentrations.

(Continued)

Table 3.3 (Continued)

References	Plant species	SBC composition	SBC concentrations	Key conclusions
[50]	*Zea mays* L. *Cucumis sativus* L. *Abelmoschus esculentus* L.	Wheat straw (cellulose), acrylic acid	6.12%	• Average number of seeds germination was increased in presence of SBC. • Root length of the germinated seed was found to be higher with SBC. • Application of SBC is soil improved the shoot height, stem diameter, and total plant leaf area.
[38]	*Arachis hypogaea* L.	Rice husk, gelatin, acrylic acid, acrylamide	15, 30, 60, and 90[a]	• Plant growth rate (both root and shoot growth) was increased with 60 kg/ha SBC application rate under water stress condition. • Significant decrease in the photosynthetic pigment content (chl a, chl b, carotenoids) in leaves due to water stress. • Noticeable increase in pigment content with the SBC amendment.
[47]	*Capsicum annuum* L.	Potato starch, gelatin, acrylic acid, urea, rice husk, 2-hydroxyethyl methacrylate	0.25%, 0.50%, 0.75%, and 1%	• Leaf area index (LAI), number of leaves per plant, and plant height was increased with 1% SBC amendment under deficit irrigation conditions. • Application of SBC save significant amount of irrigation water without affecting the crop yield.

[a] Concentration reported in kg/ha.

validated the existing root water uptake model for SBC amended soils. Such studies are required to evaluate the efficiency of SBC under water stress conditions.

3.5 Controlled Nutrient Release from SBC

Controlled nutrient release from SBC has been widely studied in the literature as it can improve the nutrient use efficiency and decrease the cost of the fertilizer application, toxicity for humans and the environment. The important and essential plant macro- and micronutrients are potassium, nitrogen, phosphorus, iron, calcium, copper, sulfur, and boron. Wu and Liu [51] reported that about 40% to 70% of nitrogen and 50% to 70% of potassium of the applied fertilizers are lost due to leaching and are not available for plant roots. Therefore, the engineered and controlled nutrient release system is an effective approach to reduce nutrient loss after application and minimize the environmental problems associated with the leaching of chemical fertilizers. Recent research has focused on the development of fertilizer-coated SBC particles (granular form) as a controlled carrier of fertilizer for agricultural applications.

Table 3.4 compares the fertilizer release capacity of different SBC with varying monomer and biopolymer compositions. It can be observed that the developed SBC has an excellent slow-release property of nutrients from its polymer network. The water molecules enter the SBC and expand the polymer network. This allows the release of nutrients through the developed channels inside the polymer structure [52]. It may be noted that the nutrient coated with the SBC could not travel from the polymer network to the external environment in the absence of water. Hence, soil moisture plays a vital role in the controlled release of nutrients from SBC. Moreover, the release of nutrients can be controlled by incorporating inorganic materials inside the polymer network. The incorporation of clay minerals (such as montmorillonite, kaolinite, attapulgite) provides tortuous pathways, which restrict the diffusion of nutrients outside the SBC network. Kenawy et al. [53] reported a decreased rate of urea release with the increasing percentage of attapulgite mineral in SBC network. The nutrient release can also be affected by the soil salinity and pH. The nutrient release capacity increases with the increase in soil salinity [52]. On the other hand, the effect of pH on nutrient release is dependent on the nutrient type. Acidic soil would enhance the release of nitrogen, whereas alkaline soil would enhance the release of phosphorous fertilizer [52].

Table 3.4 Comparison of fertilizer release amount and biodegradability of superabsorbent biopolymer composite (SBC) in soil.

SBC composition	Fertilizer release amount in soil (%)	Biodegradability by % weight loss	References
Chitosan, acrylic acid, acrylamide, NPK fertilizer	79% N (after 30 days) 62% P (after 30 days) 69% K (after 30 days)	–	[51]
Nitrogen and phosphorus coated starch, acrylic acid, and acrylamide	45% N (after 15 days) 55% P (after 15 days)	50% (after 50 days)	[54]
Maize and cassava starch, polyacrylamide, urea	70% N (after 96 hours)	–	[55]
Cellulose acetate, ethylene-diamine-tetraacetic dianhydride	–	55% (after 120 days)	[48]
Yeast, sodium alginate, polyvinyl alcohol	–	47% (after 120 days)	[15]
Sodium alginate, rice husk, acrylamide, acrylic acid, urea, potassium dihydrogen phosphate, ammonium dihydrogen phosphate	71% NPK (after 30 days)	–	[56]
Corncob, bentonite, urea, acrylic acid	44.5% N (after 30 days)	–	[57]
Maize starch, polyacrylamide, mono-ammoniumdihydrogen phosphate, di-ammonium hydrogen phosphate	85% P (after 30 days)	–	[58]
Sodium alginate, biochar, acrylic acid, acrylamide	95% N (after 30 days) 83% P (after 30 days) 72% K (after 30 days)	–	[59]
Maize straw and rice straw-biochar, urea, polyvinyl alcohol	65.28% N (after 22 days)	33.88% (after 120 days)	[60]
Maize bran, acrylic acid, acrylamide, urea, potassium dihydrogen phosphate, ammonium dihydrogen phosphate	60% NPK/Fe (after 30 days)	–	[61]

Sugarcane bagasse, polyacrylamide, attapulgite	98% urea (after 20 days)	–	[53]
Chicken feather protein, potassium polyacrylate, polyvinyl alcohol	69.46% N (after 30 days) 65.23% P (after 30 days)	60% (after 120 days)	[52]
Starch, urea, acrylic acid	90% urea (after 30 days)	40% (after 90 days)	[6]
Lignin, urea, attapulgite, acrylic acid	71% urea (after 28 days)	35% (after 40 days)	[62]
Microcellulose, acrylic acid, sodium periodate, ethylene glycol, urea	90% urea (after 30 days)	39% (after 40 days)	[30]
Rice husk, gelatin, acrylic acid, acrylamide	–	60% (after 160 days)	[38]
Acrylic acid, urea, sodium dihydrogen phosphate, dibasic sodium phosphate, ammonium polyphosphate	76% N (after 30 days) 55% P (after 30 days)	–	[31]

3.6 Biodegradability of SBC

Biodegradation and ecocompatibility of SBC are important factors to be considered before its application in the agricultural and horticultural application as the long-term retention of SBC in soil may cause environmental concerns. Researchers have conducted biodegradation studies of SBC in two different mediums, including broth culture (with various bacterial strains) and field soil [6, 63]. For the evaluation of degradation kinetics in broth culture, a nutrient medium was prepared to culture different bacterial strains (*Pseudomonas aeruginosa* and *Bacillus subtilis*) [6]. The SBC particles were added to the nutrient medium, and the extent of biodegradation was monitored from the bacterial growth on the polymer samples. The weight loss of the SBC samples during the degradation period was measured to quantify the degradation kinetics. On the other hand, the degradation kinetics in field soil was evaluated using the soil burial method [7, 15, 30, 38, 54, 60]. In this method, a certain quantity of SBC samples was placed in a nylon bag and buried inside the soil matrix for different time durations. The degradation kinetics of SBC was measured from the weight loss of SBC particles at different time intervals. The soil burial method is more popular for the evaluation of SBC degradation kinetics as it is more representative of actual field conditions.

The degradation kinetics of SBC (measured using soil burial method) with different compositions are summarized in Table 3.4. It is quite explicit that the SBC has an excellent biodegradability inside soil matrix and the half-life of most of the SBCs was less than four months. Sarmah and Karak [6] have compared the degradation kinetics of SBC with a synthetic polymer and reported that the inclusion of natural materials, such as biopolymers (which are more susceptible to soil microorganisms), inside the polymer structure increase the degradability rate of SBC. However, the degradability of SBC is also dependent on the various soil parameters, such as soil moisture content, soil texture, soil salinity, native bacterial population, and the type of chemical fertilizers used. Soil texture and presence of chemical fertilizer are important aspects to be considered as the interaction of SBC with ionic solvent can weaken polymer chains and crosslinking structures of SBC, leading to a higher degradation rate. There are not many studies reported in the literature that explored the influence of soil-specific parameters on the degradation kinetics of SBC. Such study will be helpful to understand the life span of SBC in different field conditions and provide an accurate application frequency of SBC based on the soil properties of agricultural fields.

3.7 Summary and Future Perspectives

This chapter presents a review on the role of biopolymers and their composite for sustainable agriculture, considering the individual properties of biopolymer composite (such as compositions, swelling characteristics, biodegradability), soil (soil density, soil porosity, water retention characteristics, field capacity, and infiltration characteristics), plant (plant height, root length, biomass, stomatal conductance, and photosynthetic yield), and highlighting the interdependency among those properties. The review indicates that superabsorbent biopolymer composite (SBC), a new generation of smart and biodegradable crosslinked polymer networks, are able to absorb and retain a large quantity of water within its structure. These hydrophilic polymer composites can improve the soil-water retention capacity, reduce nutrient leaching, and conserve moisture in the plant root zone. However, the improvement in soil properties with SBC amendment was found to be dependent on soil texture and its pore structure. There are ambiguities in the literature related to the application rate of SBC, as a higher dose of SBC may not necessarily improve soil and plant properties. It is not clear from the previous studies whether all the stored water in SBC amended soil is available for plant root uptake. The retained water is trapped inside the SBC network with strong molecular bonds and hence may not be entirely extractable by plant roots with higher SBC concentrations. The controlled release of nutrients from SBC is an added advantage for its application in agriculture and to increase nutrient use efficiency. It was observed that the fertilizer-coated SBC could slowly release the essential plant nutrients from the polymer network for a prolonged duration to reduce the leaching of nutrients and excess application of fertilizer. From an ecological point of view, these SBC are biodegradable (with half-life of less than four months) and produce no harmful residual products after degradation.

Figure 3.8 summarizes the basic framework for future directions of research to achieve sustainability in agricultural and food production systems utilizing the SBC. A complete understanding of solid waste management systems is necessary to identify and isolate the suitable waste and natural biopolymers for their utilization as raw materials. These raw materials should be biorenewable, ecofriendly, and economically feasible for the production of SBC through chemical modifications. Additional efforts need to be carried out to convert various agricultural wastes to SBC using low energy-intensive processes. This process will not only contribute towards the waste valorization but also prevent environmental pollution resulting from the burning of agricultural waste. The production of crosslinked SBC

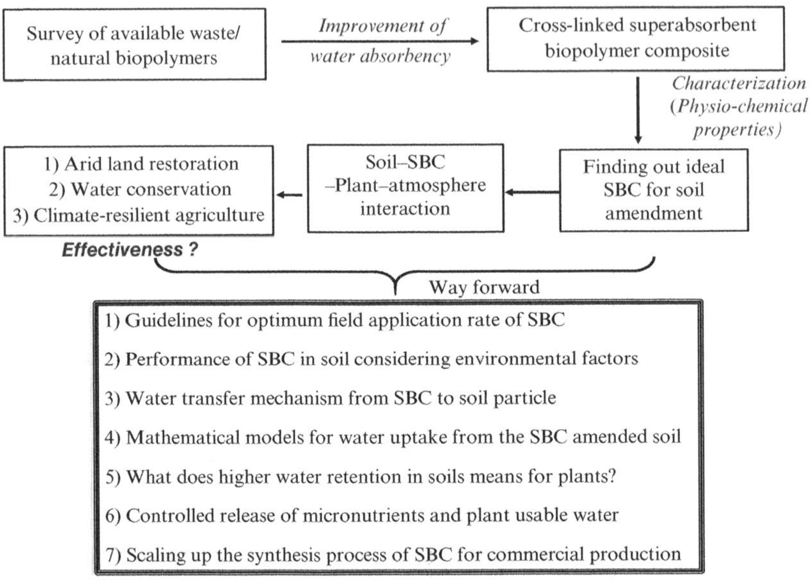

Figure 3.8 Future direction to achieve sustainability in agriculture utilizing superabsorbent biopolymer composite (SBC).

through graft polymerization has tremendously advanced over the past decade, and various waste biopolymers have been transformed into high water-absorbing SBC. However, the production of SBC is still in the laboratory stage despite its excellent biocompatibility and low production cost. Therefore, more intensive research is required to scale up the synthesis process for mass commercial production.

The role of SBC for achieving sustainability in agricultural systems needs to be thoroughly investigated through interdisciplinary approaches, considering the soil–SBC–plant–atmosphere interaction. This can be achieved through the utilization of SBC in agricultural land (with varying soil texture) under real climatic conditions with different plant species. The root water uptake and transport are sensitive to both climatic conditions and plant species. The soil-water dynamics in the plant root zone with the presence of SBC is not well understood in the past studies, and a holistic study is required to understand the moisture movement pathway from swollen polymer to plant roots (whether plant roots can directly uptake the stored water or the stored water transferred to soil for plant root uptake). This opens up an interesting area of interdisciplinary research involving geotechnical, chemical, and biotechnological experts. A wide range of non-intrusive techniques needs to be developed and explored to quantify the water movement mechanism in the soil–SBC–plant continuum.

Desertification of agricultural land is one of the major challenges for humanity, and sustainable agriculture is the best possible way to address this crisis and fulfilling sustainable development goals. The concept of sustainable agriculture was originally developed to improve the production of food and fiber to meet the requirement of the growing world population with the maximum use of natural and renewable resources. These approaches also need to be extended for ecological restoration of arid, deserted, and degraded land. Well-planned research initiatives are required to evaluate the effectiveness of various biopolymer composites for the restoration of deserted land through soil and water conservation, nutrient cycling, and microbial formulation. In addition, the life cycle assessment of the biomaterials in soil matrix is one of the key aspects to assess their environmental impacts but is largely neglected in most of the studies. Environmental impact assessment of agricultural system should adopt a broader perspective, considering the negative influence of fertilizers, pesticides (even in the case of biobased materials), and ruminate the effects of agricultural practices on the soil environment. This will pave the way for scientific implementation of new generation green materials for the restoration of arid and polluted land, drought management, and sustainable agriculture thereby contributing towards the fulfilment of various sustainable development goals laid out by the United Nations.

References

1 Saha, A., Sreedeep, S., and Manna, U. (2020). Superabsorbent hydrogel (SAH) as a soil amendment for drought management: a review. *Soil and Tillage Research* 204: 104736. doi: 10.1016/j.still.2020.104736.

2 Sreedeep, S., Gadi, V.K., Bordoloi, S. et al. (2019). Sustainable geotechnics: a bio-geotechnical perspective. In: *Frontiers in Geotechnical Engineering* (ed. G.M. Latha), 313–331. Singapore: Springer. doi: 10.1007/978-981-13-5871-5_15.

3 Saha, A., Gupt, C.B., and Sreedeep, S. (2021). Recycling natural fiber to superabsorbent hydrogel composite for conservation of irrigation water in semi-arid regions. *Waste and Biomass Valorization* 1–16. doi: 10.1007/s12649-021-01489-9.

4 Nassaj-Bokharaei, S., Motesharezedeh, B., Etesami, H., and Motamedi, E. (2021). Effect of hydrogel composite reinforced with natural char nanoparticles on improvement of soil biological properties and the growth of water deficit-stressed tomato plant. *Ecotoxicology and Environmental Safety* 223: 112576. doi: 10.1016/j.ecoenv.2021.112576.

5 Mazloom, N., Khorassani, R., Zohury, G.H. et al. (2020). Lignin-based hydrogel alleviates drought stress in maize. *Environmental and Experimental Botany* 175: 104055. doi: 10.1016/j.envexpbot.2020.104055.

6 Sarmah, D. and Karak, N. (2020). Biodegradable superabsorbent hydrogel for water holding in soil and controlled-release fertilizer. *Journal of Applied Polymer Science* 137 (13): 48495. doi: 10.1002/app.48495.

7 Thombare, N., Mishra, S., Siddiqui, M.Z., et al. (2018). Design and development of guar gum-based novel, superabsorbent and moisture retaining hydrogels for agricultural applications. *Carbohydrate Polymers* 185: 169–178. doi: 10.1016/j.carbpol.2018.01.018.

8 Narjary, B., Aggarwal, P., Singh, A. et al. (2012). Water availability in different soils in relation to hydrogel application. *Geoderma* 187: 94–101. doi: 10.1016/j.geoderma.2012.03.002.

9 Saha, A., Rattan, B., Sreedeep, S., and Manna, U. (2020). Quantifying the interactive effect of water absorbing polymer (WAP)-soil texture on plant available water content and irrigation frequency. *Geoderma* 368: 114310. doi: 10.1016/j.geoderma.2020.114310.

10 Yang, L., Yang, Y., Chen, Z. et al. (2014). Influence of super absorbent polymer on soil water retention, seed germination and plant survivals for rocky slopes eco-engineering. *Ecological Engineering* 62: 27–32. doi: 10.1016/j.ecoleng.2013.10.019.

11 George, A., Sanjay, M.R., Srisuk, R. et al. (2020). A comprehensive review on chemical properties and applications of biopolymers and their composites. *International Journal of Biological Macromolecules* 154: 329–338. doi: 10.1016/j.ijbiomac.2020.03.120.

12 Raj, S.N., Lavanya, S.N., Sudisha, J., and Shetty, H.S. (2011). Applications of biopolymers in agriculture with special reference to role of plant derived biopolymers in crop protection. In: *Biopolymers: Biomédical and Environmental Applications* (eds S. Kalia and L. Avérous), 461–481. Hoboken: Wiley.

13 Reddy, M., Ponnamma, D., Choudhary, R., and Sadasivuni, K.K. (2021). A comparative review of natural and synthetic biopolymer composite scaffolds. *Polymers* 13 (7): 1105. doi: 10.3390/polym13071105.

14 Xu, X., Bai, B., Ding, C. et al. (2015). Synthesis and properties of an ecofriendly superabsorbent composite by grafting the poly (acrylic acid) onto the surface of dopamine-coated sea buckthorn branches. *Industrial & Engineering Chemistry Research* 54 (13): 3268–3278. doi: 10.1021/acs.iecr.5b00092.

15 Feng, D., Bai, B., Wang, H., and Suo, Y. (2017). Novel fabrication of biodegradable superabsorbent microspheres with diffusion barrier through thermo-chemical modification and their potential agriculture applications for water holding and sustained release of fertilizer. *Journal of Agricultural and Food Chemistry* 65 (29): 5896–5907. doi: 10.1021/acs.jafc.7b01849.

16 Xiao, X., Yu, L., Xie, F. et al. (2017). One-step method to prepare starch-based superabsorbent polymer for slow release of fertilizer. *Chemical Engineering Journal* 309: 607–616. doi: 10.1016/j.cej.2016.10.101.

17 Zain, G., Nada, A.A., El-Sheikh, M.A. et al. (2018). Superabsorbent hydrogel based on sulfonated-starch for improving water and saline absorbency. *International Journal of Biological Macromolecules* 115: 61–68. doi: 10.1016/j.ijbiomac.2018.04.032.

18 Amiri, F., Kabiri, K., Bouhendi, H. et al. (2019). High gel-strength hybrid hydrogels based on modified starch through surface cross-linking technique. *Polymer Bulletin* 76 (8): 4047–4068. doi: 10.1007/s00289-018-2593-6.

19 Warkar, S.G. and Kumar, A. (2019). Synthesis and assessment of carboxymethyl tamarind kernel gum based novel superabsorbent hydrogels for agricultural applications. *Polymer* 182: 121823. doi: 10.1016/j.polymer.2019.121823.

20 Zhao, C., Zhang, M., Liu, Z. et al. (2019). Salt-tolerant superabsorbent polymer with high capacity of water-nutrient retention derived from sulfamic acid-modified starch. *ACS Omega* 4 (3): 5923–5930. doi: 10.1021/acsomega.9b00486.

21 Dispat, N., Poompradub, S., and Kiatkamjornwong, S. (2020). Synthesis of ZnO/SiO2-modified starch-graft-polyacrylate superabsorbent polymer for agricultural application. *Carbohydrate Polymers* 249: 116862. doi: 10.1016/j.carbpol.2020.116862.

22 Ghobashy, M.M., Abd El-Wahab, H., Ismail, M.A. et al. (2020). Characterization of Starch-based three components of gamma-ray cross-linked hydrogels to be used as a soil conditioner. *Materials Science and Engineering: B* 260: 114645. doi: 10.1016/j.mseb.2020.114645.

23 Junlapong, K., Maijan, P., Chaibundit, C., and Chantarak, S. (2020). Effective adsorption of methylene blue by biodegradable superabsorbent cassava starch-based hydrogel. *International Journal of Biological Macromolecules* 158: 258–264. doi: 10.1016/j.ijbiomac.2020.04.247.

24 Moharrami, P. and Motamedi, E. (2020). Application of cellulose nanocrystals prepared from agricultural wastes for synthesis of starch-based hydrogel nanocomposites: efficient and selective nanoadsorbent for removal of cationic dyes from water. *Bioresource Technology* 313: 123661. doi: 10.1016/j.biortech.2020.123661.

25 Motamedi, E., Motesharezedeh, B., Shirinfekr, A., and Samar, S.M. (2020). Synthesis and swelling behavior of environmentally friendly starch-based superabsorbent hydrogels reinforced with natural char nano/micro particles. *Journal of Environmental Chemical Engineering* 8 (1): 103583. doi: 10.1016/j.jece.2019.103583.

26 Olad, A., Doustdar, F., and Gharekhani, H. (2020). Fabrication and characterization of a starch-based superabsorbent hydrogel composite reinforced with cellulose nanocrystals from potato peel waste. *Colloids and*

Surfaces A: Physicochemical and Engineering Aspects 601: 124962. doi: 10.1016/j.colsurfa.2020.124962.

27 de Vasconcelos, M.C., Gomes, R.F., Sousa, A.A. et al. (2020). Superabsorbent hydrogel composite based on starch/rice husk ash as a soil conditioner in melon (*Cucumis melo* L.) seedling culture. *Journal of Polymers and the Environment* 28 (1): 131–140. doi: 10.1007/s10924-019-01593-x.

28 Baloch, F.E., Afzali, D., and Fathirad, F. (2021). Design of acrylic acid/nanoclay grafted polysaccharide hydrogels as superabsorbent for controlled release of chlorpyrifos. *Applied Clay Science* 211: 106194. doi: 10.1016/j.clay.2021.106194.

29 Liu, Y., Zhu, Y., Mu, B. et al. (2021). Synthesis, characterization, and swelling behaviors of sodium carboxymethyl cellulose-g-poly (acrylic acid)/semi-coke superabsorbent. *Polymer Bulletin* 1–19. doi: 10.1007/s00289-021-03545-9.

30 Mohammadbagheri, Z., Rahmati, A., and Hoshyarmanesh, P. (2021). Synthesis of a novel superabsorbent with slow-release urea fertilizer using modified cellulose as a grafting agent and flexible copolymer. *International Journal of Biological Macromolecules*. doi: 10.1016/j.ijbiomac.2021.05.191.

31 Wang, W., Yang, Z., Zhang, A., and Yang, S. (2021). Water retention and fertilizer slow release integrated superabsorbent synthesized from millet straw and applied in agriculture. *Industrial Crops and Products* 160: 113126. doi: 10.1016/j.indcrop.2020.113126.

32 Zhao, C., Tian, H., Zhang, Q. et al. (2021). Preparation of urea-containing starch-castor oil superabsorbent polyurethane coated urea and investigation of controlled nitrogen release. *Carbohydrate Polymers* 253: 117240. doi: 10.1016/j.carbpol.2020.117240.

33 Hussien, R.A., Donia, A.M., Atia, A.A. et al. (2012). Studying some hydro-physical properties of two soils amended with kaolinite-modified cross-linked poly-acrylamides. *Catena* 92: 172–178. doi: 10.1016/j.catena.2011.12.010.

34 Lee, S.S., Chang, S.X., Chang, Y.Y., and Ok, Y.S. (2013). Commercial versus synthesized polymers for soil erosion control and growth of Chinese cabbage. *SpringerPlus* 2 (1): 1–10.

35 Cannazza, G., Cataldo, A., De Benedetto, E. et al. (2014). Experimental assessment of the use of a novel superabsorbent polymer (SAP) for the optimization of water consumption in agricultural irrigation process. *Water* 6 (7): 2056–2069. doi: 10.3390/w6072056.

36 Montesano, F.F., Parente, A., Santamaria, P. et al. (2015). Biodegradable superabsorbent hydrogel increaseswater retention properties of growing media and plant growth. *Agriculture and Agricultural Science Procedia* 4: 451–458. doi: 10.1016/j.aaspro.2015.03.052.

37 Hasija, V., Sharma, K., Kumar, V. et al. (2018). Green synthesis of agar/gum arabic based superabsorbent as an alternative for irrigation in agriculture. *Vacuum* 157: 458–464. doi: 10.1016/j.vacuum.2018.09.012.

38 Kenawy, E.R., Seggiani, M., Hosny, A. et al. (2021). Superabsorbent composites based on rice husk for agricultural applications: swelling behavior, biodegradability in soil and drought alleviation. *Journal of Saudi Chemical Society* 25 (6): 101254. doi: 10.1016/j.jscs.2021.101254.

39 Fredlund, D.G. and Rahardjo, H. (1993). *Soil Mechanics for Unsaturated Soils*. New York: John Wiley & Sons, Inc.

40 Saha, A. and Sreedeep, S. (2021). Estimation of unsaturated hydraulic conductivity function: implication of low to high suction measurements. *Acta Geophysica* 69 (2): 547–559. doi: 10.1007/s11600-021-00548-y.

41 Saha, A., Sreedeep, S., and Manna, U. (2020). Evaluation of capacitance sensor for suction measurement in silty clay loam. *Geotechnical and Geological Engineering* 38 (4): 4319–4331. doi: 10.1007/s10706-020-01297-3.

42 Brooks, R.H. and Corey, A.T. (1964). *Hydraulic properties of porous media*. Hydrology Paper No. 3. Civil Engineering Department, Colorado State University, Fort Collins, CO.

43 Saha, A. and Sreedeep, S. (2021). Importance of volumetric shrinkage curve (VSC) for determination of soil–water retention curve (SWRC) for low plastic natural soils. *Journal of Hydrology* 596: 126113. doi: 10.1016/j.jhydrol.2021.126113.

44 Saha, A., Rattan, B., Sreedeep, S., and Manna, U. (2020). Effect of water absorbing polymer amendment on water retention properties of cohesionless soil. In: *Advances in Computer Methods and Geomechanics* (ed. A. Prashant, A. Sachan, and C.S. Desai), 185–195. Singapore: Springer. doi: 10.1007/978-981-15-0886-8_15.

45 Saha, A., Sreedeep, S., Manna, U., and Sahoo, L. (2020). Transformation of non-water sorbing fly ash to a water sorbing material for drought management. *Scientific Reports* 10 (1): 1–16. doi: 10.1038/s41598-020-75674-6.

46 Nnadi, F. and Brave, C. (2011). Environmentally friendly superabsorbent polymers for water conservation in agricultural lands. *Journal of Soil Science and Environmental Management* 2 (7): 206–211. doi: 10.5897/JSSEM.9000040.

47 Rashad, M., Kenawy, E.R., Hosny, A. et al. (2021). An environmental friendly superabsorbent composite based on rice husk as soil amendment to improve plant growth and water productivity under deficit irrigation conditions. *Journal of Plant Nutrition* 44 (7): 1010–1022. doi: 10.1080/01904167.2020.1849293.

48 Senna, A.M. and Botaro, V.R. (2017). Biodegradable hydrogel derived from cellulose acetate and EDTA as a reduction substrate of leaching NPK compound fertilizer and water retention in soil. *Journal of Controlled Release* 260: 194–201. doi: 10.1016/j.jconrel.2017.06.009.

49 Montesano, F.F., Parente, A., Santamaria, P., et al. (2015). Biodegradable superabsorbent hydrogel increases water retention properties of growing media and plant growth. *Agriculture and Agricultural Science Procedia* 4: 451–458. doi: 10.1016/j.aaspro.2015.03.052.

50 Sasmal, P.K. and Patra, S. (2020). Effect in growth of corn plant from cellulose-based hydrogel derived from wheat straw. *Journal of the Institution of Engineers (India): Series E* 1–6. doi: 10.1007/s40034-020-00180-3.

51 Wu, L. and Liu, M. (2008). Preparation and properties of chitosan-coated NPK compound fertilizer with controlled-release and water-retention. *Carbohydrate Polymers* 72 (2): 240–247. doi: 10.1016/j.carbpol.2007.08.020.

52 Kong, W., Li, Q., Li, X. et al. (2019). A biodegradable biomass-based polymeric composite for slow release and water retention. *Journal of Environmental Management* 230: 190–198. doi: 10.1016/j.jenvman.2018.09.086.

53 Kenawy, E.R., Seggiani, M., Cinelli, P. et al. (2020). Swelling capacity of sugarcane bagasse-g-poly (acrylamide)/attapulgite superabsorbent composites and their application as slow release fertilizer. *European Polymer Journal* 133: 109769. doi: 10.1016/j.eurpolymj.2020.109769.

54 Jin, S., Wang, Y., He, J. et al. (2013). Preparation and properties of a degradable interpenetrating polymer networks based on starch with water retention, amelioration of soil, and slow release of nitrogen and phosphorus fertilizer. *Journal of Applied Polymer Science* 128 (1): 407–415. doi: 10.1002/app.38162.

55 Qiao, D., Liu, H., Yu, L. et al. (2016). Preparation and characterization of slow-release fertilizer encapsulated by starch-based superabsorbent polymer. *Carbohydrate Polymers* 147: 146–154. doi: 10.1016/j.carbpol.2016.04.010.

56 Olad, A., Gharekhani, H., Mirmohseni, A., and Bybordi, A. (2017). Synthesis, characterization, and fertilizer release study of the salt and pH-sensitive NaAlg-g-poly (AA-co-Aam)/RHA superabsorbent nanocomposite. *Polymer Bulletin* 74 (8): 3353–3377. doi: 10.1007/s00289-016-1899-5.

57 Wen, P., Han, Y., Wu, Z. et al. (2017). Rapid synthesis of a corncob-based semi-interpenetrating polymer network slow-release nitrogen fertilizer by microwave irradiation to control water and nutrient losses. *Arabian Journal of Chemistry* 10 (7): 922–934. doi: 10.1016/j.arabjc.2017.03.002.

58 Alharbi, K., Ghoneim, A., Ebid, A. et al. (2018). Controlled release of phosphorous fertilizer bound to carboxymethyl starch-g-polyacrylamide and maintaining a hydration level for the plant. *International Journal of*

Biological Macromolecules 116: 224–231. doi: 10.1016/j.ijbiomac.2018.04.182.

59 Baki, M. and Abedi-Koupai, J. (2018). Preparation and characterization of a superabsorbent slow-release fertilizer with sodium alginate and biochar. *Journal of Applied Polymer Science* 135 (10): 45966. doi: 10.1002/app.45966.

60 Chen, S., Yang, M., Ba, C. et al. (2018). Preparation and characterization of slow-release fertilizer encapsulated by biochar-based waterborne copolymers. *Science of the Total Environment* 615: 431–437. doi: 10.1016/j.scitotenv.2017.09.209.

61 Gharekhani, H., Olad, A., and Hosseinzadeh, F. (2018). Iron/NPK agrochemical formulation from superabsorbent nanocomposite based on maize bran and montmorillonite with functions of water uptake and slow-release fertilizer. *New Journal of Chemistry* 42 (16): 13899–13914. doi: 10.1039/C8NJ01947A.

62 Chen, Y., Li, W., and Zhang, S. (2021). A multifunctional eco-friendly fertilizer used keratin-based superabsorbent as coatings for slow-release urea and remediation of contaminated soil. *Progress in Organic Coatings* 154: 106158. doi: 10.1016/j.porgcoat.2021.106158.

63 Tanan, W., Panichpakdee, J., and Saengsuwan, S. (2019). Novel biodegradable hydrogel based on natural polymers: synthesis, characterization, swelling/reswelling and biodegradability. *European Polymer Journal* 112: 678–687. doi: 10.1016/j.eurpolymj.2018.10.033.

4

Biopolymers in Bioengineering and Medical Technology

*Maya Rajan, Shahena S., Vinaya Chandran, and Linu Mathew**

School of Biosciences, Mahatma Gandhi University, Kottayam, Kerala
** Corresponding author*

4.1 Introduction

Biopolymers (Figure 4.1) are derived naturally from the cells of living organisms such as fungi, yeast, plants, and bacteria. They are the polymers of organic units like peptides, sugar moiety and nucleotides [1] and made of monomeric units connected together by covalent bonds. Some of them are from living organisms and include alginate and carrageenan from seaweeds [2] and chitosan present in insects and shells of other organisms including fungi and algae [3]. Generally, biopolymers are degradable and utilized in different areas such as food industry, manufacturing, and biomedical engineering. Biopolymers are promising due to their nontoxic, biocompatible characters and abundance. Some of the examples of biopolymers are starch, cellulose, carbohydrate, protein, DNA, RNA, lipids, and collagen. Materials delivered by engineered chemistry from organic sources like vegetable oil, fats, and resins are also referred to as biopolymers [4]. Compared to synthetic polymers, biopolymers are complex nuclear assemblages and have characteristic structure and shape. Their structure and shape are important to their function. One of the examples is hemoglobin, which would not be able to carry oxygen if it did not have its quaternary structure [5]. Term "bio" implies the source of biopolymers from living creatures that are frequently biodegradable. These polymers get degraded through the activity of living organisms and soil enzymes to products such as carbon dioxide and

Applications of Biopolymers in Science, Biotechnology, and Engineering, First Edition.
Edited by Arunima Reghunadhan, Akhina H., and Sabu Thomas.

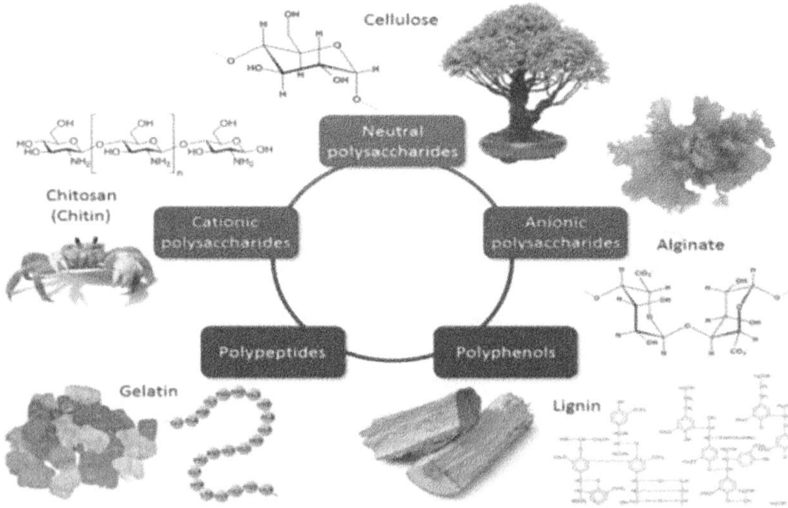

Figure 4.1 Natural and renewable biopolymers. *Source:* [9]/NIDOC.

water without harming the environment. They are regarded as renewable and biodegradable alternatives to petroleum-based plastics [6, 7].

Biopolymers are produced by microbial fermentation in special media with suitable supplements and optimal growth conditions or chemical polymerization of monomers without any harmful effect on biological systems [6]. Presently synthetic polymers such as poly lactic acid, polyvinyl alcohol, and poly glycolic acid are introduced as new innovations [8].

4.2 Different Types of Biopolymers

Biopolymers can be categorized based on their origin, number of monomeric units, and degradability. Based on the degradability they are again classified into two groups, biodegradable and nonbiodegradable or biobased and nonbiopolymers. Based on the polymer backbone they are again classified into different groups such as polysaccharides, polycarbonates, vinyl polymers and polyamides [5].

On the basis of type

- Sugar based polymers: sucrose or starch is used for preparation.
- Starch based polymers: starch act as a natural polymer, composed of glucose and mainly found in plants.

- Cellulose based biopolymer: polymer is made up of glucose obtained from natural sources.
- Synthetic materials: degradable polymers from synthetic materials.

On the basis of origin

- Natural biopolymers.
- Synthetic biopolymers.
- Microbial polymers.

On the basis of monomeric units

- Polysaccharides.
- Proteins.
- Polynucleotides.

4.3 Biopolymers Utilized as Materials

Biopolymers such as poly lactic acid (PLA) and poly-3 hydroxy butyrate can be utilized as plastics supplanting the requirement for polystyrene and polyethylene-based plastics. A few of these plastics are degradable or UV degradable when exposed to light or air. Biopolymers will breakdown and some of them used for domestic composting.

4.4 Biodegradable Polymers

These are polymers comprised of monomers connected to one another through functional groups having steady back bone [10]. They are broken down into biologically acceptable particles that are metabolized and evacuated from body through typical metabolic pathways [11]. Based on biodegradability they are divided into biodegradable polymers (collagen, poly-glycolic acid) and nonbiodegradable polymers (polyethylene) [12].

4.5 Advantages of Biopolymers

Biopolymers have various advantages:

- Controllable degradation rate.
- Reduce side effects.
- Sustained delivery of drug.
- Stabilization of drug.

4.6 Biopolymers for Medical Applications and Bioengineering Uses

Biopolymers are used in various medical applications for controlled drug delivery, occlusion, isolation, cellular proliferation, and suturing. Most medical devices are made up of natural (collagen) and synthetic (PVC, silicone) polymers, metals, alloys, glasses, and ceramics [13]. Traditional materials such as manufactured polymers, ceramics, and metals show some disadvantages like immunological rejection and biodegradation products of synthetic polymers present in the body lead to unwanted immunogenic responses. Biomaterial may be a nonviable material associated with biological systems and used in therapeutic applications. Biomaterials are used [14]:

- To help within the repair of tissue, including sutures, bone fracture plates and ligament repair contraptions.
- To supplant the tissues that are diseased, as in joint replacement, artificial heart valves and arteries.
- To supplant a portion of the work of major organs like haemodialysis, oxygenation, and insulin delivery.
- To provide drugs to the body (see Figure 4.2).

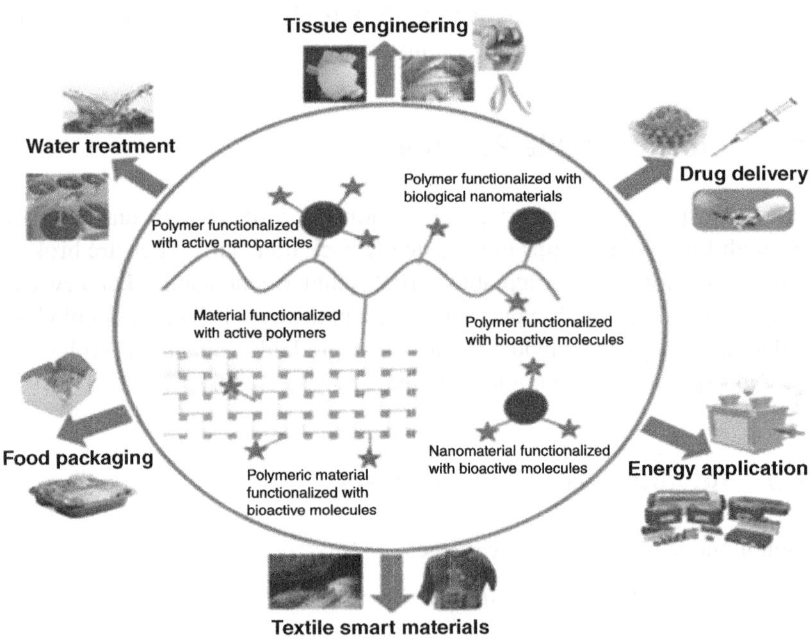

Figure 4.2 Applications of polymeric materials in different fields. *Source:* [15] / MDPI/CC BY 4.0.

The applications of biopolymers in the therapeutic field can be generally grouped into three categories:

- Drug delivery.
- Wound closure.
- Surgical implant devices and healing products.

Biodegradable capsules are used to deliver drugs into the body. Resorbable nonwovens for the replacement of human tissue, as well as sutures, staples, and clips, are available for wound healing [11].

4.6.1 Polyhydroxyalkonates

Polyhydroxyalkonates (PHA) is intracellular material of bacterial cells. It is a thermoplastic biopolymer. These polymers are currently used for medical application. The fundamental applications are controlled drug release, bandages, sutures, gauzes, wound care, and bone plates. Moreover, it is utilized as auxiliary materials in personal cleanliness products and packaging applications [16].

4.6.2 Polylactic Acid

Polylactic Acids (PLA): glycolic acid, L-lactic acid, and D-lactic acid are used to make lactide monomers for ring opening polymerization. PLA-PGA copolymers have been used to control the release of antibiotics, anticancer agents, hormones, insulin, and proteins. Polylactide sutures are the most commonly used surgical sutures. After the incision has healed, it is degraded within the body. This copolymer is used in sutures, clip, and reinforcement materials and also in orthopaedic surgery. The use of copolymer PLA-PGA prostheses has the advantage of being non-toxic and biocompatible [17].

4.6.3 Collagen

Collagen is a protein that is utilized extensively in the culinary and cosmetic sectors. Collagen is a biomaterial as well as a drug delivery system. It is easily absorbed by the body and has a low antigenicity and is non-toxic and biodegradable [18]. Collagen-based drug delivery is the most common application of collagen in biomedical sciences. Collagen sponges are useful in dressing severe wounds [19]. The use of collagen-based gel as an injectable formulation is one of the successful uses of collagen for controlled delivery [19]. Commercially available collagen-based wound healing materials include gel, films, sponges, membranes, hydrogels, and powders. Electro spinning, scaffold formation, and drug molecule interaction improve the physical, mechanical, and functional properties of biopolymers. For example, quercetin and curcumin incorporated separately into the collagen matrix demonstrated diabetic wound healing via reactive oxygen species scavenging. A nano fibrous collagen scaffold with a pattern similar to normal skin improved the healing process by accelerating fibroblasts, reducing inflammation, and promoting keratinocyte migration [20].

4.6.4 Gelatin

Gelatin is a soluble proteinaceous substance containing all essential amino acids except tryptophan [21]. It is made from various sources including cattle bone, fish, pig skin etc. Gelatin is used as a foaming, emulsifying, and wetting agent in medical and pharmaceutical applications, which is one of its most successful applications [22]. It is a polymer that is biodegradable, biocompatible, and completely absorbable. Gelatin offers a diverse range of tissue engineering applications. Gelatin-based drug delivery has demonstrated success in gene and siRNA delivery to induce therapeutic protein expression or silencing. Gelatin-based biomaterials have proven to be effective in the engineering of cardiovascular, hepatic, skin, bone, and corneal tissue [23].

4.6.5 Chitosan

Chitosan is a chitin polysaccharide derivative. It possesses excellent biodegradability, biocompatibility, permeability, and ability to form gel and film. The main applications are drug carriers for controlled release [24], antimicrobial agents [25], osteogenesis [26], and ulcer and other lesion healing [27, 28]. Chitosan promotes gene expression while also protecting DNA. Chitosan combined with some kind of anticancer agents provides anticancer effects by slow drug release in cancer tissue over time [29]. It also promotes phagocytosis, interleukin-1 (IL-1), transforming growth factor B1 (TGF-B1) production and enhancing immune response. So, it works as an anti-inflammatory agent. In bone tissue engineering, chitosan and hydroxyl apatite are potential biomaterials. A hydrogel sheet made up of alginate powder or chitosan provide a moist condition for wound healing [30]. Chitosan is primarily employed as a drug carrier molecule. Chitosan combined with polyphosphate and silver has hemostatic and antimicrobial properties [31]. The immobilization of chitosan with taurine and rutin promotes wound healing. Chitosan combined with cerium oxide nano cubes demonstrated anti-inflammatory activity [32]. Chitosan-based dressings are also available on the market e.g., Opticell used for surgical wounds, diabetic burns, leg ulcers etc.

4.6.6 Silk

Silk is used as a coating material influencing anti-coagulant properties. The biocompatible coating of silk fibroin promotes adhesion of human fibroblast on polyurethane and poly carbonate urethane surfaces. It also provides anti-thrombogenic effect and reduces the inflammatory response in wound dressing. Coating of silk fibroin with nanoparticles promote biocompatibility. Similarly, fibroin films are used for anti-bacterial effect in combination with silver or titanium dioxide nanoparticle leading to application as bone implants, wound dressing, and surgical coatings [33].

4.6.7 Cellulose

Cellulose is a glucose polymer. It is used as a drug delivery device. Bacterial cellulose is very effective in wound healing due to its high water retention capacity and purity. The most common cellulose fibers used in medical applications are natural cotton fibers and regenerated cotton fibers such as lyocell and modal [34]. Bioengineering of cellulose through surface immobilization with drugs or other therapeutic molecules have showed improved functional properties. Cellulose combined with antimicrobials prevent infection at the wound site. Cellulose combined with drugs such as vaccarin and povidone iodide provides efficient wound healing abilities [35]. In burn wounds, a hydrogel formed by UV cross-linking cellulose with acylic acid promotes neovascularization and re-epithelialization [36]. Conjugated cellulose with tungsten oxide and polydopamine increased the integrin β 1 expression with anti-inflammatory and anti-bacterial properties [37].

4.6.8 Carrageenan

It is a sulfated polygalactan with an ester-sulfate content ranging from 15% to 40%. It has anticoagulant and antithrombotic properties. Carrageenan inhibits the replication of some enveloped viruses such as human immune deficiency virus, herpes simplex virus, human rhinoviruses etc. It inhibits the entry of virions into the cell [38]. It is both suppository and topical bases. It also serves as a binder in dentifrices and prevents liquid–solid separation [39].

4.6.9 Hyaluronic Acid

Is a naturally occurring polymer seen in vertebrate tissue. It also found in some bacteria. It is biocompatible and biodegradable. The first human application of hyaluronan was as a vitreous substitution or replacement during eye surgery in the 1950s. It is used for the delivery of drugs to eyes

[40]. Other medical application of hyaluronic acid is wound healing, epithelial regeneration, scaffold material for ear surgery, protect corneal tissue and identifies the presence of tumors [41]. It's an excellent polymer because it is a natural product that degrades easily into simple sugars. The significant application of this polymer is eye surgery, arthritis treatment and wound healing processes. In diabetic ulcers, hyaluronan cross-linked with epoxy demonstrated angiogenesis and dermal restoration [42]. Molded hydrogels containing hyaluronic acid, bisphosphonate, and silver have effective antimicrobial properties [43].

4.7 Major Biomedical Applications of Biopolymers

Table 4.1 shows various biomedical applications of biopolymers.

Table 4.1 Biomedical application of biopolymer.

Sl.No.	Biopolymer	Shape	Applications
1	Poly(α-hydroxy acid),collagen, polyglyconate	Suture, clip	Binding materials
2	PLLA, polyglactin	Rod, screw, pin	Bone setting materials
3	Gelatin	Jelly, spray, sheet	Antiadhesion materials
4	Chitin, poly-L-leucine	Fiber	Wound cover attributes
5	All degradable polymers	Needle, microsphere, micro capsule	Drug delivery system

4.8 Conclusion

Biopolymers are naturally occurring polymers found in green plants, bacteria, fungi, and animals. Biopolymers, particularly those of carbohydrate origin, have a promising effect in the field of medical application in a variety of forms. This is due to their low toxicity, stability, biodegradability, and renewable nature. Wool, silk, and cellulose are examples of traditional polymers. Biotechnological approaches such as genetic engineering, recombinant DNA, nanotechnology, and fermentation now provide new specific properties to natural polymers and open up new applications for uncommon polymers and their derivatives. Biopolymers have unique properties. Due to their unique properties and structures biopolymers have many applications.

References

1 Kumari, P., Lal, S., and Singhal, A. (2021). Advanced applications of green materials in catalysis applications. In: *Applications of Advanced Green Materials* (ed. S. Ahmed), 545–571. Amsterdam: Elsevier.

2 Ghada, E.A.A., Hala, R.W., Abeer, A.A., and Mohamed, E.H. (2017). A novel alginate–CMC gel beads for efficient covalent inulinase immobilization. *Colloid and Polymer Science* 295: 495–506.

3 Abeer, A.A., Faten, A.M., Mohamed, E.H. et al. (2017). Covalent immobilization of Alternaria tenuissima KM651985 laccase and some applied aspects. *Biocatalysis and Agricultural Biotechnology* 9: 74–81.

4 Hernandez, N., Williams, R.C., and Cochran, E.W. (2014). The battle for the "green" polymer. Different approaches for biopolymer synthesis: bioadvantaged vs. bioreplacement. *Organic and Biomolecular Chemistry* 12: 2834–2849.

5 Mohan, S., Oluwafemi, O.S., Kalarikkal, N. et al. (2016). Biopolymers–application in nanoscience and nanotechnology. *Recent Advances in Biopolymers* 1 (1): 47–66.

6 Muratore, G., Nobile, D., Buonocore, G.G. et al. (2005). The influence of using biodegradable packaging films on the quality decay kinetic of plum tomato (Pomodorino Datterino). *Journal of Food Engineering* 67 (4): 393–399.

7 Rhim, J.W., Park, H.M., and Ha, C.S. (2013). Bio-nanocomposites for food packaging applications. *Progress in Polymer Science* 38 (10–11): 1629–1652.

8 Tirrell, J.G. and Tirrell, D.A. (1996). Synthesis of biopolymers: proteins, polyesters, polysaccharides, and polynucleotides. *Current Opinion in Solid State and Materials Science* 1 (3): 407–411.

9 Hassan, M.E.S., Bai, J., and Dou, D.Q. (2019). Biopolymers; definition, classification, and applications. *Egyptian Journal of Chemistry* 62 (9): 1725–1737.

10 Modjarrad, K. and Ebnesajjad, S. (eds.) (2013). *Handbook of Polymer Applications in Medicine and Medical Devices*. Amsterdam: Elsevier.

11 Pachence, J.M. and Kohn, J. (2000). Biodegradable polymers. *Principles of Tissue Engineering* 22: 263–277.

12 Agarwal, S. (2020). Biodegradable polymers: present opportunities and challenges in providing a microplastic-free environment. *Macromolecular Chemistry and Physics* 221 (6): 2000017.

13 Rebelo, R., Fernandes, M., and Fangueiro, R. (2017). Biopolymers in medical implants: a brief review. *Procedia Engineering* 200: 236–243.

14 Ikada, Y. and Tsuji, H. (2000). Biodegradable polyesters for medical and ecological applications. *Macromolecular Rapid Communications* 21 (3): 117–132.

15 El-Ghoul, Y., Alminderej, F.M., Alsubaie, F.M. et al. (2021). Recent advances in functional polymer materials for energy, water, and biomedical applications: a review. *Polymers* 13 (24): 4327.

16 Onar, N. (2004). Usage of biopolymers in medical applications. 3rd Indo-Czech Textile Conference, Istanbul, Turkey.

17 Martin, D.P. and Williams, S.F. (2003). Medical applications of poly-4-hydroxybutyrate: a strong flexible absorbable biomaterial. *Biochemical Engineering Journal* 16 (2): 97–105.

18 Yadav, P., Yadav, H., Shah, V.G. et al. (2015). Biomedical biopolymers, their origin and evolution in biomedical sciences: a systematic review. *Journal of Clinical and Diagnostic Research* 9 (9): ZE21.

19 Lee, C.H., Singla, A., and Lee, Y. (2001). Biomedical applications of collagen. *International Journal of Pharmaceutics* 221 (1–2): 1–22.

20 Sahana, T.G. and Rekha, P.D. (2018). Biopolymers: applications in wound healing and skin tissue engineering. *Molecular Biology Reports* 45 (6): 2857–2867.

21 Mariod, A.A. and Adam, H.F. (2013). Review: gelatin, source, extraction, and industrial applications. *Acta Scientiarum Polonorum Technologia Alimentaria* 12 (2): 135–147.

22 Lobo, L. (2002). Coalescence during emulsification. Effect of gelatin on rupture and coalescence. *Journal of Colloid and Interface Science* 254: 165–174.

23 Nikkhah, M., Akbari, M., Paul, A. et al. (2016). Gelatin-based biomaterials for tissue engineering and stem cell bioengineering. In: *Biomaterials from Nature for Advanced Devices and Therapies* (eds. N.M. Neves and R.L. Reis), 37–62. New York: Wiley.

24 Hamman, J.H. (2010). Chitosan based polyelectrolyte complexes as potential carrier materials in drug delivery systems. *Marine Drugs* 8: 1305–1322.

25 Kong, M., Chen, X.G., Xing, K., and Park, H.J. (2010). Antimicrobial properties of chitosan and mode of action: a state-of-the-art review. *International Journal of Food Microbiology* 144: 51–63.

26 Tanodekaew, S., Prasitsilp, M., Swasdison, S. et al. (2004). Preparation of acrylic grafted chitin for wound dressing application. *Biomaterials* 25: 1453–1460.

27 Maia, R.C.C., Franco, L.O., Stamford, T.C.M. et al. (2006). Chitin produced by *Cunninghamella elegans* (IFM 46109) and applied to wound healing. *Asian Chitin Journal* 2: 11–20.

28 Mori, T., Murakami, M., Okumura, M. et al. (2005). Mechanism of macrophage activation by chitin derivatives. *Journal of Veterinary Medical Science* 67: 51–56.

29 Wang, M. (2003). Developing bioactive composite materials for tissue replacement. *Biomaterials* 24: 2133–2151.

30 Di Martino, A., Sittinger, M., and Risbud, M.V. (2005). Chitosan: a versatile biopolymer for orthopedic tissue-engineering. *Biomaterials* 26: 5983–5990.

31 Moura, L.I., Dias, A.M., Leal, E.C. et al. (2014). Chitosan-based dressings loaded with neurotensin—an efficient strategy to improve early diabetic wound healing. *Acta Biomaterialia* 10: 843–857.

32 Huang, X., Li, L.D., Lyu, G.M. et al. (2018). Chitosan-coated cerium oxide nanocubes accelerate cutaneous wound healing by curtailing persistent inflammation. *Inorganic Chemistry Frontiers* 5: 386–393.

33 Kundu, B., Kurland, N.E., Bano, S. et al. (2014). Silk proteins for biomedical applications: bioengineering perspectives. *Progress in Polymer Science* 39 (2): 251–267.

34 Drury, J.L. and Mooney, D.J. (2003). Hydrogels for tissue engineering: scaffold design variables and applications. *Biomaterials* 24 (24): 4337–4351.

35 Qiu, Y., Qiu, L., Cui, J., and Wei, Q. (2016). Bacterial cellulose and bacterial cellulose-vaccarin membranes for wound healing. *Materials Science and Engineering C* 59: 303–309.

36 Mohamad, N., Amin, M.C., Pandey, M. et al. (2014). Bacterial cellulose/acrylic acid hydrogel synthesized via electron beam irradiation: accelerated burn wound healing in an animal model. *Carbohydrate Polymers* 114: 312–320.

37 Fawal, G.F., Abu-Serie, M.M., Hassan, M.A., and Elnouby, M.S. (2018). Hydroxyethyl cellulose hydrogel for wound dressing: fabrication, characterization, and in vitro evaluation. *International Journal of Biological Macromolecules* 111: 649–659.

38 Clark, D.T., Gazi, M.I., Cox, S.W. et al. (1993). The effects of *Acacia* arabica gum on the in vitro growth and protease activities of periodontopathic bacteria. *Journal of Clinical Periodontology* 20 (4): 238–243.

39 Onishi, T., Umemura, S., Yanagawa, M. et al. (2008). Remineralization effects of gum arabic on caries-like enamel lesions. *Archives of Oral Biology* 53 (3): 257–260.

40 Dechert, T.A., Ducale, A.E., Ward, S.I., and Yager, D.R. (2006). Hyaluronan in human acute and chronic dermal wounds. *Wound Repair and Regeneration* 14: 252–258.

41 Leach, J.B. and Schmidt, C.E. (2004). *Hyaluronan Encyclopedia of Biomaterials and Biomedical Engineering*. 779–789. New York: Marcel Dekker.

42 Zhu, J., Li, F., Wang, X. et al. (2018). Hyaluronic acid and polyethylene glycol hybrid hydrogel encapsulating nanogel with hemostasis and sustainable antibacterial property for wound healing. *ACS Applied Materials & Interfaces* 10 (16): 13304–13316.

43 Shi, L., Zhao, Y., Xie, Q. et al. (2018). Moldable hyaluronan hydrogel enabled by dynamic metal–bisphosphonate coordination chemistry for wound healing. *Advanced Healthcare Materials* 7: 1700973.

5

Biopolymers and Composites in Tissue Engineering

Trinath Biswal

Department of Chemistry, VSS University of Technology, Burla, India

5.1 Introduction

Towards the middle of the 20th century, humans were fully dependent upon the industry of synthetic polymeric materials and composites for the manufacture of various product materials and devices. Although, the commercialization and fabrication of synthetic petroleum-based polymeric materials and their composites are cost effective, their excessive use without any recycling and management is highly hazardous to our society. Today, this is an important global issue, because most of the parts of our globe, including the marine environment, is affected by nonbiodegradable plastic materials and composites. Therefore, in order to solve this global problem, researchers give emphasis to the development of new biobased biodegradable materials, mainly biopolymers of different kinds and their composites, which serves as an alternative to the synthetic petroleum-based polymeric materials [1, 2]. The development of biopolymer composites is now receiving more attention because of their high-performance and multifunctional properties such as abundant availability in nature, low or no toxicity, excellent biocompatibility, light weight, renewable and ecofriendly to nature with almost no adverse impact on the environment. Currently, many biopolymer composite materials are substituting for synthetic polymeric materials and composites in many sectors such as medical implant

Applications of Biopolymers in Science, Biotechnology, and Engineering, First Edition.
Edited by Arunima Reghunadhan, Akhina H., and Sabu Thomas.

materials in tissue engineering, packaging, optical applications, cosmetics, automobile sector and biochemistry [3, 4]. In recent years, the biodegradable polymer composites and some biopolymers have attracted scientists in the development of some suitable biomaterial devices, which can be used successfully in tissue engineering applications. The sustainable biocomposite and green composite possesses excellent recyclable and renewable potential, which is necessary for application in the biomedical sector [5]. The protection of our ecosystem is an inspiring impulse behind the production of such kinds of biopolymer composite. The recycling of the used waste plastic materials is expensive and difficult, therefore the accumulation of these waste plastic products around the globe is continuously increasing and is now at a level that is dangerous to the biological community of the world. The use of biopolymer composite materials solves the problem of recycling the synthetic polymers and specially reduces the use of fossil fuel (petroleum oil source) [6, 7]. The design and development of light weight, stronger, durable, and multifunctional composites of biobased polymeric materials is essential for application in the biomedical fields. The modified biopolymers contain the monomers, which are obtained from the biopolymers such as polypeptides, polysaccharides, and other biopolymers. The biomaterials used effectively for tissue engineering applications includes polymer composites, hydrogels, nanocomposites, biocompatible pristine and bioceramics. The scaffolds of biopolymer are now used effectively for tissue engineering applications such as bone, cartilage, blood vessel, ligament, skin, nerve, tendon and dermis [8, 9].

5.2 Biopolymer

The polymers derived from natural biological origin of living resources including plants, microorganisms and animals are termed natural polymers or biopolymers. The different biopolymers include arabinoxylan, chitin, alginate, chitosan, cellulose, starch, pectin, zein keratin, gelatine and polyhydroxyalkanoates (PHAs) [10]. In the composite of biopolymers either or both the matrix phase and reinforcing material are from natural resources. The composition of a biopolymer depends upon its functional properties, whereas the functional behavior of the material depends upon crystallinity of the material (whether amorphous or crystalline in nature) [11, 12]. For example, cellulose is an important structural polymer, the features of it partially depends upon its property of crystallinity [13]. The physical, biological, and chemical properties of the material make it as a valuable multifunctional material for different kind of applications.

Modified polysaccharides are potentially used in various applications, including paint, adhesives, cement, antibacterial coating, cosmetics, tissue engineering, medicines, and manufacturing of many important products [14]. The different classification of biopolymers is shown in Figure 5.1

Chitosan, a popular biopolymer, belongs to the family of carbohydrate polysaccharides and is used extensively in different sectors but mostly in the biomedical field because of some of its unique properties including biosustainability, non-toxicity biodegradability and compatibility to human tissue. The major source of chitin is from marine animals such lobsters, shrimp, crabs, etc., which, on deacetylation, produces chitosan and can be effectively used for the development and design of biocomposite materials. Arabinoxylan is an important biopolymer of the carbohydrate family and is used in a number of medicinal applications [15]. The commercialization of biosustainable polymers and their composites is a challenging issue because of some technical and fabrication problems. Another important hurdle to the wide use and industrialization of biopolymers and their composites is their functional properties, which can be unfavorable e.g., high production cost, low availability, and slower rate of synthesis than the petroleum-based synthetic polymers. The difficulty of wide use of biopolymers and their composites can be solved in a number of ways including grafting, blending, mixing, and reinforcing with suitable ceramics and polymeric materials. Some of the novel biomaterial synthesized from biopolymer incudes rubber,

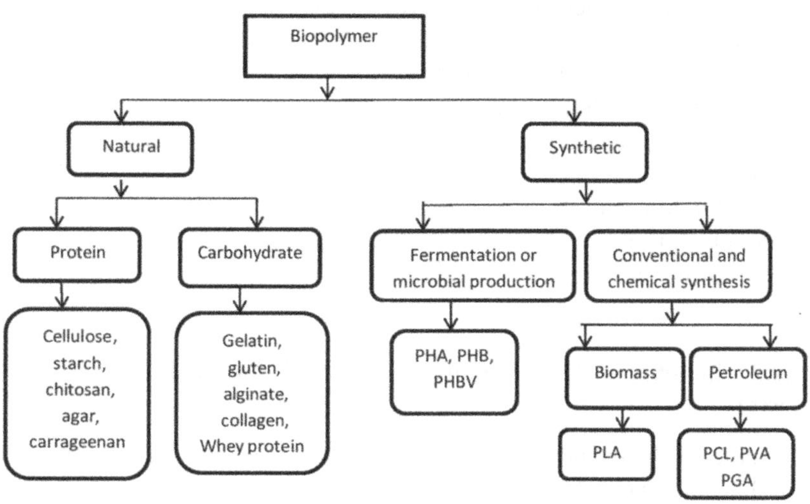

Figure 5.1 Classification of biopolymer. PHA, polyhydroxyalkanoates; PHB, polyhydroxybutyrate; PHBV, poly(3-hydroxybutyrate-co-3-hydroxyvalerate); PLA: polylactic acid; PCL, polycaprolactone; PVA, polyvinyl alcohol; PGA, polyglycolide.

cellophane, linoleum, and celluloid. The biomaterials cellophane and cellulose are synthesized from cellulose, which is the most naturally abundant biopolymer and organic substance on our globe. From the middle of the 20th century the modified biopolymeric materials and their composites have been of interest to the industrial sector, which is continuously replacing the synthetic polymeric materials [16, 17].

5.2.1 Properties of Biopolymers

The biopolymers are usually polymers derived from natural biological origin and the monomer units of the biopolymer are covalently bonded and form macro structures. On the basis of monomeric units, biopolymers are classified into three categories.

- Polynucleotides, which are long chain biopolymer having 13 or more than that nucleotide monomer units.
- Polypeptides, which are normally short chain biopolymers of amino acids as monomer units.
- Polysaccharides, which are usually straight chain polymeric materials having carbohydrate structure [18, 19].

Biopolymers are used in medical sectors, especially in tissue engineering applications, because of the following interesting properties:

- Obtained from renewable natural resources.
- Sustainable in nature.
- Biodegradable and biocompatible.
- Bioabsorbable to body fluid.
- Non-carcinogenic.
- Non-toxic.
- Non-immunogenic.
- Carbon neutral.
- Non-thrombogenic [20].

5.2.2 Benefits of Using Biopolymers in the Biomedical Sector

The different biomedical applications of biopolymers and their composites are shown in Figure 5.2.

Because of their many essential properties, there are many advantages of using biopolymers in medical applications:

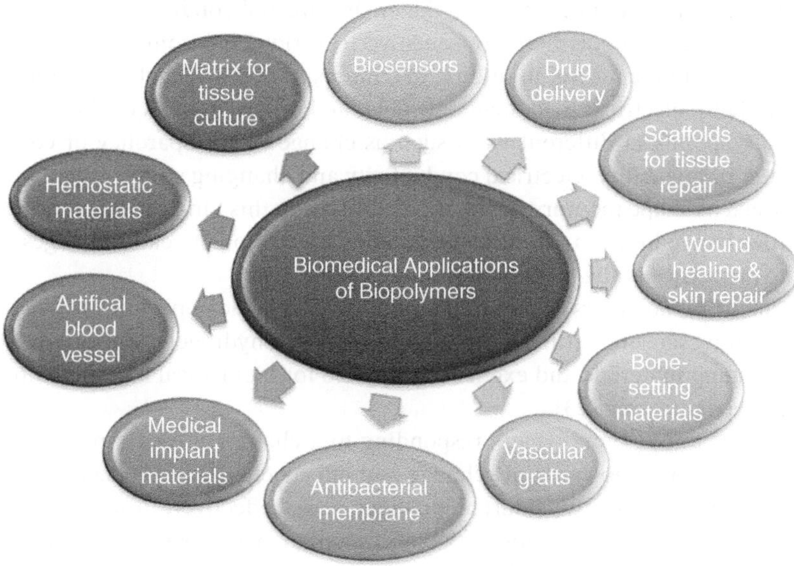

Figure 5.2 Different biomedical applications of biopolymers and their composites.

- Used in localized drug delivery system.
- Sustained delivery of drugs.
- Stabilization of the used drug.
- Reduction in the frequency of doses.
- Fewer side effects.
- Improved compliance of the patients.
- Easy to control the rate of degradation.
- Ecofriendly to the environment [21, 22].

5.2.3 Smart Biopolymers

The biopolymers are normally defined as biologically degradable polymers obtained from living organisms, otherwise known as polymeric biomolecules. The smart or intelligent biopolymer is more attractive in biomedicine, packaging, tissue engineering and biotechnology and plays many important functions in drug delivery. The smart biopolymers are renewable and sustainable biomaterials. The smart biopolymers are also known as functional biopolymers or stimuli-responsive biopolymers, which are normally polymeric materials exhibiting high performance and which can change its function according to the environmental conditions. The smart or intelligent biopolymers are sensitive to various environmental factors including temperature, humidity, pH, intensity or wavelength of the incident light, electrical field, magnetic field, and different kinds of chemicals, and responds in different ways such as change in transparency or color, water permeability, electrical conductivity and changing shape, commonly termed as shape memory polymeric materials. In this kind of biopolymer, a little change in the environmental condition leads to noticeable changes in the properties of the biopolymer.

Smart biopolymers are used in a number of specific applications such as actuators, sensors, artificial muscles, synthesis of hydrogels, biodegradable packaging materials and extensively applied for biomedical sectors mainly tissue engineering [23].

For example, biopolymers responding to a change in conformation are highly sensitive towards pH change and effectively used in drug delivery. The humidity-sensitive smart biopolymer is used effectively in self-adaptive wound dressing applications, which spontaneously balances moisture around the wound area [24].

The smart biopolymers contribute a multiple responsibility and offer delicate chemical signals to the cells of our body to guide their properties and behavior. These kinds of materials facilitate the elasticity and hydration of the skin, lubricate joint areas and the gastrointestinal tracts, and prevents against pathogens through accumulating into the mucus gel, which covers our respiratory tract and eyes. Biopolymers are considered smart biopolymers and adaptable biomaterials with living origin, due to their structural features, and can be continuously manipulated either through enzymes throughout various stages of the life cycle of the organisms or that are highly responsive towards the change in climate. The smart biopolymers are also known as environmentally sensitive biopolymers or stimuli-responsive polymers. The stimuli, which promotes the changes in the physical properties of the biopolymers, are varied in nature, and include temperature, light intensity, pH, and ionic composition of solvent, electric field, and addition of specific ions [25].

In the last two decades the applications of smart biopolymer or functional biopolymers have grown at a rapid rate. These biopolymers and their composites are responding in a favorable way with change in different factors such as in temperature, moisture, pH, pressure electric or magnetic fields and many other climatic factors. Recently, many smart biopolymers have been designed and developed for suitable applications in medicine and biotechnology. Different kinds of smart biopolymers have been developed for a variety kind of applications and mainly classified into three kinds on the basis of their functionality and stimuli. These are as follows:

- pH-sensitive smart polymers.
- Stimuli responsive smart polymers.
- Thermosensitive smart polymers [26].

5.2.4 Kinds of Biopolymers

The biopolymers and synthetic polymers are categorized across a broad spectrum, which is represented in Figure 5.3. The polymer composites and biocomposites are obtained from wide range of sources, where the biopolymer of natural origin such as cellulose, microbial polyesters and proteins are normally more significant for the biological function and structure. The biopolymers or the biopolymer-based composites are classified into three major classes according to the basis of their source of occurrence:

- Polymers derived from biomass products such as cellulose, starch, arabinoxylan, collagen and keratin.
- Polymers derived by standard techniques of chemical processing using monomers of the biopolymers obtained from renewable natural resources

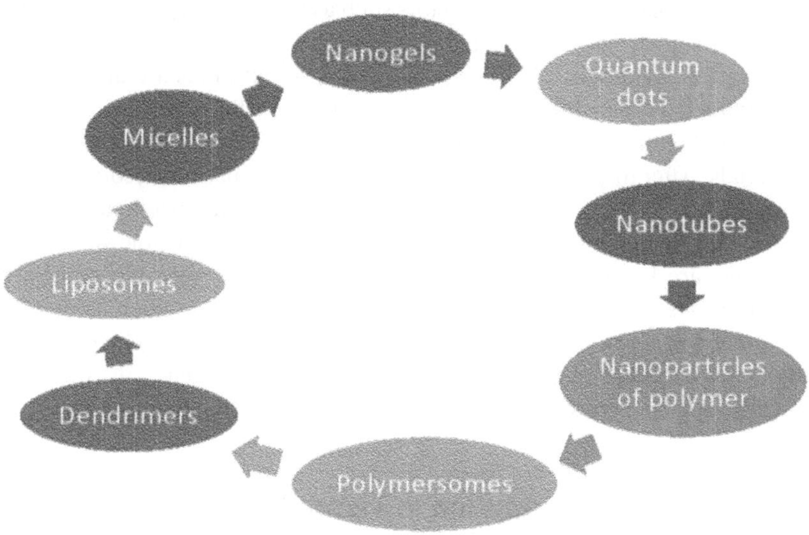

Figure 5.3 Classification of broader spectrum of these polymeric materials on the basis of source.

such as polylactic acid (PLA), polyglycolic acid (PGA), cellulose acetate (CA), etc., which are normally synthesized from carbohydrate feedstock through the process of fermentation.
- Polymers obtained by processing from microorganisms such as PHAs, whereas the bacterial cellulose is now still under development [27, 28].

5.2.5 Biocomposites

A composite material is generally formed from two or more different kinds of materials or polymers having exclusively noticeable difference in physical or chemical properties. The composites are synthesized by imparting some new useful properties or enhancing some required properties to the material, which cannot be achieved from a single substance. If the polymeric composite possesses definite biological phases, then it is known as a bio-composite. The biocomposites are normally prepared from bioderived or natural polymers such as chitin, chitosan, zein, alginate, pectin, PHAs, arabinoxylan, PLA, etc. The sustainable and effective biocomposites synthesized from natural or biopolymeric materials and degradable inorganic materials as fillers are commonly termed green composites. Now these composites are the focus of researchers considering the ecofriendliness to the environment. Biocomposite materials are synthesized and developed on the basis of the targeted applications in various sectors such as biomedical, packaging, computer parts, pharmacy, medical implant, agriculture textile, paper, automobiles and many others. The biopolymers of different kinds are quite stable and extracted from different natural resources, easily recyclable, ecologically, and commercially viable, and therefore categorized as biosustainable materials. After disposal of these biomaterials, they can start decomposition with the help of different ecological, climatic factors and microbial communities. The biocomposite materials, which are both biocompatible and biodegradable, are known as green biocomposites. Various green biocomposite materials are synthesized by using several inorganic fillers such as titania, alumina, and silica, etc. Hence, the development of bioplastic materials produces a wide range of ecofriendly and feasible product materials, which is already replacing the synthetic polymeric materials in the market [29, 30].

5.3 Biopolymers Applied in Tissue Engineering

There are number of biopolymers available in nature, which in modified form or directly or in composite form can be used effectively in different biomedical applications. The tissue engineering applications of some important biopolymers are as follows.

5.3.1 Collagen

Collagen is a crucial protein of extracellular matrix (ECM), which normally supports different tissues such as tendons, bones, skin, ligaments, cartilage, and cornea tissue. Hence, collagen is treated as an appropriate scaffold or matrix for tissue engineering applications. Collagen undergoes interaction with the different connective tissue cells, transduces vital signals for the effective regulation of cell differentiation, proliferation, adhesion, survival, and migration. Although, collagen is of different kinds, the most suitable collagen used for tissue engineering is type-I collagen [31]. This kind of collagen possesses some essential unique properties such as low antigenicity, adequate mechanical strength, better biodegradability, capability of being cross-linking and good biocompatibility, which makes type-I collagen a novel biomaterial for application in tissue engineering. Now a new scaffold has been developed by Laiva et al. having pro-angiogenic gene-activated characteristics, which influences angiogenesis, because of the synthesis of paracrine angiogenic factors, which is required for wound healing. It was synthesized by the combination of collagen-chondroitin sulphate scaffold with the polyethyleneimine nanoparticles having stromal derived factor-1 alpha (SDF-1α) gen [32].

The nervous system plays a crucial role in metabolic activity in the human body. It modulates the activity of every organ system along with a number of different physiological techniques in both indirect and direct pathways. The collagen-based scaffold offers an effective platform for the repair and regeneration of nerves. It was reported that the 3D scaffolds of collagen nanofibers synthesized by electrospun process are highly effective for spinal cord injury (SCI). The bone tissue normally consists of HA and collagen type-I with a small amount of collagen type-V. Therefore, collagens on blending with the nano-inorganic filler materials are popularly used for synthesis of scaffolds, which is the mimic of natural ECM and can be effectively used for repairing bone tissue [33]. The 3D scaffolds of collagen-$Ca_3(PO_4)_2$ optimizes the mechanical strength and cytocompatibility through modifying the scaffolds by using some specific concentration of H_3PO_4-based solutions and is effectively used in bone tissue engineering. Ligaments and tendons are normally fibrous connective tissues, which consist of 70%–80% of the dry weight collagen. Both the ligament and tendon possess extremely weak capability of spontaneous regeneration and fully recovery is very difficult. Collagen scaffolds offers an outstandingly effective pathway of regeneration and repair of ligaments and tendons. A fibrous scaffold of PCL/collagen was developed using an electrospinning process, which possesses the capability of resisting flow and high pressure for a longer time period and offers a suitable environment for the vascular cell growth. Because of the excellent property of biocompatibility scaffold of PCL/collagen on coupling with the vascular cells it can be effectively used for the formation of tissue-created blood vessels. Skin is the largest organ in our whole body and consists of epidermis, hypodermis, and dermis layers. The wound developed over skin is a potent and old problem associated with surgery. The substitutes related to skin suppresses the synthesis of fibrotic scar, but the frequently used skin substitutes, like autografts and allografts, never solve the full thickness or massive skin loss, which may cause skin dysfunction. After extensive research, it was found that the scaffolds based on collagen shows a potential benefit in the tissue engineering, where the collagen type-I synthesizes up to 70%–80% of the dermis in dry weight. Again, it was found that scaffolds of glutaraldehyde (**GA**)-treated chitosan/collagen possess excellent cytocompatibility for effective cell proliferation and infiltration. In case of open wounds the fibrous scaffold of collagen increases desertion of the proliferation of fibroblasts, surface tissue debris and the new capillaries in the initial phase of the wound healing [34, 35]. The chemical structure of collagen is shown in Scheme 5.1.

5.3.2 Gelatin

Gelatin is also an important biopolymer obtained by partial hydrolysis of collagen, and because of its property of biodegradability and biocompatibility in the physiological environments it can be widely used in pharmaceutical and medical fields. Gelatin shows less antigenicity than collagen, which is normally observed owing to its animal origin. Gelatin is normally a denatured protein, which is derived from collagen through the use of acid and alkaline hydrolysis. The process of the gelatin synthesis permits the flexibility of the poly-ion complexation of gelatin scaffold, hydrogel, or carrier with both positively or negatively charged cells and biomolecules. Because of many essential properties of gelatin, it can be used for synthesis of a number of tissues and application in drug delivery. It was found that the microcapsule of alginate-gelatin shows a better microenvironment in human umbilical vein

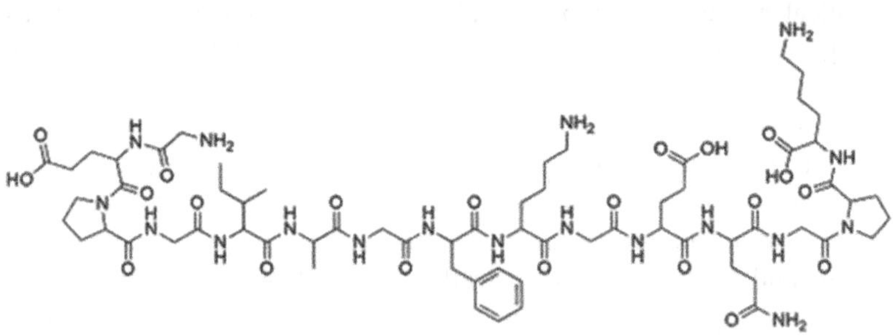

Scheme 5.1 Chemical structure of collagen.

endothelial cells (HUVECs) in only five days. Gelatin shows a number of beneficial properties like excellent biodegradability, non-toxicity, good biocompatibility, lower cost, and easy accessibility that are necessary for tissue engineering [35, 36]. The denatured form of gelatin comparatively shows less antigenic effect than collagen because its modified form of chemical structure, cell proliferation and the alternation of the cell adhesion is the cause of acceleration of the biological characteristics of the device used for tissue implantation [37]. The gelatin/ PCL blend on insertion into the matrix of gelatin serves as a good biomaterial for nerve tissue engineering. Because of the property of high biocompatibility, the gelatin nanoparticles are effectively used as a candidate material for neural tissue regeneration [38]. For example, the acetate/PLA scaffolds coated by gelatin nanoparticles shows better cell feasibility and functionality than the uncoated scaffolds [39]. The chemical structure of gelatin is shown in Scheme 5.2.

The gelatin/chitosan/PEDOT hybrid scaffolds enhance the growth of neurite and stimulates cell adhesion and proliferation. Here is a novel composite material synthesized by using biopolymer matrix reinforced with gelatin that shows adequate mechanical strength and therefore can be effectively used for tissue culture. But the major problem when using it for application in tissue engineering is poor thermal stability, mechanical strength, and a low rate of degradation.

Because of the properties of biodegradability and biocompatibility, gelatin has been more popularly used as 2D scaffolds such as micro-sheets, surface coatings and 3D scaffolds such as microparticles and hydrogels. The platform of these biomaterials enhances proliferation and assisted differentiation of different kinds of tissues and cells *in vitro* [40, 41].

5.3.3 Chitosan

Chitosan is found to be the second most common natural biopolymer on our planet. It is the polymer of linear polysaccharide group and is obtained by the deacetylation of the natural polysaccharide chitin, which is present

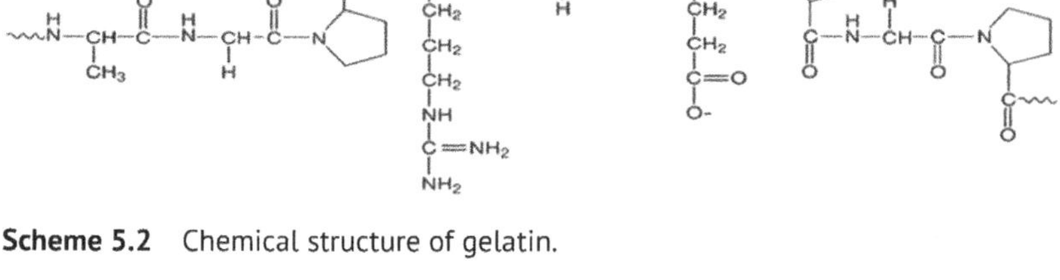

Scheme 5.2 Chemical structure of gelatin.

in high concentration at shellfish and crustaceans. Chitosan exhibits some essential properties such as capability of forming gel, higher capability of adsorption, adequate biodegradability, non-cytotoxic, non-antigenic, biologically renewable, and excellent biocompatibility along with better biological properties including antibacterial, antifungal, and antitumor activity. The structure of chitin and chitosan is shown in Scheme 5.3.

The chitosan in hydrogel form is found to be a highly effective candidate material in the application of tissue engineering owing to its excellent properties of cell adhesion, cell survival, interaction, and neurite outgrowth. Again, the 3D porous scaffold of chitosan on coupling with nerve growth factor (NGF) demonstrates the synergistic effect of neural stem cells (NSC) and possesses the ability of regenerating the CNS and PNS. Although chitosan is a suitable material for tissue engineering applications, the major problems of chitosan are its stiffness, brittle characteristics, and low mechanical strength. The degree of deacetylation of chitosan normally in the range of 70%–95%, and the molecular weight is usually in between 10 to 1000 kDa. The biopolymer chitosan is applicable effectively in variety of tissue engineering such as cartilage, vascular grafts, bone, skin regeneration, wound healing, and cell culture. The hydrogel of chitosan/gelatin is thermo-sensitive in nature and effective for angiogenic use through sustained release of stem cells derived from adipose, which causes the gradual degradation of gelatin [42, 43].

5.3.4 Silk

Silk is a protein biopolymer and is synthesized from spiders, mites, scorpions, silkworms, and flies. Spider silk is an amazing material, which is virtually weightless, very strong and elastic in nature and the mechanical properties of spider silk are better than any other modern silk. It is biodegradable in nature

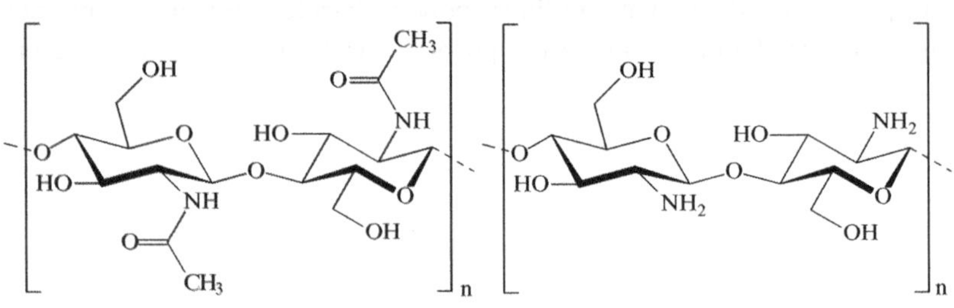

Scheme 5.3 Chemical structure of chitin and chitosan.

and environmentally friendly. Normally lower amounts of silk fibroin are present in natural silk because it is synthesized by silkworms. Fibroin and sericin are the key components of spider silk. Fibroin is a fibrous protein, which produces the core part of the silk and is consist of fibroin heavy chain, fibrohexamerin and a light chain. Because of improved mechanical properties, slow rate of degradability makes this biomaterial suitable for tissue engineering applications. The major advantage of using silk in tissue engineering is the excellent transparency potential [44]. The chain polymeric structure of silk is shown in Scheme 5.4.

The porous scaffold of the immobilized silk fibroin having basic fibroblast growth factor bFGF was synthesized, which is found to be highly effective for skin regeneration. The human skin functionally and anatomically consists of two primary layers:

- The outer epidermis layer, which mainly consist of keratinized stratified squamous epithelium and mainly contains keratinocytes.
- The dermis, which is normally an irregular high density connective tissue that basically contains fibroblasts, which interdigitates and underlies along with the epidermis.

The dermal fibroblasts proliferate and spread over the scaffolds and fibroin coatings lacking the secretion of the pro-inflammatory interleukins. The oral keratinocytes are able to proliferate on the meshes of the woven fibroin, which can be efficiently used in the application of wound healing. The sponges of fibroin–alginate blend and fibroin films are able to accelerate the rate of wound healing *in vivo* over other clinical applicable materials. Hence, the fibroin-based biomaterial facilitates epithelialization and offers superior performance for repairing ventral hernias.

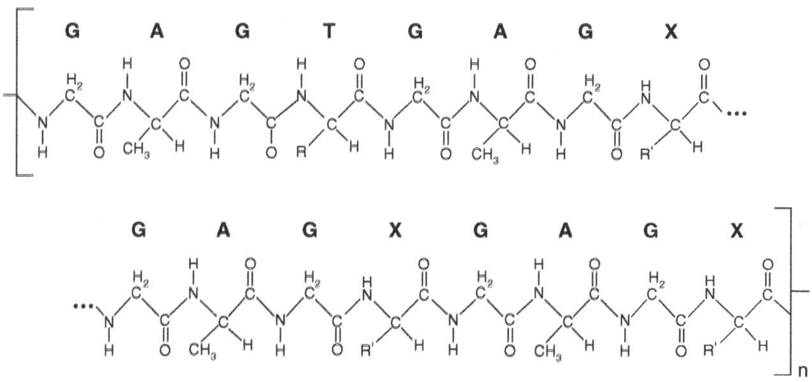

Scheme 5.4 Chain polymeric structure of silk.

The scaffolds based on fibroin show excellent performance in cartilage tissue engineering. The porous scaffolds are used to culture human mesenchymal stem cells (MSCs) and articular chondrocytes *in vitro*. The scaffolds of silk were synthesized in a serum-free environment, which is extremely significant in terms of potentiality of clinical use. But, however, presently the mechanical properties of the prepared cartilage are substandard to that of natural cartilage, therefore further modification of the culture environment is necessary.

Again, it was reported by many researchers that silk-based biomaterials or composites are useful in case of bone tissue engineering. The solution of fibroin used for synthesis of 3D porous scaffold of silk through the process of salt-leaching, is used as an effective biomaterial. Silk yarns are effective for ligament reconstruction tissue engineering. Because of adequate mechanical properties of the matrix of silk it is effective for repairing and reconstruction of MSCs and fibroblasts of anterior cruciate ligament. Recently, it was found that the film of heparinized silk fibroin possesses quite appropriate mechanical properties and hence, can be effectively applied in artificial blood vessels. The composites of collagen gel–silk filament is designed for effective use of vascular tissue engineering. The compatibility of silk fibroin with blood can be improved further by grafting the water-soluble polymers such as 2-methacryloyloxyethyl phosphorylcholine (MPC) on the surface or through blending with S-carboxymethyl kerateine. The co-culture of osteoblasts and endothelial cells on the nets of silk fibroin can form micro-vessel structure [45, 46].

5.3.5 Elastin

Because of some unique amazing properties elastin-based biomaterials are used effectively in the tissue engineering application. Elastin is a form of structural protein, which possesses some specific unique properties such as good stability, self-assembly, elasticity and excellent biological activity. Elastin transports elasticity to different organs and tissues mainly ligament, tendons, skin, blood vessels, and lungs, where the property of elasticity maintenance is a key aspect. Hence, in order to improve the property of elasticity in some tissues, mainly soft tissue such as ligament, skin, tendon, blood vessels, the polypeptides elastin can be effectively used [47].

The insertion of elastin like polypeptides (ELPs) in the vicinity of any biomaterials is highly important, where the impact of elastic property in part of our body is essential mainly in the areas of soft tissue and also capable of targeting the brain. For example, ELPs possesses the capability of restoring the thermally passive and receptive of some specific part of CNS in the treatment of the neurodegenerative defects [48]. The chemical structure of elastin is shown in Scheme 5.5.

Elastin is a magnificent biopolymer mainly effective for soft tissue engineering, normally ligaments, skin, lungs and extensively in the blood

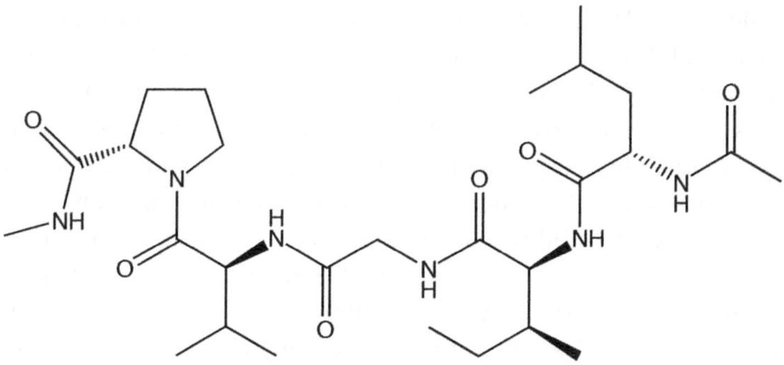

Scheme 5.5 Chemical structure of elastin.

vessels, which have the capability of relaxing and stretching almost billion times or more than that during the lifetime of an animal or human. Tropoelastin is a pioneer protein of elastin and in the peptide group of these protein molecules, which are present in the form of an intermolecular crosslinking arrangement. It is insoluble in water having half-life period ($t_{1/2}$) of about 70 years and is the most stable form of protein. Elastin is a sporadic amino acid containing almost 75% hydrophobic residues such as alanine, valine, and glycine. It is an almost insoluble protein because of the cross-linking bonds within the adjacent chains. The amazing protein tropoelastin is normally composed of three different varieties of domains established in distinct exons [49]. The first domain is hydrophobic in nature and consists of residues of alanine, valine, proline, and glycine and repeatedly occurs in the replication of several amino acids corresponding Gly-Val-Pro-Gly-Val, Gly-Val-Gly-Val-Pro and Gly-Val-Gly-Val-Ala-Pro. The second domain is hydrophilic in nature and consists of residues of alanine and lysine, which are responsible for cross-linking in between the adjacent chains. The residues of the amino acid lysine are repeatedly scattered between the residues of alanine e.g., Ala-Ala-Ala-Lys-Ala-Ala-Lys-Ala-Ala. The third domain of elastin is hydrophobic in nature and the lysine residues present in it are normally responsible for cross-linking the adjacent chain molecules. The hydrophobic domains neighboring both the crosslinking domains are promoting the process of self-assembly, hence elastin is considered as an appropriate biopolymer for application in soft tissue engineering [50].

5.3.6 Hyaluronic Acid

Hyaluronic acid (HA) is another important biopolymer, which is otherwise known as hyaluronan. It is a non-sulphated linear chain anionic glycosaminoglycan whose structure is composed of the repeating units of the disaccharides β-1, 4-D-glucuronic acid and β-1,3-N-acetyl-D-glucosamide. It functions as a lubricating material in the extracellular tissues of various parts of human body. HA is a suitable biopolymer for tissue engineering applications due to its astonishing properties such as biodegradability, non-toxicity, bioresorbability, easy availability, biocompatibility and easily forms hydrogel. HA has almost the same content of water present in human tissue and has excellent compatibility with human tissue and performs a significant role in facilitating the proliferation and differentiation of seed cells. Since the natural HA has small reservation time because of quick degradation through hyaluronidase, therefore proper modification, design and cross-linking of the active groups present in HA is highly essential. After proper modification and design, HA is widely used individually or in composite form as suitable candidate materials for tissue engineering applications. The HA-based scaffolds are highly effective for application in tissue engineering, because of some of its interesting properties such as necessary rate of biocompatibility, controllable rate of degradation through regulating the extent of cross-linking, required porosity for encapsulation of cells, excellent proliferation and differentiation [51, 52]. It is most effectively used in cartilage tissue. Because of the presence of a number of –OH and –COOH groups in the chemical structure of HA, it is found to be highly hydrophilic and capable of forming gel-like structure in aqueous medium owing to the columbic forces of attraction in between the adjacent chain of the polymer. The properties of good ionic exchange capacity, permeability and high rate of diffusion makes HA physiologically active for tissue engineering applications. Because of gel-like structure it functions as a lubricant at joint areas in our body, protects the penetration of microorganisms and toxic materials into the body. Due to its viscoelastic nature and that it is transparent aqueous medium it can be used effectively as filler materials for eye structures. The presence of HA serves as a vital role in the division of cells, angiogenesis, regeneration of tissues, wound healing, and cell migration. HA undergoes interaction with the receptors present at the surface of the stem cells such as transduction, intracellular signal and influences the function of the cell including cell movement, proliferation, survival, and differentiation. Hence, the HA based biomaterials have been used extensively as injectable hydrogels to fulfil the body fluid at the joint parts of bone. The chemical structure of HA is shown in Scheme 5.6.

The scaffold of HA is designed in such a manner that it should exhibit adequate cytocompatibility, biodegradability, biocompatibility, appropriate immune response, facilitate cell growth, and possess suitable mechanical strength at the site of implantation. It maintains the equilibrium between the porosity and mechanical strength and enhances the rate of degradation,

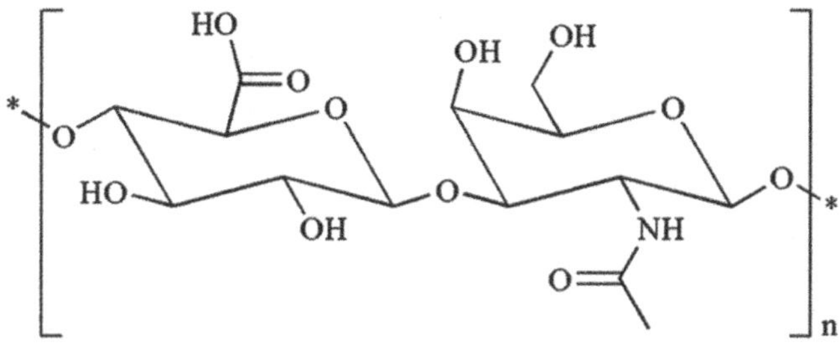

Scheme 5.6 Chemical structure of hyaluronic acid.

solubility, required biological properties, and viscosity. The use of the hydrogel form of HA enhances proliferation and also the longevity of neural precursors, therefore used for treatment of damaged nerve and CNS. The combination of HA and collagen is found to be highly effective for neural stem cells. The HA and chitosan blend can be effectively used for recovery of damaged peripheral nerves [53, 54].

5.3.7 Alginate

Alginate is an anionic biopolymer and mainly obtained from brown seaweed. The monomer of alginate contains two molecules such as α-L-guluronic acid (G) and β-D-mannuronic acid (M) bonded through (1–4)-linkage. The M and G molecules form three blocks: consecutive M residues, G residues and the alternating MG residues. Alginate can be easily processed into 3D scaffold materials like hydrogels, sponges, microcapsules, fibers microspheres and foams. The biomaterial based on alginate can be successfully utilized as cell carriers for application in tissue engineering. Alginate is an interesting biopolymer used widely owing to its unique properties such as good biocompatibility, excellent non-antigenicity, low cost, low toxicity, better capability of chelating, and properties of adequate gelation. The major source of natural alginate is marine resources, which contain toxic polyphenolic substances, proteins, heavy metals, and endotoxins, which is the major downside of natural alginate. The gel state of alginate is particularly highly effective in tissue engineering, and which offers the regeneration of the defective blood vessels, bone, cartilage, pancreas, liver, muscle and repair of certain peripheral nerves [55, 56]. The structure of alginate is shown in Scheme 5.7.

The application of alginate gel improves wound area and prevents pathogens entering it. The use of gauze is again the cause of some secondary

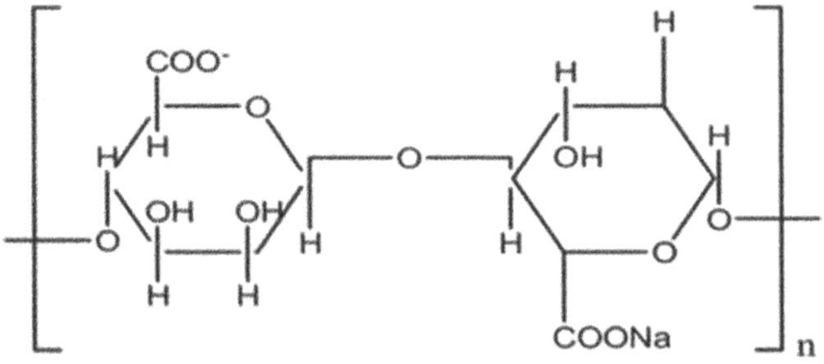

Scheme 5.7 Chemical structure of alginate.

injury during peeling off, so improved quality of novel wound dressing substances has been developed and designed, which produce a moist occlusive environment to facilitate the healing process. Many categories of dressing materials based on alginate, such as gel, sponge, semi-occlusive or occlusive, have been designed and which are highly effective. The dressing based on alginate such as sponges, electrospun mats and hydrogels is designed in such a way that it behaves as a promising material for wound healing and provides many beneficial characteristics such as homeostatic ability and capability of gel formation, which absorbs wound exudate fluids. Alginate has many desirable properties required for wound dressing, which include excellent water absorptivity, optimum rate of water vapor transmission, conformability and slight antiseptic properties, and necessary biodegradability and nontoxicity. Some specific alginate dressing materials such as Kaltostat, accelerate the healing of the wound area through stimulation of monocytes and increases the level of cytokines, specifically tumor necrosis factor-α and interleukin-6. The alginate hydrogel is designed in such a way that its physical properties tally with the properties of articular cartilage and the mechanical properties comparable with the native tissue. Hence, gel-microspheres, injectable hydrogels and solids have been utilized successfully in the regeneration and production of cartilage.

The formation of new bone at damaged sites is a great challenge for doctors in the reconstructive surgery. The major downside of bone tissue is when tumors are removed or trauma sites, and the necessary strategy is *in situ* introduction of osteogenesis. The injectable scaffold of hydrogel uses stem cells, which can differentiate to synthesize bone tissue. The scaffold of alginate on coupling with mesenchymal stem cells (MSCs) is capable of synthesizing required bone tissue. Alginate can be utilized for creating tissue in gels, which exhibit both angiogenic and osteogenic properties [57, 58].

5.4 Applications of Biopolymer Composite for Tissue Engineering

The biopolymers and their composites have many significant benefits over synthetic polymeric materials. The biopolymer-based composites in different forms used in tissue engineering applications are shown in Figure 5.4.

Polysaccharide-based composite materials are a proper alternative candidate material of metals, alloys, and ceramics for application in tissue engineering, because of many surprising properties such as it having good mechanical stability, and it being multi-functional, sustainable, ecofriendly, biocompatible, and biodegradable. Hence biopolymer composite material is considered an ideal sustainable candidate material for therapeutic effectiveness, wound cure, and some other tissue engineering [59].

5.4.1 Application in Bone Tissue Engineering

Bone tissue engineering (BTE) is a challenging technique, where the 3D bone scaffolds contain bioactive molecules and viable cells. Bone functions in numerous ways, stimulating human physiologic processes such as homeostasis, metabolic demand, development, and reproduction. The load-bearing skeleton is supported by bone and also keeps our internal organs safe. Bone retains the crucial electrolytes homeostasis by storing

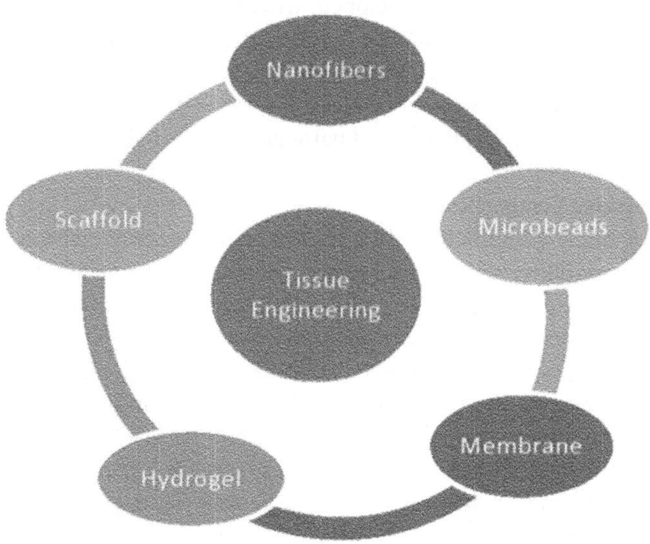

Figure 5.4 Different forms of biopolymer composite used in tissue engineering.

PO_4^{3-} and Ca^{2+} ions, absorbing the toxic metals into its porous matrix. Bone is a vital tissue, which undergoes continuous reconstruction and resorption via the exchange of fluids and chemicals. Bone is an extremely complex tissue on the basis of architectural and functional diversity. The extracellular matrix (ECM) of bone specially consists of inorganic mineralized constituent and organic non-mineralized matrix. The bone tissue engineering is normally meant for designing scaffolds to facilitate angiogenesis, which includes growth factors with the porous structure that are required for vascular growth [60].

5.4.1.1 Collagen-Based Composite

The composite of type I collagen and $Ca_3(PO_4)_2$ is found to be ideal material for fabrication of suitable bone scaffold. The type I collagen and nano HA composite that was developed, exhibits micro- and macroporous structure (1–5 μm and 50–100 μm respectively) owing to multilevel freezing. The composite formed by using nanoparticles of bioactive glass (nBG) and collagen with phosphatidylserine (PS) binds Ca^{2+} ions with greater affinity and supports repairing of tissues. The microcomposite of collagen on combination with microparticles (mHA) produces a scaffold that stimulates osteogenesis and angiogenesis *in vivo*. The same composite, if synthesized through freeze-drying process, shows enhanced mineralization, bioactivity, and completely removed the defect *in vivo*. The scaffold made of the composite of glycosaminoglycan (GAG), collagen and cobalt merged bioactive glass (CoBG) of microparticles was developed, which stimulates vascularization, angiogenesis and induces osteoblast proliferation. But, however, the major problem of using collagen composite is the generation of allergic reactions. The major source of natural collagen is porcine or bovine, which is the cause of increasing the probability of pathogen transmission [61, 62].

5.4.1.2 Gelatin-Based Composites

Gelatin possesses many benefits including biocompatibility, easy commercialization, biodegradability, low cost, and removes the immunogenicity associated with it and prevents the transmission of pathogen related with collagen. The gelatin-based composite can be synthesized in a number of ways on combination with suitable polymers or ceramic materials. The process of synthesis of composite by electrospraying of nHA and electrospinning of gelatin possesses similar structure to extracellular bone matrix. The scaffold of Gel/nHA and gelatin synthesized by electrospun process increases the cell proliferation and mineralization. The mechanical properties of the Gel/nHA scaffold increases by addition of gellan gum and polycaprolactone (PCL). The gellan gum raises the elastic modulus of synthesized scaffold, whereas the presence of PCL reduces the final strain, but, however, the composite shows adequate cell proliferation and adhesion. A nanofibrous scaffold synthesized through electrospinning gelatin and polyvinyl alcohol (PVA) along with nanoparticles of β-caryophyllene (BCP) exhibits high tensile strength, good cell adhesion and accelerate the bone creation *in vivo*. The porous scaffold synthesized through freeze drying process using gelatin and nano–submicro octacalcium phosphate (OCP), which serves as an appropriate material for the bone apatite crystals. It is extremely biodegradable and nearly fully replaces the newly synthesized bone *in vivo* but exhibits lower elastic modulus of about 250 MPa. A new and novel scaffold of gelatin–silicon doped with hydroxyapatite microparticles (mSiHA) synthesized by rapid prototyping process shows compression behavior almost same as trabecular bone [63].

5.4.1.3 Chitosan-Based Composites

Chitosan is a well-known polysaccharide and the composite synthesized by using chitosan exhibits many benefits such as low immunogenicity, appreciable antibacterial properties, good biodegradability, and biocompatibility. The composite of chitosan is effective towards bone tissue because it stimulates the deposition of mineral rich matrix and enhances the growth of osteoblasts. The scaffolds of porous CS–nHA synthesized by *in situ* precipitation process shows enhanced adhesion and proliferation of osteoblast cells and almost six times more alkaline phosphatase (ALP) activity compared to scaffolds of pure chitosan. The CS–nBG films synthesized by solvent casting technique shows improved mineralization and cell differentiation as compared to only chitosan scaffold. The incorporation of gelatin to the CS–nBG composite enhances the cell attachment and viability, hence in order to remove the difficulties of release of more ions, a new technique known as micro-contact printing is used, which controls and regulates the distribution of the density of nBG at the surface of chitosan. The biological response of CS–nHA scaffold can be enhanced by blending with gelatin and the mechanical properties are improved by addition of PCL. The addition of pectin/gelatin and functionalized multiwalled carbon nanotube (CNT) also increases both biological response and mechanical properties of CS–nHA scaffold. The thin film synthesized by using HA and CS through the technique of solvent casting shows improved tensile strength and elastic modulus. The paste made by the composite of β-TCP and CS with microparticles of CaO and ZnO enhances the bone growth rate *in vivo*. The rate of growth of the woven bone tallies with the rate of degradation of the composite material. The major downside of using chitosan-based composite material in bone implants is bacterial infection, which can be improved by addition of some metal nanoparticles such as nanoparticles of Ag, Zn, or Cu [64, 65].

5.4.1.4 Natural Polymers Synthesized Through Bacteria

Some of the natural polymers extracted by bacteria normally exhibit good biocompatibility and physical properties, which are similar to that of synthetic polymeric materials. Polyhydroxyalkanoates (PHA) is the most important class of the biopolymer synthesized through bacteria. The biopolymer poly (3-hydroxybutyric acid) (P3HB) is a member of PHA family, which stimulates the bone growth owing to its properties of piezoelectricity. P3HB on decomposition generates 3-hydroxybutyric acid, which is one of the constituents of human blood, therefore enhances the calcium influx to the cells and also prevents apoptosis, which makes it extremely non-toxic and biocompatible. The composite of P3HB–nHA synthesized by solvent-casting technique exhibits noticeable cell differentiation and proliferation and greater ALP activity when compared to polymer only. The appropriate percentage of chitosan addition to this composite material has the property of crystallinity of PHB resulting the formation of a soft, porous scaffold, and can be designed to any shape and size. Although the composite of nBG and P3HB shows excellent bioactivity and hardness with enhanced ALP activity, an increase in the concentration of nBG retards the cell proliferation. Polyhydroxybutyrate-co-(3-hydroxyvalerate) (PHBV) is an important copolymer of polyhydroxyvalerate and P3HB, which exhibits higher degradation time as compared with P3HB. The PHBV–nHA nanofibrous scaffold synthesized through the technique of electrospinning by utilizing silk fibroin exhibits enhanced cell attachment and bioactivity as compared to the only PHBV scaffold. But with an increase in silk fibroin and HA the elastic modulus of the scaffolds decreases [66].

5.4.1.5 Composite of Some Other Biopolymers

The composite of β and α chitin with nHA on hydrogel form performs good cell attachment, proliferation and exhibits improved mechanical strength of many fibroblast cell lines as compared to any individual polymer. Chitin on combination with nBG and nano-silica shows excellent bioactivity and enhanced cell viability because of the presence of silica. Since silk fibroin (SF) possesses excellent mechanical strength and biocompatibility, the composite of nHA with SF proves itself to be a good candidate material for bone tissue engineering. Its properties are similar to those of the natural bone of the human body and stimulate quick bone recovery within the cavity. The composite of the SF reinforced with microparticles of HA synthesized by porogen leaching and solvent casting technique shows eight times more Young's modulus, if refined with human mesenchymal stem cells (hMSCs), because of the increase in the structural connectivity associated with the scaffold. The porous scaffold synthesized by using wollastonite

($CaSiO_3$) and SF shows excellent degradability, bioactivity and enhanced mechanical properties. The porous scaffold of it having porosity of 81%–85%, exhibits enhanced hydrophilicity, if the percentage of ceramic phase increases, then the scaffold cell proliferation increases. The nano-needles of titania (TiO_2) on incorporation with alginate shows a lower rate of degradation, improved cell attachment, and controlled swelling compared with a scaffold of alginate. An injectable form of hydrogel synthesized by $Ca_3(PO_4)_2$ with the biopolymer alginate performs as a negligibly hostile bone filling material [67, 68].

5.4.2 Skin Tissue Engineering

Skin is treated as the most sensitive and vital organ in human body, because of its multifunctional activity such as protection and safety of internal organs, regulating and controlling the body temperature, and functioning as a sense organ. The skin of our body is composed of three different layers: epidermis, hypodermis, and dermis.

- The outer layer is waterproof and is called the epidermis. It serves as a protective barrier against foreign substances and pathogens and plays a vital role in the regulation of humidity and controlling body temperature. The cells of the epidermis are mainly composed of keratinocytes (90%). Most of the portions of the epidermal cell are populated by melanocytes, Merkel cells and Langerhans cells.
- Almost 90% of the weight of skin is due to the dermis component, and it serves as the base of our skin. The dermis layer is soft tissues consisting of ECM of many kinds of cells including hair follicles and glands, and which is fully vascularized by blood vessels. It also contains nerve endings. Fibroblasts are the major dermal cells, which consist of elastin and collagen and provide good mechanical strength and elasticity to our skin.
- The deepest layer, having elastic fibrous tissue along with the mucous properties, is known as the hypodermis. This layer stores deposited fat and contains nerves and blood vessels. Traumas are mainly fire hazards, poisoning, chronic disease, damage due to surgery and physical penetration. The hypodermis protects against injury, dehydration, and infection [69, 70].

Currently there are no major prototypes of skin that possesses adequate mechanical and biochemical characteristics to control and regulate non-antigenic, disinfection, non-toxic and inflammatory effects. The 3D nanofibrous structures of biopolymer composite material offers many benefits such as excellent mechanical strength, and greater surface area to volume ratio. Various kinds of hydrogel have been developed for the regeneration of skin by using either natural biopolymers such as alginate, cellulose, collagen, hyaluronic acid (HA), silk, and chitosan, or synthetic polymeric materials such as PGA and PVA. The fibrous scaffolds based on collagen on combination with Integra, Terudermis and Biobrane are commercial products used for skin treatment. Recently, modified HA in combination with chondroitin sulphate, has been used effectively for skin proliferation and regeneration. The alginate-based thin films are highly effective for implantation of burn injury skin. The stem cells entrenched in epidermal layers stimulates the regeneration and repair of the epidermis. Both allogeneic and fibrin dermis epidermal stem cells sustain their wound healing capability. Hence, implantation of such epidermal stem cells is the cause of deficiency in appropriate skin regeneration. Therefore, the modification in the skin morphogenesis and improvement in the cell biology are highly essential for skin regeneration, supporting the normal aesthetics and practicality of healthy skin [71].

5.4.3 Wounds Healing Application

Wounds are normally uneven skin deformities or breakage; puncture owing to the physical or thermal damage, or because of some chronic diseases.

On the basis of healing techniques, usually wounds may be classified as either acute or chronic. Chronic wounds are normally tissue cuts or scratches that heal fully within 8 to 12 weeks, whereas the acute wound continues to recur having the recovery time often of more than 12 weeks. Chronic wounds include venous or ischemic or leg ulcers. The injuries or damage affect the deep layers of epidermis, muscles, follicles, muscles, and soft tissue. In addition to the dermal and epidermis surface, wounds are also predominantly associated with deep tissue or hypodermic fat. The physical and biological process of wound regeneration include the coordinated cooperation between various biological systems [72]. The use of biomaterials and biocomposites synthesized from chitin, silk, hyaluronate, cellulose, collagen, alginate, and gelatin are found to be more popular and widely used because of bioactivity, and antimicrobial, angiogenic, biocompatibility, and immunomodulatory properties. Now, the bioactive glass on combination with some other biopolymers like gelatin, collagen, and chitosan can be suitably used in the blood clotting and wound healing applications. The thermal and mechanical properties of various biopolymers are considerably enhanced by addition of kaolin. A multifunctional hemostatic composite of kaolin–polyurethane (PU) foam was synthesized by adding kaolin in the robust PU foam by a one-step process. The mechanical properties of kaolin are controlled by the concentration of kaolin. The presence of 5% wt. or less offers good elasticity property and extraordinary resilience with > 140% strain at break. The 5%–10% kaolin foam stimulates the hemostatic ability. The concentration of kaolin is independent from the absorption property of the PU foams. This foam can be effectively utilized in multifunctional wound dressings, because it possesses excellent properties of drug release, is mechanically robust, cytocompatibility, and has good absorption characteristics [73]. The composites of collagen, chitosan, zeolite, and alginate have been used to cure localized clotting owing to their inherent hemostatic characteristics. TiO_2 exhibits amazing antibacterial properties, therefore using TiO_2 as filler material or reinforcing material in the scaffolds of the biopolymer promotes the cell growth. The membrane of CS/TiO_2 composite possesses excellent properties of porosity, mechanical strength, flexibility, and crystallinity, therefore it can be used effectively in the treatment of wound healing survival of fibroblast cells, proliferation and improving mouse fibroblast cells. Since the composite delivers remarkable antibacterial activity against *Staphylococcus aureus*, it is a good candidate material for wound healing [74]. Banana peel possesses good properties, mainly higher percentage of minerals, antioxidants, phenolic structure, and antimicrobial characteristics against bacteria, fungi, and yeast. Hence, recently, a new, impressive membrane of nanocomposite was synthesized from chitosan and banana peel powder (BPP), where the BPP functions as both cross-linking agent and reinforcing material for chitosan. The membrane of CS–BPP with 10% by wt. of BPP shows remarkable sensitivity and activity towards *Candida albicans*. Recently, the particles of nanocellulose (NC) have been utilized effectively as wound-care dressings due to its excellent mechanical strength, elasticity, conformability, and capacity for holding water. The particles of NC show high surface to volume ratio, which facilitates the cell migration and proliferation. Now, a new promising bionanocomposite have been synthesized by using poly(vinyl pyrrolidone) (PVP), using NC as the filler material and chitosan as matrix, which shows excellent performance for wound dressings because of improved swelling capacity and tensile strength. It was observed that the addition of 5% wt. of NC exhibits outstanding cytotoxicity, blood compatibility and extraordinary antibacterial activity against gram-positive and gram-negative bacteria. Silk sericin is an important bioprotein obtained from silkworms that possesses remarkable antibacterial properties, oxidation resistance and UV resistance, with good capability of moisture release and absorption [75].

Another interesting composite material (hydrogel) synthesized by using semi-interpenetrating polymer network (semi-IPN) on cross-linking with nanoclay lithium magnesium silicate hydrate (LMSH) with silk sericin, which exhibits highly effective performance in every kind of wound healing or dressing [76].

5.4.4 Oral Tissues

Oral tissue delivers a vital function in human body. Dental caries and periodontitis, and some other factors, are causes of injury to oral tissues. The constituents present within the oral cavity may be destroyed by corrosion due to exposure to body fluids within the mouth, and the synthesis of cytotoxic substances. A structure of tooth along with dentin and enamel was first produced from porcine tooth buds. A composite was synthesized by utilizing poly (lactic-co-glycolic) acid (PLGA) and polyglycolic acid (PGA), which was used for cell cultivation of the roots of the teeth and used for the growth of tooth tissue. The scaffolds of biodegradable polymer normally used for design of 3D salivary glandular tissue are used for dental tissue engineering. Another new composite was developed by using PLA and HA, which is used to generate temporomandibular joints and for implantation of some specific cells to form vascular bones along with osteochondral tissue [77].

5.4.4.1 Biopolymer Composite on Dental Application

Teeth are found to be the hardest biological material present in the human body. The tooth of human beings composed of three kinds of tissue: enamel, dental pulp, and dentin.

- Enamel is the outer or external layer of the tooth and is found to be most highly mineralized and hardest material present the body. Enamel creates an impenetrable and tight joint over the dentin, which is synthesized through the ameloblast cells of ectodermal origin. The composition of the enamel is approximately mineral of 97 wt.%, water 2 wt.%, and organic material 1 wt.%. The organic material is composed of enameline, insoluble and soluble form of proteins with low percentages of lipids and carbohydrates.
- The dentin is produced by cementoblast or odontoblast depending on whether the dentin is situated in the tooth root or crown. The dentin is present in between the pulp, enamel and with both these two structures produces the functional material termed as tooth. Dentin is a highly complex structure and divided into three layers. The primary layer of dentin has lower mineral tissue when compared to enamel. Dentin contains hydroxyapatite of 70 wt.%, organic material of 20 wt.% and water of 10 wt.%. The matrix structure of it is synthesized by using type I collagen and also

collagen-associated substances with assimilation of the crystals of hydroxyapatite in a haphazard way. This kind of dentin forms the key part of the roots and crown and is synthesized earlier as the tooth enters into the occlusion. The primary dentin consists of one external layer and one interior layer, where the interior layer is more mineralized as compared to the external layer, because the external layer contains larger collagen tubes having size 150 μm. The primary dentin lies over the secondary one. The process of formation of this kind of dentin originates after the crown comes to the occlusion. The pulp chamber becomes squeezed at the time of synthesis of secondary dentin. When the tooth is damaged and becomes irregular structure, only tertiary dentin is produced. The resin-based composite materials have to be applied as restorative systems because of the need for aesthetic restorations [78, 79].

5.4.4.1.1 Bio-Composites of Silver Nanoparticles

The use of conventional antibiotics in membranes enhances the resistance to bacterial risk. Hence, because of them having fewer tendencies to promote resistance to antibiotics, silver nanoparticles are incorporated to the membranes to stimulate antibacterial activity. The membrane synthesized by using Ag nanoparticles with collagen in colloidal solution is suitable for periodontal regeneration therapy [80].

5.4.4.1.2 Chitosan-Based Composites

The membrane, developed by using composite synthesized by using poly(ethylene oxide) (PEO) and chitosan through the technique of electrospinning with addition of nano-Ag particles, exhibits excellent biological performance and antibacterial potential. Another membrane synthesized by using chitosan and polyurethane (PU) on addition with nano-Ag particles also exhibits very strong antibacterial activity and promotes tissue regeneration efficiently [81].

5.4.4.1.3 Hydroxyapatite-Based Composites

The major components of dentine are hydroxyapatite (HA) and collagen. Hence, hydroxyapatite is taken as a suitable selection for dental and bone regeneration. The composite synthesized by using chitosan matrix reinforced with HA powder shows enhanced mechanical properties, therefore the membranes of this composite are highly suitable for dental application. The presence of HA influences the rate of degradation and the mechanical strength of the prepared composite membranes, therefore indicating potential application for dental treatment. The microspheres based on nanohydroxyapatite/chitosan composite synthesized by a solution casting technique shows remarkable results. A novel blend of gelatin/chitosan membrane with titania nanoparticles and HA inhibit oral pathogens and microorganisms. The membrane developed exhibits the properties of biodegradability, osteoconductivity and biocompatibility, which are necessary for dental tissue engineering [82].

5.4.4.1.4 Composites Based on Calcium Phosphate

The biocomposite biopolymer matrix reinforced with $Ca_3(PO_4)_2$ releases the phosphate and calcium ions into the oral environment, therefore its deposition in the teeth serves as an apatitic substance in tooth remineralization. The hybrid composite material synthesized by using PLA matrix with carbon nanotubes and $Ca_3(PO_4)_2$ facilitates the stimulation of the osteoblastic differentiation and is found to be a remarkable material for hard tissue implants specially in the dental field. The membranes of PLA have been incorporated with the $Ca_3(PO_4)_2$ to improve mechanical strength and other properties that are required for dental treatment. Tricalcium phosphate (TCP) is a well-known resorbable and bioactive ceramic material found in crystalline forms β-TCP and α-TCP and the chemical composition of it is similar to the mineral phase of the bone tissue. HA also contains high percentage of $Ca_3(PO_4)_2$, which fulfils most of the inorganic materials of the bone tissue and impacts the bioabsorption and bioactivity properties. The composite synthesized by using chitosan and β-TCP with collagen matrix forms membranes for GBR with adequate mechanical properties [83].

5.4.4.1.5 Composites of Bioactive Glass

The composite synthesized by using different common biopolymers with PCL, PLLA, PLGA and PEG with bioactive glass exhibits glassy structure with excellent properties of bioactivity and biocompatibility, therefore is used as a suitable candidate material for dental restoration. The scaffold of PCL and gelatin synthesized by electrospin techniques with the incorporation of nanoparticles of bioactive glass is an amazing novel material for dental tissue engineering. This novel material, on loading with dexamethasone (odontogenic drug), promotes odontogenesis [84].

5.4.4.1.6 Hybrid Films

Instead of using single bioactive membranes, there are a number of benefits of using multiphasic and hybrid membranes, which are found effective for periodontal tissue engineering. The different multiphasic membranes are developed and designed with variation in the porosity, structure, composition, and technique of fabrication. The hybrid membranes are synthesized by the optimal combination of various biopolymers. The membranes of chitosan /collagen blend after synthesis were cross-linked through γ-ray irradiation treatment. The membrane produced possesses high mechanical strength with improved cell proliferation. The membrane synthesized by using elastin/gelatin/sodium hyaluronate is a suitable material for bio-ink. The most interesting membrane designed is the central layer of chitosan sandwiched by two membranes of collagen, which exhibits outstanding structural and mechanical properties to serve as barriers in guided dental and bone tissue regeneration [85].

5.4.5 Blood Vessels

Repairing of damaged blood vessels is a challenge for medical science because many of the artificial scaffolds are found to be unsuitable for this application and require a longer time for the recovery of the patient. To mimic the morphology and function of the electrospinning, ECM have been applied widely for designing 2D and 3D fibers by utilizing some specific artificial and natural polymers. The electrospun tubular scaffolds, having small size, are synthesized by using different synthetic polymeric materials with vascular proteins to address these difficulties. Bacterial nanocellulose (BNC) fabricated by the dimers such as β-1,4 linked D-glucose offers some novel properties like excessive water holding capability, ultrafine nanofiber matrix, adequate crystallinity, proper chemical purity, outstanding biocompatibility, and wet tensile strength. The tubes of BNC composite on doping with PVA function as artificial blood vessels. The tensile strength and water permeability were raised after BNC tubes were blended with PVA. The composite of D-BNC/PVA shows very good tensile strength as compared to S-BNC/PVA composite [86, 87].

It was found recently that using the bilayer composite of chitosan/gelatin it is possible to mimic the tensile strength and structural properties of the natural biological macroporous blood vessels. The macroporous layers provide a huge surface area for cell adhesion and proliferation. The external nonporous layer provides higher elasticity, cell protection ability and flexibility. The scaffolds of both random and associated nanofiber stimulated the feasibility and the cardiomyocyte attachment. The PCL on blending with gelatin fiber forms a novel interesting material, which exhibits enhanced interactions of cell–matrix as blood vessel scaffolds. The composite of P7G3-C (70% PCL and 30% gelatin) is treated as a potential template for the scaffolds of blood vessel tissue [88].

5.5 Conclusion

Among the various kinds of materials, polymer is considered a promising material for different kinds of biomedical implants and tissue engineering. The effort must be to balance the industrial productivity and capability of the planet to generate resource materials with the proper neutralization and management of wastes to nullify the extra pollution load to our environment. The materials used in the healthcare or pharmaceutical industry generate a high volume of biomedical wastes, which largely contributes to increasing the overall pollution load of our globe and creates negative impact not only to human health, but also our whole ecosystem. Nature provides us many important materials, such as biopolymers, to maintain environmental sustainability. Hence, in order to protect depletion of valuable stock resources, biopolymers, having the properties of biocompatibility and biodegradability, are used as a promising material for different medical applications such as tissue engineering and drug release. These days many biopolymer composites are developed according to the requirement of clinical therapy, which possesses high tensile strength and mechanical properties as compared to the single biopolymer. The poor dispersion and mechanical properties are the key downsides of biopolymers and biopolymer-based composite materials. The various kinds of biopolymers such as keratin, chitosan, alginate, HA, gelatin, elastin collagen, and synthetic biomaterials such as PGA, PLA, PCL, bio-PET, PEG, PLGA etc., and their composites have been successfully utilized for various soft and hard tissue engineering including wound healing, bone tissue regeneration, cartilage repairing, blood vessel generation and dental tissue engineering. In spite of several advantages of using the biopolymers and its composites, these are still not fully commercialized as they are unable to replace the synthetic polymers, ceramics, metals and their composites in tissue engineering applications.

References

1 Biswal, T. (2020). Biopolymers for tissue engineering applications: a review. *Materials Today: Proceedings* 41 (2): 397–402. doi: 10.1016/j.matpr.2020.09.628.

2 Reddy, M., Sai, B., Ponnamma, D. et al. (2021). A comparative review of natural and synthetic biopolymer composite scaffolds. *Polymers* 13: 1105–1155. doi: 10.3390/polym13071105.

3 Khan, M.U.A., Razak, S.I.A., Arjan, W.S. et al. (2021). Recent advances in biopolymeric composite materials for tissue engineering and regenerative medicines: a review. *Molecules* 26 (3): 619–637. doi: 10.3390/molecules26030619.

4 Aaliya, B., Sunooj, K.V., and Lackner, M. (2021). Biopolymer composites: a review. *International Journal of Biobased Plastics* 3 (1): 40–84.

5 Koyyada, A. and Orsu, P. (2021). Natural gum polysaccharides as efficient tissue engineering and drug delivery biopolymers. *Journal of Drug Delivery Science and Technology* 63: 1–19. doi: 10.1016/j.jddst.2021.102431.

6 Kohane, D. and Langer, R. (2008). Polymeric biomaterials in tissue engineering. *Pediatric Research* 63: 487–491. doi: 10.1203/01.pdr.0000305937.26105.e7.

7 Salerno, A. and Pascual, C.D. (2015). Bio-based polymers, supercritical fluids, and tissue engineering. *Process Biochemistry* 50 (5): 826–838. doi: 10.1016/j.procbio.2015.02.009.

8 Nemati, S., Kim, S., Shin, Y.M., and Shin, H. (2019). Current progress in application of polymeric nanofibers to tissue engineering. *Nano Convergence* 6 (1): 36–51. doi: 10.1186/s40580-019-0209-y.

9 Nathanael, A.J. and Oh, H.T. (2020). Biopolymer coatings for biomedical applications. *Polymers* 12: 3061–3086.

10 George, A., Sanjay, M.R., Srisuk, R. et al. (2020). A comprehensive review on chemical properties and applications of biopolymers and their composites. *International Journal of Biological Macromolecules* 154: 329–338. doi: 10.1016/j.ijbiomac.2020.03.120.

11 Kumaran, S.K., Chopra, M., Oh, E., and Choi, H.-J. (2020). Biopolymers and natural polymers. In: *Polymer Science and Nanotechnology* (ed. R. Narain), chapter 11, 245–256. Amsterdam: Elsevier. doi: 10.1016/b978-0-12-816806-6.00011-x.

12 Aggarwal, J., Sharma, S., Kamyab, H., and Kumar, A. (2020). The realm of biopolymers and their usage: an overview. *Journal of Environmental Treatment Techniques* 8 (2): 1005–1009.

13 Seddiqi, H., Oliaei, E., and Honarkar, H. (2021). Cellulose and its derivatives: towards biomedical applications. *Cellulose* 28: 1893–1931.

14 Agnihotri, S. and Dhiman, N.K. (2017). Development of nano-antimicrobial biomaterials for biomedical applications. In: *Advances in Biomaterials for Biomedical Applications*. Advanced Structured Materials, 66 (ed. A. Tripathi and J. Melo), 479–545. Singapore: Springer. doi: 10.1007/978-981-10-3328-5_12.

15 Jiménez-Gómez, C.P. and Cecilia, J.A. (2020). Chitosan: a natural biopolymer with a wide and varied range of applications. *Molecules* 25 (17): 3981–4023. doi: 10.3390/molecules25173981.

16 Nosrati, H., Pourmotabed, S., and Sharifi, E. (2018). A review on some natural biopolymers and their applications in angiogenesis and tissue engineering. *Journal of Applied Biotechnology Report* 5 (3): 81–91. doi: 10.29252/jabr.05.03.01.

17 Aisyah, H.A., Tahir, P.M., Sapuan, S., and Abdan, M. (2020). A review of biocomposites in biomedical application. In: *Composites in Biomedical Applications*, 31–48. CRC Press. doi: 10.1201/9780429327766-2.

18 Bharath, K.N. and Basavarajappa, S. (2016). Applications of biocomposite materials based on natural fibers from renewable resources: a review. *Science and Engineering of Composite Materials* 23 (2): 1–11. doi: 10.1515/secm-2014-0088.

19 Francis, R., Sasikumar, S., and Gopalan, G.P. (2013). Synthesis, structure, and properties of biopolymers (natural and synthetic). In: *Polymer Composites* (ed. S. Thomas, K. Joseph, S.K. Malhotra et al.), 11–107. Weinheim: Wiley-VCH. doi: 10.1002/9783527674220.ch2.

20 Rebelo, R., Fernandes, M., and Fangueiro, R. (2017). Biopolymers in medical implants: a brief review. *Procedia Engineering* 200: 236–243. doi: 10.1016/j.proeng.2017.07.034.

21 Wróblewska-Krepsztul, J., Rydzkowski, T., Michalska-Pożoga, I., and Thakur, V.K. (2019). Biopolymers for biomedical and pharmaceutical applications: recent advances and overview of alginate electrospinning. *Nanomaterials* 9 (3): 404–4226. doi: 10.3390/nano9030404.

22 Ulery, B.D., Nair, L.S., and Laurencin, C.T. (2011). Biomedical applications of biodegradable polymers. *Journal of Polymer Science Part B: Polymer Physics* 49 (12): 832–864. doi: 10.1002/polb.22259.

23 Pattanashetti, N.A., Heggannavar, G.B., and Kariduraganavar, M.Y. (2017). Smart biopolymers and their biomedical applications. *Procedia Manufacturing* 12: 263–279. doi: 10.1016/j.promfg.2017.08.030.

24 Hendi, A., Umair Hassan, M., Elsherif, M. et al. (2020). Healthcare applications of pH-sensitive hydrogel-based devices: a review. *International Journal of Nanomedicine* 15: 3887–3901. doi: 10.2147/ijn.s245743.

25 Altomare, L., Bonetti, L., Campiglio, C.E. et al. (2018). Biopolymer-based strategies in the design of smart medical devices and artificial organs. *The International Journal of Artificial Organs* 41 (6): 337–359. doi: 10.1177/0391398818765323.

26 Wei, W., Zhu, M., and Wu, S. (2020). Stimuli-responsive biopolymers: an inspiration for synthetic smart materials and their applications in self-controlled catalysis. *Journal of Inorganic and Organometallic Polymers and Materials* 30: 69–87.

27 Hassan, M.E., Bai, J., and Dou, D.-Q. (2019). Biopolymers; definition, classification, and applications. *Egyptian Journal of Chemistry* 62 (9): 1725–1737.

28 Van de Velde, K. and Kiekens, P. (2002). Biopolymers: overview of several properties and consequences on their applications. *Polymer Testing* 21 (4): 433–442. doi: 10.1016/s0142-9418(01)00107-6.

29 Priyadarsini, M., Biswal, T., and Dash, S. (2019). Sustainable biocomposite its manufaturing processes and applications. *Egyptian Journal of Chemistry* 62 (4): 1151–1166.

30 Gurunathan, T., Mohanty, S., and Nayak, S.K. (2015). A review of the recent developments in biocomposites based on natural fibres and their application perspectives. *Composites Part A: Applied Science and Manufacturing* 77: 1–25.

31 León-Mancilla, B.H., Araiza-Téllez, M.A., Flores-Flores, J.O., and Piña-Barba, M.C. (2016). Physico-chemical characterization of collagen scaffolds for tissue engineering. *Journal of Applied Research and Technology* 14 (1): 77–85.

32 Laiva, A.L., Raftery, R.M., Keogh, M.B., and O'Brien, F.J. (2018). Pro-angiogenic impact of SDF-1α gene-activated collagen-based scaffolds in stem cell driven angiogenesis. *International Journal of Pharmaceutics* 544 (2): 372–379. doi: 10.1016/j.ijpharm.2018.03.032.

33 Glowacki, J. and Mizuno, S. (2008). Collagen scaffolds for tissue engineering. *Biopolymers* 89 (5): 338–344. doi: 10.1002/bip.20871.

34 Liu, Y., Ma, L., and Gao, C. (2012). Facile fabrication of the glutaraldehyde cross-linked collagen/chitosan porous scaffold for skin tissue engineering. *Materials Science and Engineering: C* 32 (8): 2361–2366. doi: 10.1016/j.msec.2012.07.008.

35 Dong, C. and Lv, Y. (2016). Application of collagen scaffold in tissue engineering: recent advances and new perspectives. *Polymers* 8 (2): 42–51.

36 Echave, M.C., Saenz del Burgo, L., Pedraz, J.L., and Orive, G. (2017). Gelatin as biomaterial for tissue engineering. *Current Pharmaceutical Design* 23 (24): 3567–3584.

37 Nikkhah, M., Akbari, M., Paul, A. et al. (2016). Gelatin-based biomaterials for tissue engineering and stem cell bioengineering. In: *Biomaterials from Nature for Advanced Devices and Therapies*, 37–62. Wiley-Blackwell. doi: 10.1002/9781119126218.ch3.

38 Gautam, S., Dinda, A.K., and Mishra, N.C. (2013). Fabrication and characterization of PCL/gelatin composite nanofibrous scaffold for tissue engineering applications by electrospinning method. *Materials Science and Engineering: C* 33 (3): 1228–1235. doi: 10.1016/j.msec.2012.12.015.

39 Liu, S., Qin, S., He, M. et al. (2020). Current applications of poly(lactic acid) composites in tissue engineering and drug delivery. *Composites Part B: Engineering* 108238. doi: 10.1016/j.compositesb.2020.108238.

40 Agarwal, T., Narayan, R., Maji, S. et al. (2016). Gelatin/Carboxymethyl chitosan-based scaffolds for dermal tissue engineering applications.

International Journal of Biological Macromolecules 93: 1499–1506. doi: 10.1016/j.ijbiomac.2016.04.028.
41 Bello, A.B., Kim, D., Kim, D. et al. (2020). Engineering and functionalization of gelatin biomaterials: from cell culture to medical applications. *Tissue Engineering Part B: Reviews*. doi: 10.1089/ten.teb.2019.0256.
42 Islam, M.M., Shahruzzaman, M., Biswas, S. et al. (2020). Chitosan based bioactive materials in tissue engineering applications-a review. *Bioactive Materials* 5 (1): 164–183. doi: 10.1016/j.bioactmat.2020.01.01.
43 Sultankulov, B., Berillo, D., Sultankulov, K. et al. (2019). Progress in the development of chitosan-based biomaterials for tissue engineering and regenerative medicine. *Biomolecules* 9: 470–485. doi: 10.3390/biom9090470.
44 Vepari, C. and Kaplan, D.L. (2007). Silk as a biomaterial. *Progress in Polymer Science* 32 (8–9): 991–1007. doi: 10.1016/j.progpolymsci.2007.05.013.
45 Bandyopadhyay, A., Chowdhury, S.K., Dey, S. et al. (2019). Silk: a promising biomaterial opening new vistas towards affordable healthcare solutions. *Journal of the Indian Institute of Science* 99: 445–487. doi: 10.1007/s41745-019-00114-y.
46 Tandon, S., Kandasubramanian, B., and Yakout, S.M. (2020). Silk-based composite scaffolds for tissue engineering applications. *Industrial & Engineering Chemistry, Research* 59: 40–58. doi: 10.1021/acs.iecr.0c02195.
47 Yeo, G.C., Mithieux, S.M., and Weiss, A.S. (2018). The elastin matrix in tissue engineering and regeneration. *Current Opinion in Biomedical Engineering* 6: 27–32. doi: 10.1016/j.cobme.2018.02.007.
48 Nettles, D.L., Chilkoti, A., and Setton, L.A. (2010). Applications of elastin-like polypeptides in tissue engineering. *Advanced Drug Delivery Reviews* 62 (15): 1479–1485. doi: 10.1016/j.addr.2010.04.002.
49 Daamen, W., Veerkamp, J., Vanhest, J., and Vankuppevelt, T. (2007). Elastin as a biomaterial for tissue engineering. *Biomaterials* 28 (30): 4378–4398.
50 Kowalczyk, T., Hnatuszko-Konka, K., Gerszberg, A. et al. (2014). Elastin-like polypeptides as a promising family of genetically engineered protein-based polymers. *World Journal of Microbiology and Biotechnology* 30: 2141–2152. doi: 10.1007/s11274-014-1649-5.
51 Rao, N.V. (2020). Hyaluronic acid–based hydrogels for tissue engineering. In: *Biomaterials for Organ and Tissue Regeneration* (ed. N.E. Varana, H. Knopf-Marques and J. Bathes), 551–565. doi: 10.1016/b978-0-08-102906-0.00014-3.
52 López-Ruiz, E., Jiménez, G., Álvarez de Cienfuegos, L. et al. (2019). Advances of hyaluronic acid in stem cell therapy and tissue engineering, including current clinical trials. *European Cells and Materials* 37: 186–213. doi: 10.22203/ecm.v037a12.
53 Dovedytis, M., Liu, Z.J., and Bartlett, S. (2020). Hyaluronic acid and its biomedical applications: a review. *Engineered Regeneration* 1: 102–113.

54 Zhu, Z., Wang, Y.M., Yang, J., and Luo, X.S. (2017). Hyaluronic acid: a versatile biomaterial in tissue engineering. *Plastic and Aesthetic Research* 4: 219–227.

55 Sun, J. and Tan, H. (2013). Alginate-based biomaterials for regenerative medicine applications. *Materials* 6 (4): 1285–1309. doi: 10.3390/ma6041285.

56 Sahoo, D.R. and Biswal, T. (2021). Alginate and its application to tissue engineering. *SN Applied Sciences* 3: 30–48. doi: 10.1007/s42452-020-04096-w.

57 Hernández-González, A.C., Téllez-Jurado, L., and Rodríguez-Lorenzo, L.M. (2019). Alginate hydrogels for bone tissue engineering, from injectables to bioprinting: a review. *Carbohydrate Polymers* 229: 1–52. doi: 10.1016/j.carbpol.2019.115514.

58 Lee, K.Y. and Mooney, D.J. (2012). Alginate: properties and biomedical applications. *Progress in Polymer Science* 37 (1): 106–126. doi: 10.1016/j.progpolymsci.2011.06.003.

59 Velema, J. and Kaplan, D. (n.d.). Biopolymer-based biomaterials as scaffolds for tissue engineering. *Advances in Biochemical Engineering/Biotechnology* 187–238. doi: 10.1007/10_013.

60 Yunus Basha, R., T.S. S. Kumar, and Doble, M. (2015). Design of biocomposite materials for bone tissue regeneration. *Materials Science and Engineering: C* 57: 452–463. doi: 10.1016/j.msec.2015.07.016.

61 Li, J., Baker, B.A., Mou, X. et al. (2013). Biopolymer/calcium phosphate scaffolds for bone tissue engineering. *Advanced Healthcare Materials* 3 (4): 469–484. doi: 10.1002/adhm.201300562.

62 Turnbull, G., Clarke, J., Picard, F. et al. (2018). 3D bioactive composite scaffolds for bone tissue engineering. *Bioactive Materials* 3 (3): 278–314. doi: 10.1016/j.bioactmat.2017.10.001.

63 Chen, K.-Y. and Yao, C.-H. (2011). Repair of bone defects with gelatin-based composites: a review. *BioMedicine* 1 (1): 29–32. doi: 10.1016/j.biomed.2011.10.005.

64 Venkatesan, J. and Kim, S.-K. (2010). Chitosan composites for bone tissue engineering—an overview. *Marine Drugs* 8 (8): 2252–2266. doi: 10.3390/md8082252.

65 Aguilar, A., Zein, N., Harmouch, E. et al. (2019). Application of chitosan in bone and dental engineering. *Molecules* 24 (16): 30093025. doi: 10.3390/molecules24163009.

66 Donnaloja, F., Jacchetti, E., Soncini, M., and Raimondi, M.T. (2020). Natural and synthetic polymers for bone scaffolds optimization. *Polymers* 12 (4): 905. doi: 10.3390/polym12040905.

67 Wang, Y., Cao, X., Ma, M. et al. (2020). A GelMA-PEGDA-nHA composite hydrogel for bone tissue engineering. *Materials* 13 (17): 3735. doi: 10.3390/ma13173735.

68 Urruela-Barrios, R., Ramírez-Cedillo, E., Díaz de León, A. et al. (2019). Alginate/gelatin hydrogels reinforced with TiO2, and β-TCP fabricated by microextrusion-based printing for tissue regeneration. *Polymers* 11 (3): 457. doi: 10.3390/polym11030457.

69 Silva, S.S., Rodrigues, L.C., Fernandes, E.M., and Reis, R.L. (2020). Biopolymer membranes in tissue engineering. *Biopolymer Membranes and Films* 141–163. doi: 10.1016/b978-0-12-818134-8.00006-7.

70 Bose, S., Koski, C.M., and Vu, A. (2020). Additive manufacturing of natural biopolymers and composites for bone tissue engineering. *Materials Horizons* 7: 2169–2170. doi: 10.1039/D0MH00277A.

71 Sahana, T.G. and Rekha, P.D. (2018). Biopolymers: applications in wound healing and skin tissue engineering. *Molecular Biology Reports*. doi: 10.1007/s11033-018-4296-3.

72 Yahya, E.B., Jummaat, F., Amirul, A.A. et al. (2020). A review on revolutionary natural biopolymer-based aerogels for antibacterial delivery. *Antibiotics* 9 (10): 648–664.

73 Lundin, J.G., McGann, C.L., Daniels, G.C. et al. (2017). Hemostatic kaolin-polyurethane foam composites for multifunctional wound dressing applications. *Materials Science and Engineering: C* 79: 702–709.

74 Anaya-Esparza, L.M., Ruvalcaba-Gómez, J.M., Maytorena-Verdugo, C.I. et al. (2020). Chitosan-TiO_2: a versatile hybrid composite. *Materials* 13 (4): 811–821. doi: 10.3390/ma13040811.

75 Kamel, N.A., Abd El-messieh, S.L., and Saleh, N.M. (2017). Chitosan/banana peel powder nanocomposites for wound dressing application: preparation and characterization. *Materials Science and Engineering: C* 72: 543–550. doi: 10.1016/j.msec.2016.11.104.

76 Yang, C., Xue, R., Zhang, Q. et al. (2017). Nanoclay cross-linked semi-IPN silk sericin/poly(NIPAm/LMSH) nanocomposite hydrogel: an outstanding antibacterial wound dressing. *Materials Science and Engineering: C* 81: 303–313. doi: 10.1016/j.msec.2017.08.008.

77 Díez-Pascual, A.M. (2019). Synthesis and applications of biopolymer composites. *International Journal of Molecular Sciences* 20 (9): 2321–2328. doi: 10.3390/ijms2009232.

78 Furtos, G., Silaghi-Dumitrescu, L., Lewandowska, K. et al. (2016). Biocomposites for orthopedic and dental application. *Key Engineering Materials* 672: 261–275. doi: 10.4028/www.scientific.net/kem.672.261.

79 da Silveira, J.V.W., Meireles, A.B., and Ferreira, E. de P. (2020). Biopolymer membranes for dentistry applications. *Biopolymer Membranes and Films* 243–272. doi: 10.1016/b978-0-12-818134-8.00010-9.

80 Gherasim, O., Puiu, R.A., Bîrcă, A.C. et al. (2020). An updated review on silver nanoparticles in biomedicine. *Nanomaterials* 10 (11): 2318. doi: 10.3390/nano10112318.

81 Tang, G., Tan, Z., Zeng, W. et al. (2020). Recent advances of chitosan-based injectable hydrogels for bone and dental tissue regeneration. *Frontiers in Bioengineering and Biotechnology* 8: 1–15.

82 Sharma, S. (2014). Biomaterials in tooth tissue engineering: a review. *Journal of Clinical and Diagnostic Research* 8 (1): 309–315. doi: 10.7860/jcdr/2014/7609.3937.

83 Ho, K.-N., Salamanca, E., Chang, K.-C. et al. (2016). A novel HA/β-TCP-collagen composite enhanced new bone formation for dental extraction socket preservation in beagle dogs. *Materials* 9 (3): 191–205. doi: 10.3390/ma9030191.

84 Hwang, P.T.J., Murdock, K., Alexander, G.C. et al. (2016). Poly(ε-caprolactone)/gelatin composite electrospun scaffolds with porous crater-like structures for tissue engineering. *Journal of Biomedical Materials Research Part A* 104 (4): 1017–1029. doi: 10.1002/jbm.a.35614.

85 Rokaya, D., Srimaneepong, V., Sapkota, J. et al. (2018). Polymeric materials and films in dentistry: an overview. *Journal of Advanced Research* 14: 25–34. doi: 10.1016/j.jare.2018.05.001.

86 Kakisis, J.D., Liapis, C.D., Breuer, C., and Sumpio, B.E. (2005). Artificial blood vessel: the Holy Grail of peripheral vascular surgery. *Journal of Vascular Surgery* 41 (2): 349–354. doi: 10.1016/j.jvs.2004.12.026.

87 Tang, J., Bao, L., Li, X. et al. (2015). Potential of PVA-doped bacterial nano-cellulose tubular composites for artificial blood vessels. *Journal of Materials Chemistry B* 3 (43): 8537–8547. doi: 10.1039/c5tb01144b.

88 Awad, N., Niu, H., Ali, U. et al. (2018). Electrospun fibrous scaffolds for small-diameter blood vessels: a review. *Membranes* 8 (1): 15–40. doi: 10.3390/membranes8010015.

6

Biopolymers, Composites, Nanocomposites, and Gels in Biotechnology

*Ayush Bhandari, Karishma Shetty, Anubhav Wadhwa, and Khushwant S. Yadav**

Shobhaben Pratapbhai Patel School of Pharmacy & Technology Management, SVKM'S NMIMS (Deemed to be University), Mumbai, India
** Corresponding author*

6.1 Biopolymers

Biopolymers are the biomolecules that are made of monomeric units that are covalently bonded together to form larger molecules. The prefix "bio" is related to materials that are associated with living creatures [1]. They comprise materials made from biological sources such as fats, vegetable oils, resins, proteins, sugars, and amino acids using synthetic chemistry.

Biopolymers are categorized in a variety of ways. Biopolymers based on their degradability can be either biodegradable or nonbiodegradable. They can be biobased and nonbiobased. When they are nonbiodegradable, they are known as bioplastics. Based on the polymer backbones, biopolymers can be classified into five types: polyesters, polysaccharides, polycarbonates, polyamides, and vinyl polymers. Depending on their use, biopolymers are categorized as biosurfactants, biodetergents, bioadhesives, bioflocculants, bioplastics, and other applications.

However, biopolymers have certain limitations that are not seen in synthetic polymers. Synthetic polymers outperform in terms of mechanical characteristics and thermal stability. They may be formed into a variety of shapes, whereas biopolymers can only be formed into a limited number of shapes [2, 3]. Apart from these drawbacks that biopolymers possess, they

Applications of Biopolymers in Science, Biotechnology, and Engineering, First Edition.
Edited by Arunima Reghunadhan, Akhina H., and Sabu Thomas.

are sustainable resources, which synthetic polymers are not. Biopolymers are also more cost-effective than synthetic polymers. Hence, systems developed using biopolymers are considered to be sustainable and are useful candidates for various applications [4].

Biopolymers can be used in their diluted and concentrated form. If used as dilute biopolymer solutions then they are extremely pseudo-plastic, have low adsorption, and have negligible visco-elasticity and yield stress, resulting in good injectivity and flow in porous media. If used as concentrated solutions then they have a higher apparent yield, which can be enhanced by controlled crosslinking.

6.1.1 Cellulose and Nanocellulose

The polysaccharide cellulose is a vast polymeric raw material with intriguing structure and characteristics. The highly functionalized, linear stiff chain homopolymer is characterized by its hydrophilicity, chirality, biodegradability, broad chemical modifying capacity, and production of varied semicrystalline fiber morphologies. It is formed by the repetitive connecting of d-glucose building blocks [5]. In cellulose, the monomer unit is β–1, 4 linked glucan chains, with hydrogen bonds forming between hydroxyl groups and oxygen atoms both inside a single glucose chain and between adjacent chains (Figure 6.1). To produce cellulose microfibrils, hydrogen bonding and van der Waals forces bring glucan chains together side by side [1].

Nanocellulose materials have evolved over the last decades as a fascinating new class of nanomaterials. They have unique properties, quantity, and renewability. Nanocellulose as a cellulosic appears to be a promising, novel, green biomaterial [6]. The term nanocellulose refers to a group of materials that include cellulose nanocrystals, bacterial nanocellulose (BNC) and cellulose nanofibrils (CNF).

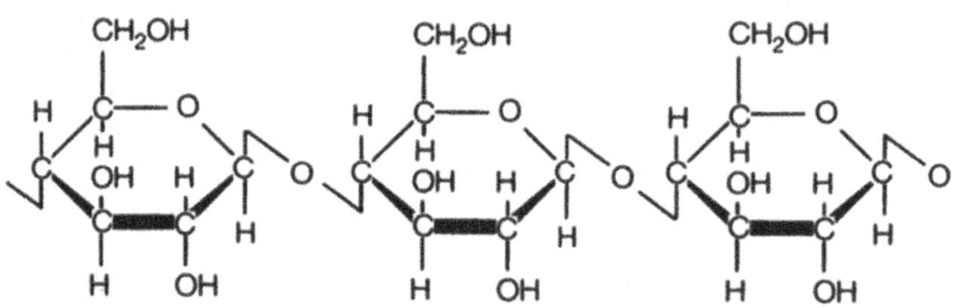

Figure 6.1 Structure of cellulose.

Cellulose nanocrystals (CNC) are cellulose derivatives generated from acid hydrolysis of cellulose, which comprises reacting cellulose with sulfuric acid at a regulated temperature and for a regulated time [7]. The CNC's diameter and length are determined by the cellulose source material and preparation parameters. The diameter, however, is measured in nanometres, while the length ranges from tenths of nanometres to some micrometers. Nanocellulose components and most amorphous cellulose sections are removed from the raw material during acid treatment, resulting in pure CNCs. Acids such as hydrochloric, phosphoric, and hydrobromic acids have also been used to isolate CNCs in addition to sulfuric acid [8].

Bacterial nanocellulose (BNC) is a promising natural biopolymer produced as an exopolysaccharide of -D glucopyranose by specific bacterial species. Even though it contains 99% water, BNC has excellent mechanical qualities [9]. Because of its strong tensile qualities, low density, and low toxicity, bacterial nanocellulose (BNC) is a promising material for the creation of high-performance renewable composites [10].

Cellulose nanofibrils (CNF) are produced by mechanical grinding and high-pressure fluidization of cellulose to eliminate its lignin content. CNF is made up of lengthy (several m) and very thin (5–20 nm) fibrils with a high aspect ratio.

Nanocellulose is used as nano reinforcement for polymer matrices in the food industry [11]. The customizable surface chemistry and mechanical/physical qualities made cellulose scaffolds an ideal material for 3D nerve cell growth and differentiation [12]. The compressive strength, compressive modulus, porosity, and biocompatibility of cellulose nanofibers can be used to mimic real bone [13]. Nanocellulose can be used to print and change film surfaces as bio-ink [14].

6.1.2 Gelatin

Gelatin derives its name from the Latin word "gelata", which refers to its most distinctive property: the ability to form gels in water [15]. It is a type of protein that is derived from animal products, which are the heterogeneous mixtures of different sized proteins such as bovine or skins [1, 16]. It is a polyampholyte that contains cationic, anionic, and hydrophobic groups. The triple helical structure is due to the repeated sequences of glycine, proline, and alanine amino acid triplets [17]. Figure 6.2 displays the basic structure of gelatin. Collagen, which is obtained from animal skin, bones, and tissues, is partially hydrolyzed to form gelatin [1].Collagen is made up of protein chains that are linked together. Collagen's tertiary structure is broken up by hydrolysis, resulting in gelatin, which shares a primary structure with up to 20 distinct amino acids in varying proportions [18].

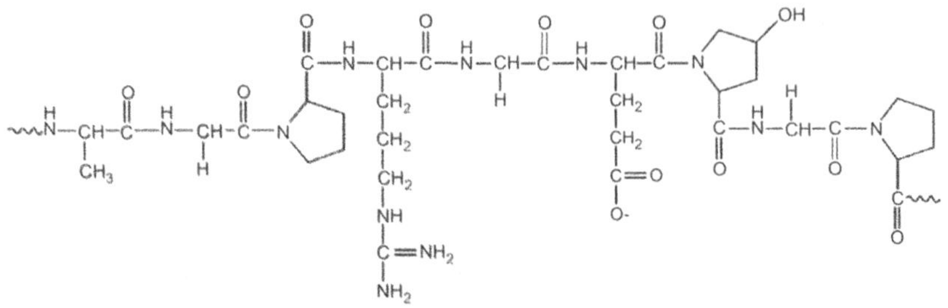

Figure 6.2 Basic structure of gelatin.

Gelatin is classified as Type-A or Type-B depending on the manufacturing procedure. Type-A gelatin is made from collagen that has been treated with acid and has an isoelectric point (pI) of 7.0 to 9.0. Type-B gelatin, on the other hand, is made by alkaline hydrolysis of collagen and has a pI of 4.8 to 5.0 [19].

Gelatin can be used as a biodegradable packaging material for food products [20]. Gelatin can be used as nanofibers for controlled release of drugs [21]. Photocrosslinkable gelatin is used to engineer the epidermis for the skin [22]. When used as a matrix, gelatin promotes therapeutic cell adhesion without compromising cell morphologies, and when used as a porogenic agent, it provides intermediate structural support when mixed with other materials. Gelatin–hydroxyapatite scaffolds, along with graphene oxide nanoflakes, intensifies osteogenic differentiation of human mesenchymal stem cells.

6.1.3 Chitosan

Chitin is a nitrogen containing high molecular weight polymer related chemically to cellulose that produces a semi-transparent horny substance and is a major component of the exoskeletons of insects, crustaceans, and arachnids [1]. After cellulose, chitin is the most abundant polymer on the planet in terms of annual production. Chitin is found in nature as structured crystalline microfibrils that constitute structural components of arthropod exoskeletons and fungal and yeast cell walls [23]. Chitosan is a cationic polysaccharide that can form polyelectrolyte complexes with other polysaccharides. It is a de-acetylated derivative of chitin. It is composed of N-acetyl-D-glucosamine and -(1 4)-linked D-glucosamine (Figure 6.3).

The viability of employing a chitosan derivative, N-carboxymethyl chitosan, to increase the mechanical performance of a fly ash-based geopolymer has been investigated [24].

With the ability to produce structures with predictable pore sizes and degradation rates, chitosan is a promising bone transplant material in orthopedic treatments.

Chitosan can bind anionic molecules like transforming growth factor (TGF), deoxyribonucleic acid (DNA), and glycosaminoglycans (GAG) [25].

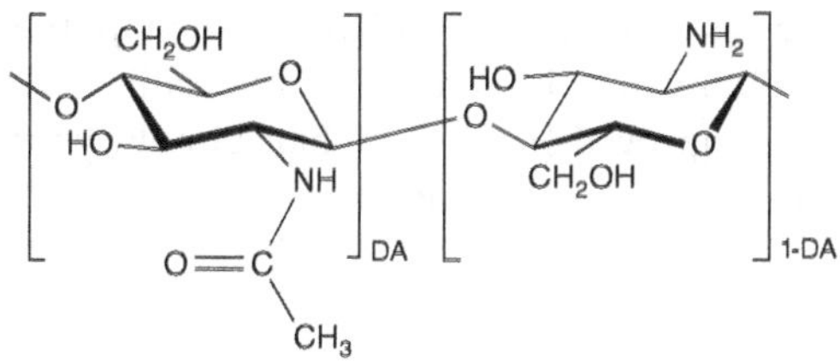

Figure 6.3 Structural unit of chitosan.

Chitosan, having mucoadhesive properties, can be kept in the nasal cavity for extended periods, allowing for better medication absorption on the nasal mucosa by delaying mucociliary clearance [26]. Chitosan scaffolds containing silver nanoparticles have antibacterial activity against *Staphylococcus aureus* and *Escherichia coli* and excellent blood coagulation ability, so silver NP can be used to cure patients with deep burns and wounds.

6.1.4 Mucin

Mucin belongs to the class of large glycoprotein (~2–3 MDa) polymers that are both membrane-tethered molecules and a primary component of the mucus gel when it is released by the epithelium's goblet cells [27]. These mucins are important components of tears, saliva, lungs, and gastrointestinal mucus, but they also serve as the foundation for the cervical mucus [28]. Mucin is extensively glycosylated, containing around 80% carbohydrates, notably N-acetylgalactosamine, N-acetylglucosamine, fructose, galactose, and sialic acid with minor amounts of mannose and sulfate. Mucins are polypeptide chains that are mainly unfolded and have a strongly O-glycosylated core region. The 5–15 monomer oligosaccharide chains show substantial branching and are connected to the protein core via O-glycosidic linkages to the hydroxyl side chains of serine and threonine, forming a "bottle brush" around the protein core. The protein core (200–500 kDa) accounts for the other 20% of molecular mass [29]. There are three basic ways to create synthetic mucins. These include (1) glycosylated amino acid synthesis in the solid-state, (2) solution polymerization from glycomonomers, and (3) glycosylation after polymerization [30]. Mucins tend to aggregate and form gels. At concentrations greater than 14 mg/mL, human tracheobronchial mucin gels below 30 °C [31]. Mucoadhesivity, or the ability for other substances to stick to mucin, is unsurprising considering that this glycoprotein has electrostatic, hydrophobic, and H-bonding interactions.

The mucin coating method and modified cell lines have the potential to boost single-cell growth and viability in bioreactors [32]. Snails produce a lot of mucins, which is also known as slime. Wound healing is aided using mucin in a mucoadhesive gel mixture enriched with honey [33]. A novel mucin/lectin film that is employed as a sacrificial multilayer system for the triggered release of polymer thin films can be both robust and versatile [34]. Mucin is now being researched for improved uses in nanobiotechnology for the treatment of serious disorders such as breast cancer.

6.1.5 Polycaprolactone

It is a semicrystalline aliphatic polyester made from caprolactone via ring-opening polymerization. Figure 6.4 shows a basic structural unit of polycaprolactone. Polycaprolactone is waxy or solid based upon the molecular weight. The molecular weight of more than 20,000 Da is a solid polymer. A molecular weight below several thousand is a waxy solid [35]. It is entirely degradable due to the activity of enzymes [36]. It is a biodegradable polymer with a glass transition temperature of -60 °C. The melting point is around 60 °C. As a result, it is frequently mixed with other polymers including polycarbonate (PC), polyethylene oxide (PEO), and starch to create composites with desired qualities [37, 38]. In physiological environments of the human body, polycaprolactone is damaged by hydrolysis of its ester bonds, and as a result, it has attracted a lot of interest for usage as an implantable biomaterial. It has excellent processability and great molecular chain flexibility. It can be spun into fibers or can be blown into the films without thermal deterioration under 200 °C.

Applications of polycaprolactone include:

- Use as an implantable biomaterial.
- Encapsulation of various drugs to achieve sustained or controlled release.
- Microcellular foams are used for tissue regeneration and cell transplantation.
- Intraocular drug delivery [39].
- To deliver and target drugs, genes, and proteins at target site efficiently, polycaprolactone is used in its nanoform. PCL nanofibers are formed along with other polymers to deliver the drugs in a sustained manner.

6.2 Polymer Composites and Nanocomposites

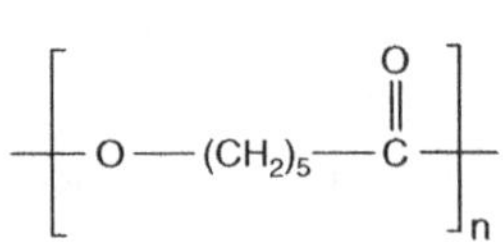

Figure 6.4 Basic structural unit of polycaprolactone.

Composites are the materials that are made up of a polymer and modifying elements such as stabilizers, plasticizers, lubricants, fillers, etc., that change the way they work. Out of which at least one of the components is in the particulate form or fibers/films with a diameter of 5 to 100 nm in polymer nanocomposites. Natural and synthetic polymers can be employed as binders in the production of polymer nanocomposites (PNC) and polymer composite materials (PCM). This happens during the manufacturing and processing of PCM and PNC, which are made of thermoplastics (materials that reversibly viscous-flow when heated) or thermosets (materials that lose their capacity to viscous-flow when heated and/or in the presence of a curing agent) [40]. Many major application areas in aircraft, land transportation, marine, sports products, and a variety of other industrial sectors rely on these. These materials, on the whole, are mature and are well-developed. Depending on the desired purpose, composites and nanocomposites can be made from a range of materials.

A nanocomposite is a system in which at least one of the components has a dimension on the nanoscale range (<100 nm) [41]. They are made by combining two or more separate materials that have diverse chemical or physical properties. These composites have qualities that are unique to the component materials, which stay separate and distinct inside the completed construction and are not linked together by formal chemical bonds. Composites are becoming increasingly essential as a means of enhancing our quality of life [42]. Out of all the composite systems used, polymeric composites combined with glass or carbon fibers are by far the most widely used composite materials in the industry [43]. These nanocomposites have a varied range of applications in tissue engineering, biotechnology, drug delivery, cancer therapy, wound healing, and other commodity applications as well. Figure 6.5 displays polymer composites and nanocomposite systems.

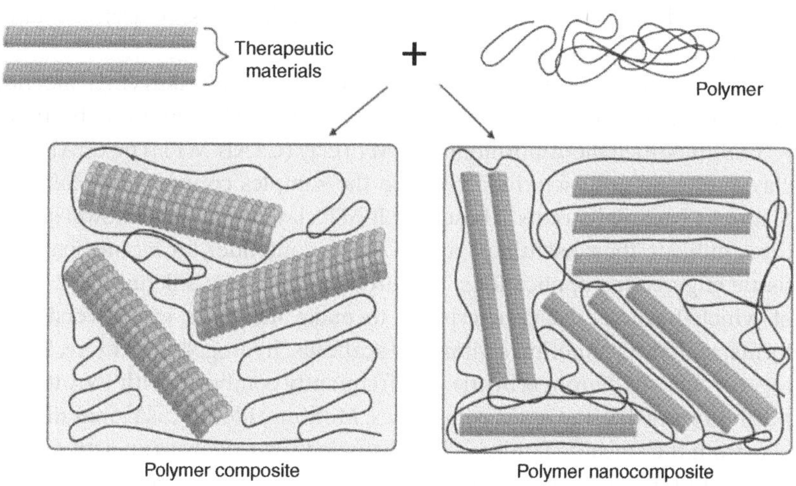

Figure 6.5 Polymer composite and nanocomposite.

Polymer composites consists of larger sized materials than those present in polymer nanocomposites. These materials are incorporated into the polymer/polymer scaffolds that forms a base for holding these materials. These therapeutic materials can be multifunctional systems, any type of clay, graphene, or graphene derivatives, etc.

6.2.1 Multifunctional Polymeric Nanocomposites

Multifunctional nanocomposites, as the name implies, have multiple applications in biomedical engineering. These nanocomposites have a varied range of applications, are considered smart, and are innovative in many engineering applications. They possess a wide surface area, a narrow pore size distribution and high drug loading capacity that helps to prevent big protein molecules from being taken up. They can be employed for drug delivery, targetability, diagnostics, stem cell treatment, and other applications by creating a single nanocomposite. Multifunctional nanomaterial was employed as a foundation in one investigation to contain multifunctional entities. The preparation was done with the SBA and MCM series of mesoporous molecular sieves. They are considered good template material for making dual-targeting entities. Antibody and magnetic targeting systems were created in these nanostructured composites. The multifunctional nanocomposites were shown to have low or no toxicity due to their extremely selective targeting characteristics. These materials have already been characterized for several functions, including delivery of drugs and other therapeutic applications [44]. As a result, multipurpose nanocomposites can be created and used in a variety of biological applications, as well as being investigated for other purposes.

Electrospinning is used to create tasar silk fibroin-polyvinyl alcohol blend-based 3D nanowebs doped with cetyl trimethyl ammonium bromide (CTAB) modified montmorillonite (MMT) clay (CTAB-MT). The CTAB-MT clay conferred antibacterial activity to the samples counter to *E. coli* and *S. aureus* bacterial strains. These CTAB-MT clay-doped 3D nanowebs' features indicate their usefulness as a possible contender for biomedical and tissue engineering applications. Researchers have used silk and PVA, both of which have great characteristics, to make materials in a diversity of forms, including nanofibers, films, 3D scaffolds, hydrogels, and more. Since regenerated silk fibroin is brittle, it is frequently combined with PVA to add flexibility and develop novel composites. Montmorillonite, a layered silicate clay from the smectite family, is being studied intensively as an active component in polymer nanocomposite biopolymers for tissue engineering and biomedical application. MMT's unique qualities, including its high surface area, aspect ratio, biocompatibility, and charged layer structure, allow it to interact with a wide range of polymers and other organic compounds, including therapeutics [45].

6.2.2 Graphene-Based Nanocomposites

Achieving targetability of delivery systems has gained tremendous attention not only in treating cancers but also to treat other ailments. In one such study, graphene-based nanocomposites were developed and fabricated with targeting ligand, polyethylene glycol (PEG). Graphene-based nanocomposites overcome the drawback of reproducibility that is majorly faced by nanobiotechnology. The number of advantages that graphene and graphene derivative systems offer, makes them a potential delivery system to track, target, and deliver the therapeutics. These systems are inexpensive, have greater stability, and are easy to prepare on a large-scale [46]. Hence, graphene-based composites are extensively being explored in many different domains, especially in nanobiotechnology (gene and protein delivery) and in biomedical applications.

6.2.3 Bioactive Glass Nanocomposite

The utilization of bioactive glass nanofibers or nanoparticles in conjunction with polymeric systems allows for the development of nanocomposites that might be employed in a variety of orthopedic applications, such as tissue engineering and regenerative medicine [47]. Bioactive glasses are made up of a silicate network with varying amounts of sodium, calcium, and phosphorus. The U.S. FDA has approved the standard 45S5 bioactive glass composite known as Bioglass (content in weight %: 24.5% Na_2O, 24.5% CaO, 45% $SiO_2$and 6% P_2O_5), which is used in the medical therapy of periodontal diseases as a bone filler and also in middle ear surgery [48]. Other bioactive glass compositions do not contain sodium, and have other elements such as potassium, fluorine, zinc, magnesium, boron, strontium, iron, and silver integrated into the silicate network. These glass composites can be functionalized by sol-gel techniques or melting to attain targeted delivery systems [47]. The capacity to interact with live tissue, generating strong attachments to bone tissue, is a trait universal to all bioactive glasses, whether melt or sol-gel generated. A calcium-deficient, carbonated apatite surface layer forms on the bioactive glass surface when it comes into touch with appropriate physiological fluid or during *in vivo* treatments, and it attaches to the bone. It is now well understood that a bioactive apatite surface layer must develop at the material/bone contact to create a link with bone. All known inorganic materials utilized for orthopedic implants, bone tissue engineering scaffolds and bone replacement share the creation of these bioactive apatite layers [49].

The recent applications of bioactive glass composites and nanocomposites are described in Table 6.1.

Table 6.1 Recent applications of composites and nanocomposites.

Type of composite	Outcome	Reference
Chitosan/bacterial cellulose composite films containing nano diamonds	Chitosan-loaded nano diamonds avoids wound infection because of its antimicrobial property. Further it also enhances wound healing properties.	[55]
Poly-lactic acid nanocomposite film combined cellulose nanocrystals and reduced graphene oxide	PLA modified scaffold significantly improved mechanical characteristics of polymer film when CNC and rGO were included. Along with mechanical properties it also exhibited other characteristics such as enhanced permeability, biocompatibility, antioxidant properties that proved to be a potential composite in biotechnology and tissue engineering.	[56]
Reduced graphene oxide-loaded nanocomposites for improving angiogenesis in tissue engineering	Developed nanocomposite scaffolds that were successful in boosting angiogenesis in a CAM model by including rGO.	[57]
Laponite nanoparticle-associated silated hydroxypropyl methyl cellulose composite hydrogel	The addition of laponites compensated for the composite hydrogel's lack of mechanical qualities, resulting in a robust composite hydrogel with minimal adverse effects. It could also self-organize chondrogenic cells in a cluster and create extracellular matrix components, indicating that it could be used to cure articular cartilage abnormalities.	[58]
Electrically conductive biocompatible composite aerogel	The foundation template for incorporating nanomaterials and other polymers was created using nano clay and bacterial cellulose aerogel. Nano clay and polyaniline were combined to have a synergistic impact on cell adhesion and biocompatibility with no carcinogenic concerns. These composite aerogels, as possible biomaterials, are likely to be useful in a variety of therapeutic applications, including drug delivery and tissue engineering.	[59]

Table 6.1 (Continued)

Type of composite	Outcome	Reference
Electrospun PCL/PGS composite fibers incorporating bioactive glass particles	Silicate or borosilicate glass particles were mixed into electrospun nanofibers composite. The BG has osteogenic characteristics that are vital for bone repair. Wound healing property was also observed due to the presence positive ions of BG.	[60]
Alginate-bioactive glass containing Zn and Mg composite	Developed BG containing Zn and Mg, enhanced the antibacterial properties, mechanical properties, and biological activities of the bone composite alginate scaffold.	[61]
3D printing of mesoporous bioactive glass/silk fibroin composite scaffolds	Using 3D technology, mesoporous BG and silk fibroin were integrated into a composite scaffold. They were shown to be mechanically robust and capable of simulating natural bone. They are known for being an excellent material for bone restoration.	[62]

PLA, polylactic acid; CNC, cellulose nanocrystals; rGO, Reduced graphene oxide; PCL, poly (epsilon caprolactone); PGS, poly (glycerol – sebacate); BG, bioactive glass.

6.2.4 Clay-Based Nanocomposites

Clay-based composites have been around since the ancient period. Previously, clay was applied to the skin to treat a variety of skin ailments. Clay is presently being researched for enhanced therapeutic uses due to its great qualities of high retention capacity, swelling and rheological capabilities, and broader availability. They have also been employed as enhanced carrier systems to transport drugs more effectively, improve composite stability, and alter drug release [50]. Various clay-based systems such as halloysite, laponite, palygorskite, and many more are being explored [51].

Electro spun scaffolds with good physical properties and the capacity to enable cell adhesion and proliferation are gaining in popularity among scientists for tissue engineering applications. In one such study, inorganic nanomaterials such as nano-silica and nano clay were included into electrospun biopolymeric composites to achieve the required criteria. The introduction of silica and clay into polymeric nanowires has been shown to strengthen and enhance the mechanical and osteogenic capabilities of the scaffolds. Both silica (AS) and clay (CLO) nanoparticles were able to cause a wide morphology in the cells cultivated on the composites and to drive the cells to colonize the scaffolds uniformly. AS and CLO have been identified as viable candidates to be used as fillers in electrospun PLA devices for bone tissue engineering applications [52].

Another example of clay that is used to develop composite systems is silver (Ag) clay . Coated catheters, bone prostheses, coated cardiovascular stents, surgical tools, wound dressings, topical burn ointments, contraceptive devices, wound healing creams, and other Ag-based items are used in clinical practice today [53]. Polyethylene nanocomposites containing antibacterial silver–clay hybrid nanofiller (Ag–MMT) have been investigated as a possible antimicrobial and cytocompatible polymeric material. Melt compounding was used to distribute Ag NPs coated MMT in high density polyethylene (HDPE) in the presence of a polyethylene grafted with maleic anhydride (PE-g-MA) to create antibacterial polyethylene nanocomposites. The lattice polymer was chosen because it offers a wide range of biomedical uses, including tubing, syringes, catheters, and device packaging. The antimicrobial polyethylene/silver–clay hybrid nanocomposite produced shows high antibacterial action as well as cytocompatibility, making it a good option for biomedical applications [54]. See Table 6.1. for recent applications of clay-based composites used in engineering and biotechnology.

6.3 Nanogels

Nanogels are the nanoscopic drug carriers used for site specific and time-controlled delivery of active pharmaceutical ingredients (API) [63]. These nanogels possess various outstanding features like high stability, excellent permeation, drug loading capacity and response to environmental stimuli [64]. Hydrogen bonding and electrostatic interactions between the polymers are responsible for crosslinking within nanogels [65]. They can hold a large amount of water and biological fluids within their structure while maintaining their arrangement. Therefore, these nanogels can overcome the challenges like drug carrying capacity and easy passage through various physiological barriers. These nanogels can be classified based on their responsive behavior as well as based on the linkage they possess. These nanogels have numerous applications in the field of biotechnology ranging from tissue engineering to gene delivery. Nanogels cover a wide plethora in the fields of biotechnology that we discusse next.

6.3.1 Types of Nanogels

Nanogels can be classified either based on their responsive behavior or on the basis of linkage they possess.

6.3.1.1 Based on Responsive Behavior

i) Stimuli responsive nanogels

Stimuli responsive nanogels can be defined as the crosslinked colloidal particles that have the ability to swell by absorption when exposed to large amount of solvent. They do not dissolve because of their complex polymer network [66]. These stimuli responsive nanogels can be pH responsive, temperature responsive, bioreduction responsive or light responsive [67]. The preparation of stimuli responsive nanogels require careful selection of the monomer that responds to the desired stimuli. Stimuli responsive nanogels serve many applications in diagnostic and therapeutic fields. They can be used for intercellular drug delivery as well as intracellular delivery of nucleic acids. Besides these, stimuli responsive nanogels are also deployed for bioimaging and photothermal therapy [68, 69].

ii) Multi-responsive nanogels

Multi-responsive nanogels are polymer based nanogel systems that can respond to a variety of external stimuli. These can provide triggered drug release along with enhanced therapeutic efficacy [70]. Multi-responsive nanogels are usually of two types: A and B type. These multi-responsive nanogels can be prepared by either microemulsion copolymerization or by simple monomer polymerization [71]. These are widely used in the market for various applications like tissue engineering [72], biosensors [73], and gene delivery [74]. At the time of writing, many experiments have been carried out for the preparation of multi-responsive nanogels and different combinations and unique responsiveness in each case reinforces them as potential candidates for drug delivery applications.

6.3.1.2 Based on Linkages

i) Hybrid nanogels

Hybrid nanogels are prepared by simply dispersing nanogel particles in organic or inorganic media [75]. The purpose of preparing hybrid nanogels is the protection of loaded drug in order to perform the designated task [76]. These hybrid nanogels have capabilities like carrying anticancer drugs and insulin very efficiently. Due to characteristics like stable interior network and large surface area, these nanogels are useful in medical imaging and are widely used in X-ray crystallography and magnetic resonance imaging [77]. Hybrid nanogels respond to any change in their environment, which makes them potentially useful for combined sensing and imaging of tissues and organs.

ii) Micellar nanogels

The main purpose behind the creation of micellar nanogels is improving the solubility and stability of hydrophobic drugs [78]. To prove this Chen et al. carried out a series of experiment where they incorporated paclitaxel in the micellar nanogel system which showed a 70% release for 70h [79]. Polylactic acid, poly lactic-co-glycolic acid, polyhydroxy butyrate, and poly caprolactone are some of the biocompatible polymers that are used to form the core of these micellar nanogels [80]. The major sources for making polymer micellar nanogels are supramolecular self-assembly of graft copolymers in water [81]. These micellar nanogels have the capability to show sol-gel transition at 37 °C.

6.3.2 Preparation of Nanogels

As of the time of writing several approaches have been investigated for the preparation of nanogels. A few of them will be discussed in depth here.

i) Physical self-assembly of interacting polymer chains

There are various physical interactions that occur between the polymers which include hydrogen bonding, van der Waals force, electrostatic force, hydrophobic interactions etc. The preparation of nanogels via this method is carried out in mild conditions. Hydrophilic polymers that are capable of electrostatic or hydrophobic interactions are mildly aggregated in aqueous media. To control the size of the nanogel, the selection of polymer is based on adequate functional groups, pH, and ionic strength [82]. The examples of some polymers that are used in preparation of nanogels via this method are pullulan and β-cyclodextrins. To exemplify this Oh et al. prepared polysaccharides based nanogels where hydrophobic moieties tend to interact with each other thereby enabling the formation of suitable nanogels [83]. Another example of this method is the preparation of gels containing cholesterol modified pullulan for transporting insulin by Akiyoshi et al [84].

ii) Cross linking of polymers

Cross linking of polymers is an easier way of nanogel formation as it involves covalent bond formation between polymer chains [85]. This cross-linking method has been employed for the synthesis of many nanogels for transporting drugs. For the first time Vinovgradov et al. used this technique to prepare cationic nanogels for the transport of polynucleotides [86]. Amino group and disulfide groups are the most used groups for cross linking. Li et al. prepared stimuli sensitive biodegradable nanogels for the delivery of doxorubicin via disulfide based cross linking of polymers [87]. Similarly, cross linking via amino group also serves a lot of advantage as it helps in increasing the reactivity of esters and iso-thiocyanides [88].

6.3.3 Applications of Nanogels

Nanogels in the pharmaceutical field are the materials that are both promising and innovative. These can be administered via multiple routes that include nasal, intra-ocular, pulmonary, oral, parenteral etc. [89]. Beginning from diagnosis to chemotherapy and from vaccine delivery to bone regeneration these nanogels can be used in multiple domains [90]. One of the currently most important use of nanogels is cancer treatment and many drugs have been incorporated into these nanogels which include cisplatin, doxorubicin, 5-flurouracil etc. [91]. The treatment of cancer necessitates the introduction of site-specific drug delivery along with predictable low toxicity to adjacent tissues and this can be achieved through nanogels [92]. Moving apart from cancer treatment, nanogels also find a major role in the treatment of autoimmune diseases. The incorporation of various immunosuppressants into nanogels has been examined widely because of its ability to target cells responsible for autoimmune diseases [93].

Nanogels find their potential application in enhancing the performance and potency of various vaccines. Nanogels prevent the enzymatic degradation of vaccines, which often occurs when delivering via conventional ways [94]. Nanogels are also deployed for bone regeneration in the body. Since, bone regeneration requires a steady release of lithium along with other medicaments locally, nanogels incorporated into a matrix of polyhydroxybutyrate are used to achieve this [95]. Delivery of genetic material into the body is a big task because of the complex nature of the human body. However, these polymeric nanogels offer a promising way to carry these genetic materials into the human body. Due to hydrophobic nature of nanogels, they can achieve a high loading capacity of hydrophilic drugs and because of its cross-linking nature they also offer a stable complex. Lastly, because of their passive targeting nature and site-specific release action these nanogels offer a potential way to deliver genetic material into human body [96].

Besides all these applications, nanogels are also used as local anesthetics. To prove this, Tan et al. carried out a series of experiments which led to the conclusion that nanogels prepared via layer-by-layer approach cause less pain during injection and provide a longer blood circulation time [97].

6.3.4 Advantages of Nanogels

Nanogels are regarded as superior drug delivery technology because they offer many advantages.

- Higher biocompatibility, biodegradability, and precise medicine delivery [98].
- Nanogels can release the drug at target site in a controlled manner which helps in increasing the therapeutic efficacy of the formulation and lowers the adverse drug reactions [99].
- Nanogels can be delivered through various routes like parenteral, oral, mucosal, nasal, vaginal etc. [100].
- Nanogels can encapsulate both hydrophobic and hydrophilic drugs [101].
- Nanogels can load comparably high amounts of drug and provides the benefit of decreased initial outflow of encapsulated drug from the solution [102].
- Nanogels are inert in the internal aqueous environment, which means that they do not cause any immune reaction in the body [103].
- Due to their small size, they offer increased permeation capabilities helping them to cross blood–brain barrier easily [104].
- Nanogels have a great affinity towards aqueous solutions. This results in their ability to swell or de-swell when placed into an aqueous medium. This characteristic helps in making nanogels the ideal candidates for the delivery of proteins and other biomolecules [105].

6.3.5 Limitations of Nanogels

Despite offering many advantages, these nanogels also have certain limitations:

- Complete removal of surfactants and solvents becomes very difficult.
- The monomers used in the nanogel preparation can cause adverse reactions in the formulation [106].
- A firm contact between the polymer and drug reduces the hydrophilicity of the nanogels thereby causing the structure to collapse [107].
- Immunogenicity and inconsistent behavior in human models are often observed [108].
- Difficulty of scale up and mass production [109].

6.3.6 Patent Landscape

The effectiveness of nanogels in when used in the fields of biotechnology is of utmost interest in the market. Only a few patents have been filed to date (Table 6.2). The patents include delivery of si-RNA and micro-RNA through nanogels and delivery of fibroblastic growth factors via chitosan quaternary ammonium salt hyaluronic acid nanogel.

Table 6.2 Current patent scenario of nanogels.

Patent details	Title of patent	Outcome	Applicant	Reference
US11000602B2	Delivery of small interfering RNA and micro-RNA through nanogels containing hydrophobic pseudo-peptides.	Reactive phenylamine derivative was synthesized successfully to modulate hydrophobicity of pH-responsive nanogels.	Nicholas Peppas, William Liechty	[110]
CN-113332454-A	Ultra-small ferro ferric oxide loaded polyethyleneimine nanogel/siRNA compound.	si-RNA nanogel loaded with ultra-small ferro ferric oxide, was prepared successfully for gene transferring.	Cao Xueyan, Shi Xiangyang, Peng Yucheng, Yang Chaoyue	[111]
CN-113750244-A	Engineered platelet carrying chemotherapeutic drug and containing external back immune checkpoint inhibitor nanogel as well as preparation method and application of engineered platelet.	Nanogels carrying drugs for chemotherapy were prepared to bypass immune checkpoint prohibitor.	Sun Jin He Zhonggui Lu Qi Ye Hao	[112]
CN-108794684-A	Responsive to temperature type si-RNA nanogel carriers and preparation method thereof.	Temperature sensitive nanogels to carry si-RNA were prepared.	Zhang Afang, Taopei Peng, Yu Yao Huan, Fang Wenlong, Wu Jianhui	[113]

(Continued)

Table 6.2 (Continued)

Patent details	Title of patent	Outcome	Applicant	Reference
EP-2906242-A1	Delivery of small interfering RNA and micro-RNA through membrane-disruptive, responsive nanoscale hydrogels.		Nicholas Peppas, William Liechty	[114]
CN-104906073-A	Preparation method of chitosan quaternary ammonium salt hyaluronic acid nanogel coated with basic fibroblast growth factors.		Xu Quanchen, Wang Zhiguo, Xu Xiaoyan Yuan, Changqing Ji Qiuxia	[115]

6.4 Conclusion

Biopolymers are cost effective and are obtained through renewable resources along with sustainable production when compared it with synthetic polymers. Cellulose and gelatin are abundantly available biopolymers with many applications in biotechnology, tissue engineering and in the pharmaceutical industry. Composites and nanocomposites are being widely explored for various applications. Mostly, polymer-based nanocomposites are utilized for their advantages such as multifunctionality, frequent dosing and minimum side effects. Graphene oxide composites are considered as ideal carriers for gene and drug delivery. Clay-based composites initially were only used for treatment of skin-related problems. However, they are now being investigated for many other nanobiotechnological applications. Nanogels have been evolved much recently and with increasing technology newer uses of these nanogels are coming into limelight every day. Delivery of proteins to tissue engineering and many more things can be done via nanogels.

References

1 Mohan, S., Oluwafemi, O.S., Kalarikkal, N. et al. (2016). Biopolymers – application in nanoscience and nanotechnology. In: *Recent Advances in Biopolymers* (ed. F.K. Perveen), 47–66. London: IntechOpen.

2 Giusti, P., Lazzeri, L., de Petris, S. et al. (1994). Collagen-based new bioartificial polymeric materials. *Biomaterials* 15: 1229–1233.

3 Hacker, M.C., Krieghoff, J., and Mikos, A.G. (2019). Synthetic polymers. In: *Principles of Regenerative Medicine* (ed. A. Atala, R. Lanza, T. Mikos and R. Nerem), 559–590. Amsterdam: Elsevier.

4 Gowthaman, N.S.K., Lim, H.N., Sreeraj, T.R. et al. (2021). Advantages of biopolymers over synthetic polymers: social, economic, and environmental aspects. In: *Biopolymers and Their Industrial Applications*, (ed. S Thomas, S. Gopi, and A. Amalraj), 351–372. Amsterdam: Elsevier.

5 Klemm, D., Heublein, B., Fink, H.P., and Bohn, A. (2005). Cellulose: fascinating biopolymer and sustainable raw material. *Angewandte Chemie - International Edition* 44 (22): 3358–3393.

6 Barampouti, E.M., Mai, S., Moustakas, K. et al. (2021). Status and perspectives of agricultural residues in a circular and resource-efficient context. In: *Clean Energy and Resources Recovery* (ed. V. Tyagi and K. Aboudi), 49–102. Amsterdam: Elsevier.

7 Deshmukh, K., Basheer Ahamed, M., Deshmukh, R.R. et al. (2017). Biopolymer composites with high dielectric performance: interface engineering. In: *Biopolymer Composites in Electronics* (ed. K.K. Sadasivuni, D. Ponnamma, J. Kim et al.), 27–128. Amsterdam: Elsevier.

8 Usmani, M.A., Khan, I., Gazal, U. et al. (2018). Interplay of polymer bionanocomposites and significance of ionic liquids for heavy metal removal. In: *Polymer-based Nanocomposites for Energy and Environmental Applications* (ed. M. Jawaid and M. Mansoob), 441–463. Amsterdam: Elsevier.

9 Numata, Y., Sakata, T., Furukawa, H., and Tajima, K. (2015). Bacterial cellulose gels with high mechanical strength. *Materials Science and Engineering: C* 47: 57–62.

10 Lee, K.Y. and Bismarck, A. (2016). Bacterial nanocellulose as reinforcement for polymer matrices. In: *Bacterial Nanocellulose: From Biotechnology to Bio-Economy* (ed. M. Gama, F. Dourado and S. Bielecki), 109–122. Amsterdam: Elsevier.

11 Arrieta, M.P., Fortunati, E., Burgos, N. et al. (2016). Nanocellulose-based polymeric blends for food packaging applications. In: *Multifunctional*

Polymeric Nanocomposites Based on Cellulosic Reinforcements (ed. D. Puglia, E. Fortunati, and J.M. Kenny), 205–252. Amsterdam: Elsevier.

12 Innala, M., Riebe, I., Kuzmenko, V. et al. (2014). 3D Culturing and differentiation of SH-SY5Y neuroblastoma cells on bacterial nanocellulose scaffolds. *Artificial Cells, Nanomedicine and Biotechnology* 42 (5): 302–308.

13 Eftekhari, S., el Sawi, I., Bagheri, Z.S. et al. (2014). Fabrication and characterization of novel biomimetic PLLA/cellulose/hydroxyapatite nanocomposite for bone repair applications. *Materials Science and Engineering: C* 39 (1): 120–125.

14 Rees, A., Powell, L.C., Chinga-Carrasco, G. et al. (2015). 3D bioprinting of carboxymethylated-periodate oxidized nanocellulose constructs for wound dressing applications. *BioMed Research International* 2015.

15 Haug, I.J. and Draget, K.I. (2009). Gelatin. In: *Handbook of Hydrocolloids* 2e (ed. G.O. Phillips and P.A. Williams), 142–163. Amsterdam: Elsevier.

16 Won, Y.-W. and Kim, Y.-H. (2008). Recombinant human gelatin nanoparticles as a protein drug carrier. *Journal of Controlled Release* 127 (2): 154–161.

17 Fraunhofer, W., Winter, G., and Coester, C. (2004). Asymmetrical flow field-flow fractionation and multiangle light scattering for analysis of gelatin nanoparticle drug carrier systems. *Analytical Chemistry* 76 (7): 1909–1920.

18 Zhang, F., Xu, S., and Wang, Z. (2011). Pre-treatment optimization and properties of gelatin from freshwater fish scales. *Food and Bioproducts Processing* 89 (3): 185–193.

19 Kommareddy, S., Shenoy, D.B., and Amiji, M.M. (2007). Gelatin nanoparticles and their biofunctionalization. In: *Nanotechnologies for the Life Sciences* (ed. C.S.S.R. Kumar). Weinheim: Wiley-VCH.

20 Nur Hanani, Z.A., Roos, Y.H., and Kerry, J.P. (2014). Use and application of gelatin as potential biodegradable packaging materials for food products. *International Journal of Biological Macromolecules* 71: 94–102.

21 Yang, D., Li, Y., and Nie, J. (2007). Preparation of gelatin/PVA nanofibers and their potential application in controlled release of drugs. *Carbohydrate Polymers* 69 (3): 538–543.

22 Zhao, X., Lang, Q., Yildirimer, L. et al. (2016). Photocrosslinkable gelatin hydrogel for epidermal tissue engineering. *Advanced Healthcare Materials* 5 (1): 108–118.

23 Rinaudo, M. (2006). Chitin and chitosan: properties and applications. *Progress in Polymer Science (Oxford)* 31 (7): 603–632.

24 Li, Z., Chen, R., and Zhang, L. (2013). Utilization of chitosan biopolymer to enhance fly ash-based geopolymer. *Journal of Materials Science* 48 (22): 7986–7993.

25 di Martino, A., Sittinger, M., and Risbud, M.V. (2005). Chitosan: a versatile biopolymer for orthopaedic tissue-engineering. *Biomaterials* 26 (30): 5983–5990.

26 Shadab, M.D., Khan, R.A., Mustafa, G. et al. (2013). Bromocriptine loaded chitosan nanoparticles intended for direct nose to brain delivery: pharmacodynamic, pharmacokinetic and scintigraphy study in mice model. *European Journal of Pharmaceutical Sciences* 48 (3): 393–405.

27 Morrison, C.B., Markovetz, M.R., and Ehre, C. (2019). Mucus, mucins, and cystic fibrosis. *Pediatric Pulmonology* 54 (S3): S84–S96.

28 Marczynski, M., Winkeljann, B., and Lieleg, O. (2021). 6 advances in mucin biopolymer research: purification, characterization, and applications. In: *Biopolymers for Biomedical and Biotechnological Applications* (ed. B. Rehm and M.F. Moradali), 181–208.

29 Bansil, R. and Turner, B.S. (2006). Mucin structure, aggregation, physiological functions, and biomedical applications. *Current Opinion in Colloid & Interface Science* 11 (2–3): 164–170.

30 Kwan, C.S., Cerullo, A.R., and Braunschweig, A.B. (2020). Design and synthesis of mucin-inspired glycopolymers. *ChemPlusChem* 85 (12): 2704–2721.

31 Bromberg, L.E. and Barr, D.P. (2000). Self-association of mucin. *Biomacromolecules* 1 (3): 325–334.

32 Shurer, C.R., Head, S.E., Goudge, M.C., and Paszek, M.J. (2019). Mucin-coating technologies for protection and reduced aggregation of cellular production systems. *Biotechnology and Bioengineering* 116 (5): 994–1005.

33 Adikwu, M.U. and Alozie, B.U. (2007). Application of snail mucin dispersed in detarium gum gel in wound healing. *Scientific Research and Essay* 2 (6): 195–198.

34 Polak, R., Crouzier, T., Lim, R.M. et al. (2014). Sugar-mediated disassembly of mucin/lectin multilayers and their use as pH-tolerant, on-demand sacrificial layers. *Biomacromolecules* 15 (8): 3093–3098.

35 Jiang, L. and Zhang, J. (2017). 7 – Biodegradable and biobased polymers. In: *Plastics Design Library*, 2e (ed. M. Kutz), 127–143. William Andrew Publishing.

36 Tokiwa, Y. and Suzuki, T. (1977). Hydrolysis of polyesters by lipases. *Nature* 270 (5632): 76–78.

37 Tsuji, H. and Suzuyoshi, K. (2002). Environmental degradation of biodegradable polyesters 1. Poly(ε-caprolactone), poly[(R)-3-hydroxybutyrate], and poly(L-lactide) films in controlled static seawater. *Polymer Degradation and Stability* 75 (2): 347–355.

38 Dubois, P., Jacobs, C., Jerome, R., and Teyssie, P. (1991). Macromolecular engineering of polylactones and polylactides. 4. Mechanism and kinetics of lactide homopolymerization by aluminum isopropoxide. *Macromolecules* 24 (9): 2266–2270.

39 Boia, R., Dias, P.A.N., Martins, J.M. et al. (2019). Porous poly(ε-caprolactone) implants: a novel strategy for efficient intraocular drug delivery. *Journal of Controlled Release* 316: 331–348.

40 Yamamoto, H., Hoshina, W., Kurashima, H. et al. (2007). Engineering of poly(DL-lactic-co-glycolic acid) nanocomposite particles for dry powder inhalation dosage forms of insulin with the spray-fluidized bed granulating system. *Advanced Powder Technology* 18 (2): 215–228.

41 Teixeira, M. do C., Santini, A., and Souto, E.B. (2017). Delivery of antimicrobials by chitosan-composed therapeutic nanostructures. In: *Nanostructures for Antimicrobial Therapy* (ed. A. Ficai and A.M. Grumezescu), 203–222. Amsterdam: Elsevier.

42 Haghi, A.K., Oluwafemi, O.S., Jose, J.P., and Maria, H.J. (2013). *Composites and Nanocomposites*. London: Taylor & Francis.

43 Joshi, G., and Yadav, K.S. (2023). Applications of bioresins and biopolymers derived from natural resources as composites in drug delivery. In: *Green Sustainable Process for Chemical and Environmental Engineering and Science* (ed. T. Altalhi and Inamuddin), 21–34. Amsterdam: Elsevier.

44 Yiu, H.H.P., Niu, H.J., Biermans, E. et al. (2010). Designed multifunctional nanocomposites for biomedical applications. *Advanced Functional Materials* 20 (10): 1599–1609.

45 Batra, R., Purwar, R., Kulanthaivel, S., and Mishra, P. (2021). Cetyl trimethyl ammonium bromide modified montmorillonite-doped tasar silk fibroin/polyvinyl alcohol blend 3D nanowebs for tissue engineering applications. *Macromolecular Materials and Engineering* 306 (11): 2100450.

46 Yang, K., Feng, L., Hong, H. et al. (2013). Preparation and functionalization of graphene nanocomposites for biomedical applications. *Nature Protocols* 8: 2392–2403.

47 Boccaccini, A.R., Erol, M., Stark, W.J. et al. (2010). Polymer/bioactive glass nanocomposites for biomedical applications: a review. *Composites Science and Technology* 70 (13): 1764–1776.

48 Fiume, E., Barberi, J., Verné, E., and Baino, F. (2018). Bioactive glasses: from parent 45S5 composition to scaffold-assisted tissue-healing therapies. *Journal of Functional Biomaterials* 9: 24.

49 Rezwan, K., Chen, Q.Z., Blaker, J.J., and Boccaccini, A.R. (2006). Biodegradable and bioactive porous polymer/inorganic composite scaffolds for bone tissue engineering. *Biomaterials* 27 (18): 3413–3431.

50 Viseras, C., Carazo, E., Borrego-Sánchez, A. et al. (2019). Clay minerals in skin drug delivery. *Clays and Clay Minerals* 67: 59–71.

51 Mishra, D.K., Yadav, K.S., Prabhakar, B., and Gaud, R.S. (2018). Nanocomposite for cancer targeted drug delivery. In: *Applications of Nanocomposite Materials in Drug Delivery* (ed. A.M. Asiri Inamuddin, and A. Mohammad). Amsterdam: Elsevier.

52 Lopresti, F., Pavia, F.C., Ceraulo, M. et al. (2021). Physical and biological properties of electrospun poly(d,l-lactide)/nanoclay and poly(d,l-lactide)/nanosilica nanofibrous scaffold for bone tissue engineering. *Journal of Biomedical Materials Research – Part A* 109 (11): 2120–2136.

53 Roy, A., Joshi, M., Butola, B.S., and Srivastava, A.K. (2018). Silver-loaded HDPE/clay nanocomposites with antibacterial property: a potential replacement for commodity polyethylene plastic. *Polymer Composites* 39 (S1): E366–E377.

54 Roy, A., Joshi, M., Butola, B.S., and Ghosh, S. (2020). Evaluation of biological and cytocompatible properties in nano silver-clay based polyethylene nanocomposites. *Journal of Hazardous Materials* 384: 121309.

55 Ostadhossein, F., Mahmoudi, N., Morales-Cid, G. et al. (2015). Development of chitosan/bacterial cellulose composite films containing nanodiamonds as a potential flexible platform for wound dressing. *Materials* 8 (9): 6401–6418.

56 Pal, N., Banerjee, S., Roy, P., and Pal, K. (2019). Reduced graphene oxide and PEG-grafted TEMPO-oxidized cellulose nanocrystal reinforced poly-lactic acid nanocomposite film for biomedical application. *Materials Science and Engineering C* 104: 109956.

57 Chakraborty, S., Ponrasu, T., Chandel, S. et al. (2018). Reduced graphene oxide-loaded nanocomposite scaffolds for enhancing angiogenesis in tissue engineering applications. *Royal Society Open Science* 5 (5): 172017.

58 Boyer, C., Figueiredo, L., Pace, R. et al. (2018). Laponite nanoparticle-associated silated hydroxypropylmethyl cellulose as an injectable reinforced interpenetrating network hydrogel for cartilage tissue engineering. *Acta Biomaterialia* 65: 112–122.

59 Salehi, M.H., Golbaten-Mofrad, H., Jafari, S.H. et al. (2021). Electrically conductive biocompatible composite aerogel based on nanofibrillated template of bacterial cellulose/polyaniline/nano-clay. *International Journal of Biological Macromolecules* 173: 467–480.

60 Luginina, M., Schuhladen, K., Orrú, R. et al. (2020). Electrospun PCL/PGS composite fibers incorporating bioactive glass particles for soft tissue engineering applications. *Nanomaterials* 10 (5): 978.

61 Zamani, D., Moztarzadeh, F., and Bizari, D. (2019). Alginate-bioactive glass containing Zn and Mg composite scaffolds for bone tissue engineering. *International Journal of Biological Macromolecules* 137: 1256–1267.

62 Du, X., Wei, D., Huang, L. et al. (2019). 3D printing of mesoporous bioactive glass/silk fibroin composite scaffolds for bone tissue engineering. *Materials Science and Engineering C* 103: 109731.

63 Hamidi, M., Azadi, A., and Rafiei, P. (2008). Hydrogel nanoparticles in drug delivery. *Advanced Drug Delivery Reviews* 60 (15): 1638–1649.

64 Pinelli, F., Pizzetti, F., Ortolà, Ó.F. et al. (2020). Influence of the core formulation on features and drug delivery ability of carbamate-based nanogels. *International Journal of Molecular Sciences* 21 (18): 1–15.

65 Ekkelenkamp, A.E., Elzes, M.R., Engbersen, J.F.J., and Paulusse, J.M.J. (2018). Responsive crosslinked polymer nanogels for imaging and therapeutics delivery. *Journal of Materials Chemistry B* 6 (2): 210–235.

66 Ramos, J., Forcada, J., and Hidalgo-Alvarez, R. (2014). Cationic polymer nanoparticles and nanogels: from synthesis to biotechnological applications. *Chemical Reviews* 114 (1): 367–428.

67 Preman, N.K., Barki, R.R., Vijayan, A. et al. (2020). Recent developments in stimuli-responsive polymer nanogels for drug delivery and diagnostics: a review. *European Journal of Pharmaceutics and Biopharmaceutics* 157: 121–153.

68 Wu, H.Q. and Wang, C.C. (2016). Biodegradable smart nanogels: a new platform for targeting drug delivery and biomedical diagnostics. *Langmuir* 32 (25): 6211–6225.

69 Wang, H., Qian, J., and Ding, F. (2017). Recent advances in engineered chitosan-based nanogels for biomedical applications. *Journal of Materials Chemistry B* 5 (34): 6986–7007.

70 Basak, S. (2020). The age of multistimuli-responsive nanogels: the finest evolved nano delivery system in biomedical sciences. *Biotechnology and Bioprocess Engineering* 25 (5): 655–669.

71 Qiao, Z.Y., Zhang, R., Du, F.S. et al. (2011). Multi-responsive nanogels containing motifs of ortho ester, oligo(ethylene glycol) and disulfide linkage as carriers of hydrophobic anti-cancer drugs. *Journal of Controlled Release* 152 (1): 57–66.

72 Hashimoto, Y., Mukai, S-.A., Sasaki, Y., and Akiyoshi, K. (2018). Nanogel tectonics for tissue engineering: protein delivery systems with nanogel chaperones. *Advanced Healthcare Materials* 7 (23): 1–14.

73 Wu, W. and Zhou, S. (2010). Hybrid micro-/nanogels for optical sensing and intracellular imaging. *Nano Reviews* 1 (1): 5730.

74 Vashist, A., Kaushik, A.K., Ahmad, S., and Nair, M. (eds.) (2017). *Nanogels for Biomedical Applications*. London: Royal Society of Chemistry.

75 Akiyoshi, K., Kang, E.C., Kurumada, S. et al. (2000). Controlled association of amphiphilic polymers in water: thermosensitive nanoparticles formed by self-assembly of hydrophobically modified pullulans and poly(N-isopropylacrylamides). *Macromolecules* 33 (9): 3244–3249.

76 Soni, K.S., Desale, S.S., and Bronich, T.K. (2016). Nanogels: an overview of properties, biomedical applications, and obstacles to clinical translation. *Journal of Controlled Release* 240: 109–126.

77 Deng, H., Zhong, Y., Du, M. et al. (2014). Theranostic self-assembly structure of gold nanoparticles for NIR photothermal therapy and X-ray computed tomography imaging. *Theranostics* 4 (9): 904–918.

78 Liggins, R.T. and Burt, H.M. (2002). Polyether-polyester diblock copolymers for the preparation of paclitaxel loaded polymeric micelle formulations. *Advanced Drug Delivery Reviews* 54 (2): 191–202.

79 Chen, D., Yu, H., Sun, K. et al. (2014). Dual thermoresponsive and pH-responsive self-assembled micellar nanogel for anticancer drug delivery. *Drug Delivery* 21 (4): 258–264.

80 Bae, K.H., Mok, H., and Park, T.G. (2008). Synthesis, characterization, and intracellular delivery of reducible heparin nanogels for apoptotic cell death. *Biomaterials* 29 (23): 3376–3383.

81 Sultana, F., Manirujjaman, I.-U.-H., Arafat, M., and Sharmin, S. (2013). An overview of nanogel drug delivery system. *Journal of Applied Pharmaceutical Science* 3 (8 SUPPL): S95–S105.

82 Zamurovic, M., Christodoulou, S., Vazaios, A. et al. (2007). Micellization behavior of complex comblike block copolymer architectures. *Macromolecules* 40 (16): 5835–5849.

83 Oh, J.K., Lee, D.I., and Park, J.M. (2009). Biopolymer-based microgels/ nanogels for drug delivery applications. *Progress in Polymer Science (Oxford)* 34 (12): 1261–1282.

84 Akiyoshi, K., Kobayashi, S., Shichibe, S. et al. (1998). Self-assembled hydrogel nanoparticle of cholesterol-bearing pullulan as a carrier of protein drugs: complexation and stabilization of insulin. *Journal of Controlled Release* 54 (3): 313–320.

85 Cao, Q.C., Wang, X., and Wu, D.C. (2018). Controlled cross-linking strategy for formation of hydrogels, microgels and nanogels. *Chinese Journal of Polymer Science (English Edition)* 36 (1): 8–17.

86 Vinogradov, S.V., Zeman, A.D., Batrakova, E.V., and Kabanov, A.V. (2005). Polyplex Nanogel formulations for drug delivery of cytotoxic nucleoside analogs. *Journal of Controlled Release* 107 (1): 143–157.

87 Li, Y.L., Zhu, L., Liu, Z. et al. (2009). Reversibly stabilized multifunctional dextran nanoparticles efficiently deliver doxorubicin into the nuclei of cancer cells. *Angewandte Chemie - International Edition* 48 (52): 9914–9918.

88 Chacko, R.T., Ventura, J., Zhuang, J., and Thayumanavan, S. (2012). Polymer nanogels: a versatile nanoscopic drug delivery platform. *Advanced Drug Delivery Reviews* 64 (9): 836–851.

89 Zhou, M., Wang, T., Hu, Q., and Luo, Y. (2016). Low density lipoprotein/pectin complex nanogels as potential oral delivery vehicles for curcumin. *Food Hydrocolloids* 57: 20–29.

90 Kabanov, A.V. and Vinogradov, S.V. (2009). Nanogels as pharmaceutical carriers: finite networks of infinite capabilities. *Angewandte Chemie - International Edition* 48 (30): 5418–5429.

91 Garg, T. (2016). Current nanotechnological approaches for an effective delivery of bio-active drug molecules in the treatment of acne. *Artificial Cells, Nanomedicine and Biotechnology* 44 (1): 98–105.

92 Peng, J., Qi, T., Liao, J. et al. (2013). Controlled release of cisplatin from pH-thermal dual responsive nanogels. *Biomaterials* 34 (34): 8726–8740.

93 Kaur, M., Sudhakar, K., and Mishra, V. (2019). Fabrication and biomedical potential of nanogels: an overview. *International Journal of Polymeric Materials and Polymeric Biomaterials* 68 (6): 287–296.

94 Debache, K., Kropf, C., Schütz, C.A. et al. (2011). Vaccination of mice with chitosan nanogel-associated recombinant NcPDI against challenge infection with *Neospora caninum* tachyzoites. *Parasite Immunology* 33 (2): 81–94.

95 Bai, X., Gao, M., Syed, S. et al. (2018). Bioactive hydrogels for bone regeneration. *Bioactive Materials* 3 (4): 401–417.

96 Kandil, R. and Merkel, O.M. (2019). Recent progress of polymeric nanogels for gene delivery. *Current Opinion in Colloid and Interface Science* 39: 11–23.

97 Tan, J.P.K., Wang, Q., and Tam, K.C. (2008). Control of burst release from nanogels via layer-by-layer assembly. *Journal of Controlled Release* 128 (3): 248–254.

98 Shitrit, Y., Davidovich-Pinhas, M., and Bianco-Peled, H. (2019). Shear thinning pectin hydrogels physically cross-linked with chitosan nanogels. *Carbohydrate Polymers* 225 (March): 115249.

99 Neamtu, I., Rusu, A.G., Diaconu, A. et al. (2017). Basic concepts and recent advances in nanogels as carriers for medical applications. *Drug Delivery* 24 (1): 539–557.

100 Zabihi, F., Koeppe, H., Achazi, K. et al. (2019). One-pot synthesis of poly(glycerol- co-succinic acid) nanogels for dermal delivery. *Biomacromolecules* 20 (5): 1867–1875.

101 Ahmad, Z., Pandey, R., Sharma, S., and Khuller, G. (2006). Pharmacokinetic and pharmacodynamic behaviour of antitubercular drugs encapsulated in alginate nanoparticles at two doses. *International Journal of Antimicrobial Agents* 27 (5): 409–416.

102 Guerrero-Ramírez, L.G., Nuño-Donlucas, S.M., Cesteros, L.C., and Katime, I. (2008). Smart copolymeric nanohydrogels: synthesis, characterization, and properties. *Materials Chemistry and Physics* 112 (3): 1088–1092.
103 Lewis Phillips, G.D., Li, G., Dugger, D.L. et al. (2008). Targeting HER2-positive breast cancer with trastuzumab-DM1, an antibody-cytotoxic drug conjugate. *Cancer Research* 68 (22): 9280–9290.
104 Eftekhari, R.B., Maghsoudnia, N., Samimi, S. et al. (2019). Co-delivery nanosystems for cancer treatment: a review. *Pharmaceutical Nanotechnology* 7 (2): 90–112.
105 Peng, J., Yang, Q., Xiao, Y. et al. (2019). Tumor microenvironment responsive drug-dye-peptide nanoassembly for enhanced tumor-targeting, penetration, and photo-chemo-immunotherapy. *Advanced Functional Materials* 29 (19): 1–16.
106 Kataria, K., Sharma, A., Garg, T. et al. (2014). Novel technology to improve drug loading in polymeric nanofibers. *Drug Delivery Letters* 4 (1): 79–86.
107 Kardos, A., Gilányi, T., and Varga, I. (2019). How small can poly(N-isopropylacrylamide) nanogels be prepared by controlling the size with surfactant? *Journal of Colloid and Interface Science* 557: 793–806.
108 Cho, H., Jammalamadaka, U., and Tappa, K. (2018). Nanogels for pharmaceutical and biomedical applications and their fabrication using 3D printing technologies. *Materials* 11 (2).
109 Hu, Y., Wang, J., Wang, H. et al. (2012) Microfluidic fabrication and thermoreversible response of core/shell photonic crystalline microspheres based on deformable nanogels. *Langmuir: The ACS Journal of Surfaces and Colloids* 28 (49): 17186–17192.
110 US11000602B2 – Delivery of small interfering RNA and micro-RNA through nanogels containing hydrophobic pseudo-peptides – Google Patents.
111 CN113332454A – Ultra-small ferroferric oxide loaded polyethyleneimine nanogel/siRNA compound – Google Patents.
112 CN113750244A – Engineered platelet carrying chemotherapeutic drug and containing external back immune checkpoint inhibitor nanogel as well as preparation method and application of engineered platelet – Google Patents.
113 CN108794684A – Responsive to temperature type siRNA nanogel carriers and preparation method thereof – Google Patents.
114 EP2906242A1 – Delivery of small interfering rna and micro rna through membrane-disruptive, responsive nanoscalle hydrogels – Google Patents.
115 CN104906073A – Preparation method of chitosan quaternary ammonium salt hyaluronic acid nanogel coated with basic fibroblast growth factors – Google Patents.

7

Biopolymers, Blends, Composites, Gels, and Thin Films in Drug Delivery and Drug Design

Saman Noor[1], *Nazim Hussain*[2,*], *and Areej Shahbaz*[2]

[1] *Atta-ur-Rehman School of Applied Biosciences (ASAB), National University of Sciences and Technology (NUST), Islamabad, Pakistan*
[2] *Center for Applied Molecular Biology (CAMB), University of the Punjab, Lahore, Pakistan*
* *Corresponding author*

7.1 Introduction

Natural materials are more beneficial than synthetic materials as they have a less environmental impact and are cost-effective. Also, they are frequently available in nature. Scientific communities targeting the boundary between chemistry and biology strive to explore natural biopolymers. The main goal is to build a system with many applications in biomedical sciences; this includes bioimaging, targeted drug delivery, and tissue engineering. To synthesize biocompatible materials, many natural polymers and synthetic ones have been analyzed in the last few decades [1, 2]. Plant-based proteins are widely distributed on Earth and can be extensively used in various shapes for medical purposes, such as films, nanoparticles, microparticles, fibers, and hydrogels [3, 4].

Since the 1980s, there has been much attention focused on nanotechnology and many fields of engineering such as biomedical, mechanical, space engineering, and electronics have been modified by adapting nanotechnology. The contribution of nanotechnology has led towards huge progress in various fields, especially in the field of biomedical sciences, where we use nanoparticles for

Applications of Biopolymers in Science, Biotechnology, and Engineering, First Edition.
Edited by Arunima Reghunadhan, Akhina H., and Sabu Thomas.

targeted delivery of either genes or a drug [5], DNA probing to detect the target sequences, and specific site's imaging [6] and tissue engineering [7]. Nanoparticles are of great importance in the field of nanotechnology. Various therapies are based on the use of nanoparticles for treating diabetes [8], infection [9], allergy [10], inflammation [11], and cancer [12] (Figure 7.1).

The size of nanoparticles and their large surface area greatly contribute to their various applications in therapeutics. Several ligands can be displayed on their large surface, which can be controlled or modified. These modifications include layer-by-layer structure and the presence of covalent interactions between functional and surface molecules or polymers. Moreover, their characteristic feature of changing their volume and diffusion provides rapid release and absorption behavior [13]. To synthesize nanoparticles, inorganic, organic, or hybrid materials are utilized (Figure 7.2). In addition, polymeric nanoparticles have also been extensively used for therapeutical purposes, especially as a drug/gene delivery carrier [14] and as a scaffold for tissue engineering [15].

Despite having less capacity to load drugs and wide size distribution, polymeric nanoparticles have gained much interest from biologists, pharmaceutical scientists, chemists, and engineers because of their role in targeted drug/biomaterial delivery, particularly tissues, cells compartments, and cells [16]. Compared with metal or ceramic nanoparticles, it is possible to synthesize polymeric nanoparticles in a broad range of sizes and types. These polymeric nanoparticles can uphold the localized therapeutic agent for a longer period (weeks) [17]. Among natural and synthetic biomaterials,

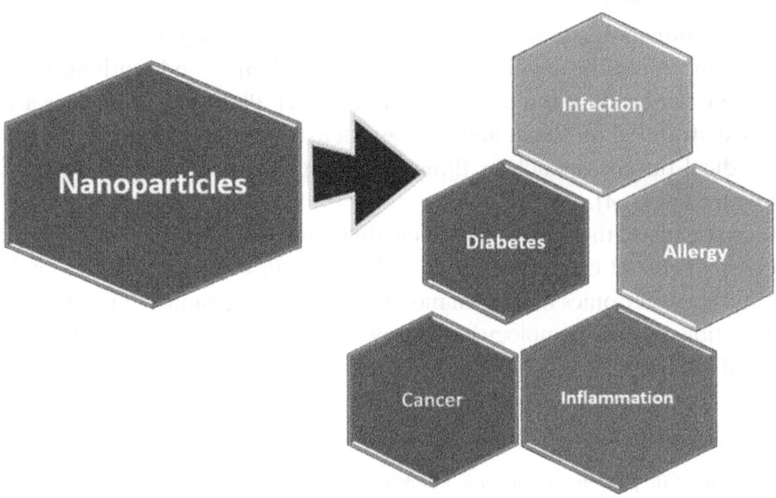

Figure 7.1 Nanoparticles in therapeutics.

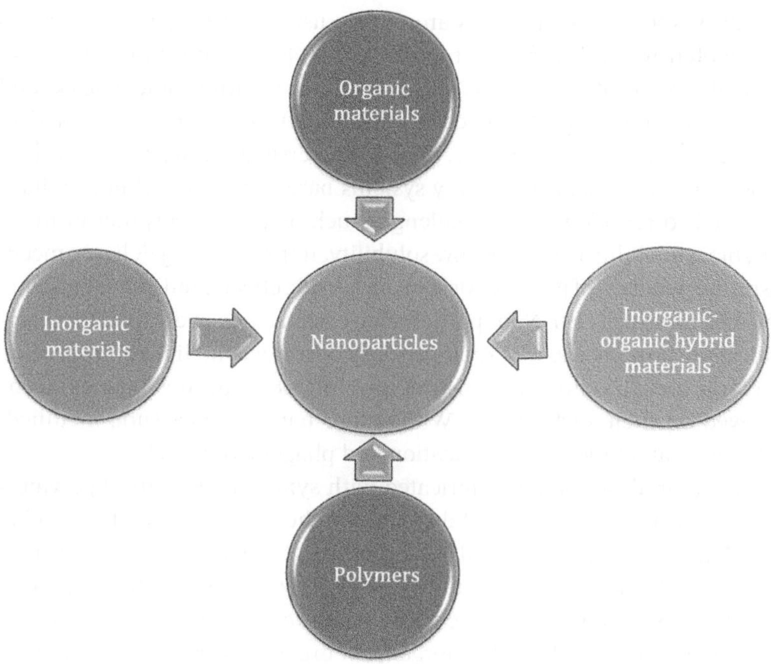

Figure 7.2 Building blocks in designing nanoparticles.

the former has the advantages of biodegradability, biocompatibility, and low immunogenicity (Figure 7.3).

Nanomaterials have numerous applications based on their mechanical, physical, electrical, magnetic, biological, and chemical characteristics.

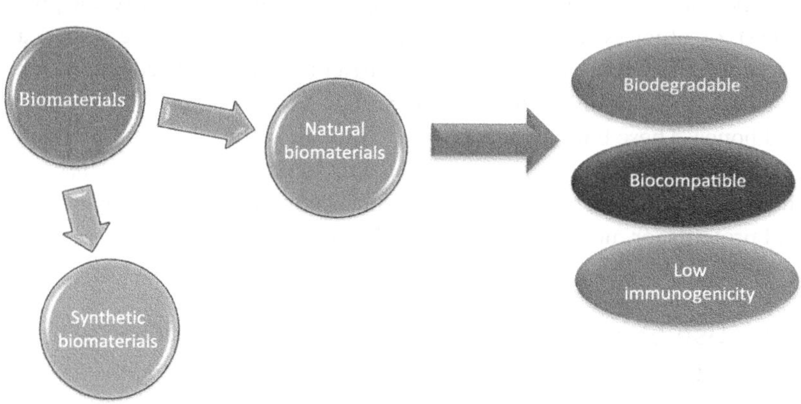

Figure 7.3 Types of biomaterials.

Nowadays, these nanomaterials are widely used in drug delivery systems and the pharmaceutical field. Drug delivery systems control the target-specific delivery of vital materials and contribute to drug rapid release and absorption, improving pharmacokinetics. A device can be formulated to regulate drug release in the body, improving persistent drug release [18]. In biomedical sciences, drug delivery systems based on nanotechnology have been used to resolve various challenges such as poor absorption in intestines, bioavailability caused by low solubility, improper drug delivery mechanism, the potential of therapeutical material, side effects, and incompatibility of plasma for drugs [19]. Nanoparticles have been extensively used for therapeutical purposes because of their property of surface modification based on their large area to volume ratio. Numerous functional groups/ligands can be displayed on their surface [20]. While other nanoparticles (non-modified) are targeted and killed by opsonization and phagocytosis [21].

Nanomaterials are largely fabricated with synthetic or natural polymers to protect from gastrointestinal degradation, hence improving the mechanism of targeted drug delivery at the target site [22]. Various methods can be used to modify or control the surface qualities size of nanoparticle and their properties based on their end-use implementation. The polymeric nanoparticles can either be modified by inorganic or organic hybrid materials. These modifications include covalent interactions between biopolymers or synthetic polymers. Nowadays, polysaccharides and protein-based biopolymers are widely used because of their antibacterial activity, biocompatibility, biodegradability, and low immunogenicity [23]. Despite having a wide size distribution and less drug/biomaterial binding ability, biopolymeric nanoparticles have been found to deliver drugs in the target cells and tissues [22]. Rapid biodegradability and improper mechanical characteristics of nanoparticles can be improved by crosslinking with genipin, polyethylene glycol (PEG), or other synthetic or biomaterials. Nanoparticles have wide-scale applications such as extensive use as a carrier in drug/gene delivery systems or as a scaffolding material for skin, cardiac, tissue engineering, cartilage, and bones. These biopolymers have been used for the treatment of allergy, inflammation, diabetes, cancer, and infections [20]. This chapter discusses the use of biopolymer nanomaterials such as polysaccharides, polypeptides, and proteins for medical use and drug delivery systems.

7.2 Use of Nanoparticles in Drug Design and Drug Delivery

By applying nanophases and nanostructures, nanotechnology has been shown to overcome the obstacles of physical and biological sciences in various fields [24]; in particular, these particles are of great importance in

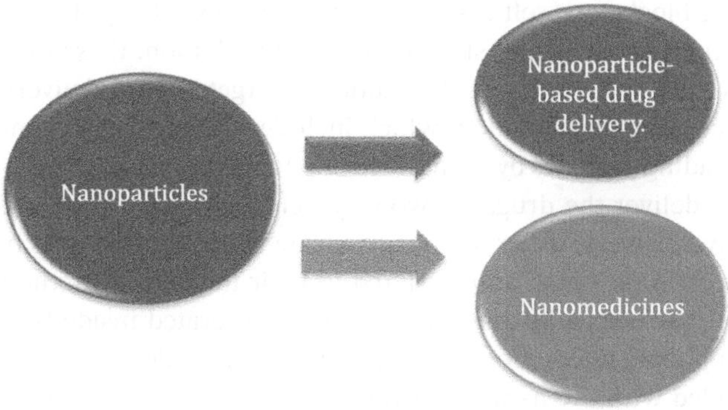

Figure 7.4 Role of nanoparticles in therapeutics.

nanomaterial-based DDS and nano-based medicines [25] (Figure 7.4). A material having a size ranging from 1 nm to 100 nm is called a nanoparticle, which can affect the boundaries of nanomedicine starting from microfluidics, biosensors, microarray tests, and drug delivery to tissue engineering [26, 27]. Nanotechnology uses therapeutic agents to produce nanomedicines at the nano level. Nanoparticles have powered the field of biomedicine, including biosensors, drug delivery, tissue engineering, and nanobiotechnology [28]. Since these nanoparticles are synthesized at the atomic or molecular level, they comprise very small nanospheres [29]. These small-sized nanoparticles are preferred over bigger molecules as they can easily move inside the body. They show distinct chemical, electrical, mechanical, structural, biological, and magnetic characteristics.

Drugs can be delivered to the target tissues or cells more accurately by encapsulating or attaching them with nanoparticles, which are used as drug delivery agents [30]. Nanomedicine is a field of science that utilizes the information and procedures of nanoscience to prevent disease and remediation and in medical biology. In this field, nano dimensional materials (nanosensors, nanorobots) are used for the diagnosis, sensory information, delivery, and activation of materials inside cells. For example, a technique based on nanoparticles has been designed combining both the imaging modes and treatment of cancer diagnosis [31]. At the nanoscale, drug designing is the most extensively studied and emerging field in the utilization of nanoparticles as these particles have the potential to modify their characteristics such as drug release patterns, solubility, diffusivity, immunogenicity, and bioavailability. This eventually leads to designing methods that have improved drug administration pathways, fewer side effects, lower toxicity, prolonged life cycle of the drug, and enhanced bioavailability and distribution [28]. Drug delivery systems (DDS) are either designed for controlled drug release or their use as targeted delivery of drugs at a distinct location. Building blocks are self-assembled spontaneously to synthesize DDS to form distinct patterns and structures [32]. In addition, these drug delivery systems must overcome the difficulties in targeted drug delivery, such as sequestration or opsonization, which includes the recognition and destruction of invading particles by mononuclear phagocyte system [33, 34].

Nanostructures deliver the drugs in two ways, either by passive or self-delivery. In passive delivery, drugs are buried inside the nanostructures by hydrophobic forces. After being administered to their targeted sites, these nanomaterials release a specific amount of drug incorporated inside their cavities via hydrophobic interactions. In self-delivery, drugs designed to be released are coupled with the nanomaterials for their convenient delivery. In this method, timing is critical for drug release. These drugs get dissociated from their carrier nanomaterial rapidly before reaching the targeted site, decreasing efficacy and bioactivity [32].

Another important purpose of using nanostructures is the targeted delivery of drugs at the particular site and is achieved in two ways, either by active or passive targeting. Active drug targeting includes conjugating peptides and antibodies with DDS, which anchor them to the surface receptors present on the target tissues or cells. While passive targeting is based on the properties like temperature, pH, shape, and molecular size, which influence the binding and interaction of drug–nanocarrier complex circulating in the bloodstream and drive them to their targeted site. Major targets of these drugs inside the body include cell surface proteins or antigens, lipid moieties of the cell membrane, and receptors expressed on the cell membrane. Nowadays, most drug delivery systems based on nanotechnology focus on the diagnosis of cancer and its treatment [35].

Recently, many polymeric nanoparticles loaded with drug/gene have been designed, and their circulatory mechanism inside bodies has been extensively studied [36]. After being injected into human bodies, these gene/drug-bearing nanoparticles need to cross the epithelial sheets and enter the blood circulatory system in order to reach their target sites. Then these nanoparticles escape from the blood circulatory system in either fenestrated or continuous tissues. In the case of continuous endothelium of healthy tissues, nanoparticles leave the blood circulatory system by intracellular pathway, paracellular process, or transmembrane transport. Conversely, in diseased tissues, fenestration sites on the fenestrated endothelium have much larger gaps (100 nm – 2 μm) than that present in healthy tissues (2 to 6 nm). These gaps increase drug penetration and accumulation in cancerous sites, known as "enhanced permeation and retention effect (EPR effect)." Fenestrations of tumor vasculature differ based on several factors, including the type of cancer, site of disease inside body, and its stage. Under numerous pathological states, modifications of these vasculatures and fenestrations can also occur [37]. For example, the progression of neovasculature is induced by tumor expansion, which is characterized by fenestrated endothelium having large gaps of 200–780 nm, thus allowing the transport of nanoparticles [38].

Harmful effects of these particles should be investigated as they interact with cells, fluids, and tissues starting from their administration to the target sites via various routes [39]. After reaching the target organ, these particles may stimulate mediators to initiate immunological or inflammatory responses. One of the most significant criteria for a drug delivery system is to engineer polymer-based nanoparticles with distinct sizes [40]. Several factors influence intercellular drug delivery, including the charge on the surface, the shape of the particle, and surface structure. These properties play a role in the internalization of various molecules via endocytosis [41]. For targeted delivery and controlled drug release, a range of factors should be considered, including particle sizes, type of polymer, biodegradability, surface properties, and solubility [42] (Figure 7.5). Stimuli sensitivity must be introduced to the nanocarriers to achieve the controlled drug release after accumulated at the target site. These drug carriers may get diffused or degraded and release the incorporated drugs in response to certain stimuli including ionic strength, pH, or temperature. pH-sensitivity is significantly effective for applying drug delivery systems as the pH difference of various tumors (6.2 to 6.9) and healthy extracellular tissues (7.2 to 7.4).

Gene therapy includes the treatment of diseases by transferring genetic materials into a particular site/cell of a diseased person to supply disease-causing abnormal genes and has been applied to various diseases, including cardiovascular diseases, AIDS, and cancer. To achieve targeted gene delivery, factors or processes affecting the nature of macromolecules, such as being recognized by the immune system and enzymatic degradation, must be avoided through nanoparticle design. In addition, genes must be encapsulated in nano-carriers to protect them from degradation by nucleases. These nano-carriers should have a small size for convenient administration, internalization into cells(endocytosis), and transport to the nucleus. They should possess the ability to escape endosome-lysosome processing [43]. For gene therapy, viral as well as non-viral vectors are used. Non-viral vectors are preferred because of their low immunogenicity compared to viral vectors, and it is easier to control and modify their properties [44]. In gene therapy, cationic polymers have a significant affinity towards complexation with DNA as non-viral vectors. Cell-specific ligands are conjugated with nanocarriers' surfaces to modify the nanoparticle surfaces, resulting in the expression of the desired targeted transgene. For target-specific gene delivery and enhanced transfer of transgene, cationic polymers and genes bearing nanoparticles are modified with various proteins, including transferrin, antigens/antibodies, or knob [45].

Materials releasing nitric oxide (NO) have emerged, exploiting enormous roles of nitric oxide in therapeutics. NO has antimicrobial properties against numerous pathogens and helps to heal wounds [46]. Several protective, harmful, and regulatory effects of NO have initiated extensive research work to design and manufacture nitric oxide-releasing materials and drugs. Nanoparticles and other nitric oxide-releasing scaffolds have a promising role based on their distinctive features, including storage and delivery of nitric oxide cargo (load) in an effective and controlled way [47]. Several nitric oxide-releasing materials have been encapsulated in the matrices of biocompatible polymers and are used as dressings of wounds, patches, on devices that contact blood and as controlled NO drug release [48]. For controlled NO drug release, dextran and chitosan have extensively been utilized as matrices [49]. Tissue engineering has now been considered as a significant cause of controlled drug delivery. The efficiency of tissue engineering can be increased by releasing therapeutic agents in a controlled manner. To enhance the efficacy of incorporated drugs, biomaterials are

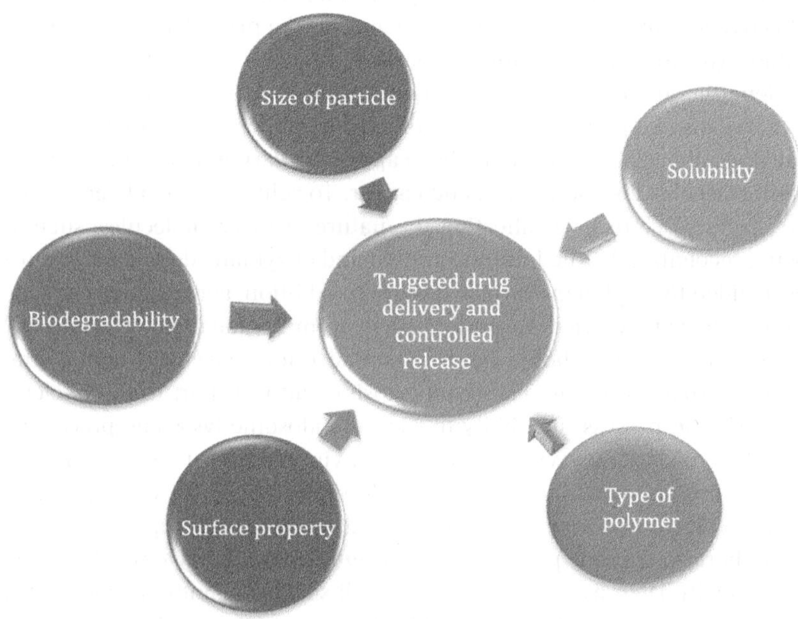

Figure 7.5 Factors affecting targeted drug delivery and controlled release.

designed at a nanoscale with controlled association [7]. Elements of gene delivery encapsulated inside scaffolds greatly increase the interactions between cells and their external environment as targeted delivery of genes triggers cues and signals to cells in spatiotemporal way for the maintenance and growth of tissues. In this way, these targeted genes increase the encapsulation of tissue construct, development, and interaction with neighboring tissues. The use of biopolymers in gene delivery makes complexes with DNA and plays a significant role in tissue engineering since being used as structural scaffolds. This association of tissue engineering and gene therapy has now been considered an advanced regenerative medicine treatment [45]. Gene-activated matrix (GAM) based localized gene delivery system combines two approaches, including their role as localized bioreactor with the expression of transgene/therapeutic gene and to serve as a template for lesion defects filling, enhancing the cell attachment, proliferation, and formation of extracellular matrix [50].

7.3 Use of Biopolymers in Drug Delivery System

Drug delivery is a collaborative field of medicine, pharmaceutical sciences, engineering, and chemistry. It is often based on drug design, routes of drug administration, and dosage form. Difficulties in conventional routes of drug administration have been overcome by the drug delivery system, which enhances the solubility of drugs, extends their duration time, increases the bioactivity of drugs, and lowers their side effects. The drug/gene delivery system currently improves the bioavailability and uptake of drugs, maintains the concentration of drugs by controlled release of therapeutic drug, and lowers the side effects by target-specific delivery of drug [51]. Several creative approaches in material chemistry have led towards the advent of biocompatible drug delivery systems (DDS), responsiveness of target, inducement, and biodegradable carriers. To overcome the barriers in intracellular and controlled drug delivery of macromolecular drugs (nucleic acids and proteins) and hydrophobic therapeutics, molecular biomaterials have been developed [52]. For efficient drug administration, the use of biological and physiological signals sensing materials is considered a promising approach for scientists and engineers to overcome many challenges. For tissue engineering and healthcare applications, both biodegradable DDS and non-biodegradable DDS such as composite polymers, bioceramics, biocompatible polymers, and hydrogels. Various significant characteristics have been exhibited by biopolymers including antibacterial properties, biocompatibility, and biodegradability. Biopolymers have a significant composition and structure similarity with macromolecules present in extracellular environment [53]. Simulation of chronic infection, side effects, and immunological responses due to synthetic polymer being administered inside the body can be reduced by using these biomaterials. These biopolymers can be modified to induce required electrical and mechanical characteristics based on their specific application and degradation rate [54].

Proteins have gained significant importance among biopolymers, exhibiting numerous beneficial roles due to having low toxicity, high nutritional value, non-antigenicity, solubility, biodegradability, and high binding ability of several therapeutic drugs, including ibuprofen and paclitaxel [55]. Their unique characteristics, including gelation, water binding ability, and emulsification, are of great interest to scientists [56]. These distinct properties distinguish them from other synthetic polymers and have a promising role in delivering gene and drugs. Since the secondary structure of proteins determines their sizes, protein-based synthesis of nanoparticles can be achieved with precision. Therapeutic nanoparticles are being synthesized by a diverse range of proteins, including albumin [57], elastin [58], silk [59], and collagen [59]. For controlled drug delivery, silk fibroin (SF) has a significant role in therapeutics because of having a number of advantageous properties, including self-assembly, malleability, ecological distribution, mechanical hardness, and biocompatibility. It can be utilized in the form of films, electrospun fibers, 3D scaffolds and hydrogels [60]. Silk fibroin and L-asparaginase were coupled to synthesize crystalline nanomaterials having diameter ranges from 50nm to 120nm [61]. These newly formed nanomaterials exhibited improved serum stability, decreased or no enzyme leakage from nanostructure particle, enhanced storage capacity and greater resistance against digestion from trypsin. For controlled drug release in SF-based nanoparticles, silk hydrogels networks and silk nanoparticles derived dual-drug release system have been formulated. Successful dual-drug release was observed by a standard drug being encapsulated in silk hydrogels and SF-based nanoparticles, resulted in a constant and rapid release from SF-derived nanoparticles [62]. Collagen has been extensively utilized as an important biomaterial because of having good biodegradability, biocompatibility, and fewer antigenic properties. It has a significant role in, and serves as a main constituent of, extracellular matrix [63]. Collagen-based synthesis of nanoparticles require additional treatment with chemicals since collagen hydrogels lack chemical crosslinking and have weaker mechanical strength. These collagen-based nanoparticles are mostly manufactured using electrostatic forces with sodium sulfate being used as a desolvating agent [64].

Recently, the curcumin-based drug delivery system has gained significant importance since it has various beneficial roles in therapeutics. It is now being encapsulated in SF-based drug delivery systems to overcome the challenges of curcumin's water solubility. To manufacture these nanofibers, a solution-enhanced dispersion technique is used, with CO_2 being crucial, resulting in enhanced water solubility and lower crystallinity [65]. Collagen is transformed into gelatin when it is partially hydrolyzed. This transformation leads to the conversion of collagen into a water soluble, irregularly structured protein that undergoes a sol-gel transition in water at lower temperature (below 35 °C) [66]. Large molecules can be delivered using gelatin-based microparticles employed as cell amplification carriers. Gelatin hydrogels can penetrate drugs between crosslinking gaps of polymer, exhibiting high surface loading capacity. During delivery of drugs, these drug molecules conveniently get diffused into blood circulatory system. Gelatin is often utilized for tissue adhesion, and its ability to form crosslinks is used for the controlled release of drugs in wound healing and pain management [67]. β-Casein is the main constituent of milk protein, and being an amphiphilic molecule, it has the ability to form micelles by hydrophobic effect, this property makes it a promising candidate for use in drug delivery system [68]. β-Casein is used in oral drug delivery systems where target-activated drug release is achieved by encapsulating various hydrophobic drugs in micelles of β-Casein. These drugs include vinblastine, docetaxel, irinotecan, paclitaxel and mitoxantrone. Human gastric N-87 tumor cells specific cytotoxic activity of paclitaxel was retained after casein was degraded with pepsin, while nanoparticles derived from β-Casein–paclitaxel were found to have no cytotoxicity in case of no prior gastric degradation [69]. β-Casein plays role in targeting stomach cancers due to its gastric digestibility [68]. In the biomedical field, zein has been considered an outstanding tool due to its ability to be handled chemically and shaped. Zein has the potential to overcome metal particles being utilized for numerous therapeutical approaches in the field of nano- and microparticles. Physiochemical properties of zein have made it a promising candidate in intracellular drug delivery, and it can replace the currently used synthetic polymers. Manufacturing strategies and use of surfactants have been investigated but important information regarding experimentation on stability and storage of zein is missing in the literature so preventing aggregation and *in vivo* trials. Generally, zein must be chemically modified for manufacturing zein-based blends in order to enhance the mechanical properties of films, hydrogels and fibers and to investigate the mechanism of drug release from the matrix of zein (*in vivo*) [70].

In drug delivery systems, cellulose and its derivatives have largely been used to modify drugs for controlled drug release. These modifications enhance drugs' gelation and solubility, making them suitable for therapeutical purposes [71]. Elseoud et al. [72] experimented on the use of chitosan nanoparticles and cellulose nanocrystals for oral repaglinide (RPG) release, which is an anti-hyperglycemic drug. Chitosan-based nanoparticles exhibited an average size range of 197 nm, on the other hand, hybrids of oxidized cellulose nanocrystals containing repaglinide (RPG) and chitosan nanoparticles showed an average diameter of 251–310 nm. Hydrogen bonding, present between the drug and cellulose based nanocrystals, plays an important role in controlled drug release. Consequently, lower drug release was observed in case of oxidized cellulose-based nanocrystals in contrast to native cellulose-based nanocrystals. Polymeric micelles-based drug targeting has been achieved by a number of methods such as enhanced ability of penetration and stimuli of holding effect, conjugation of monoclonal antibodies and the micelles or by coupling of multiple target specific ligands with the surface of micelles [73]. Polymeric micelles have been found beneficial for ocular delivery of drugs [74] as well as drug delivery against tumors [75]. Nanoparticles derived from dextran are also suitable for gene/drug delivery systems. Chemical modifications of dextran have extensively been studied for its improved activity in therapeutics. It can be modified chemically by grafting bile acids comprised of an amphiphilic steroid nucleus having a hydrophilic alpha side chain, hydrophobic beta side chain or lauryl chains [76].

7.4 Biopolymer Thin Films in Drug Delivery

Biopolymer thin films are usually adjustable and thin layers that may or may not contain plasticizer [77]. Being flexible and having thin size make them more suitable for patients [78]. These thin films are polymers matrices, having several advantages when using them as a drug delivery system [79]. Unlike liquid drugs or tablets, thin films can efficiently target sensitive sites [80]. These biopolymer thin films have many advantages, such as enhancing the efficiency of drugs, lowering the frequency of drug dose, and improving the onset of drug mechanism [79]. They have also been effectively used to prevent proteolytic degradation and eliminate the side effects of therapeutic drugs [81]. Thin films should exhibit some distinctive properties, including rapid dissolution rate, enhanced drug loading ability, adequate manufacturing stability, and extended residence duration at the administration site. They are required to show biodegradability, biocompatibility, and non-toxicity

[82]. Several polymers are being used as a backbone for manufacturing thin film [83]. In order to acquire the desired features of thin films, a single polymer or combination of various polymers can be utilized. It should be kept in mind that the polymers being used are non-irritating and have no toxicity and impurity. Water-soluble polymers have been largely used to manufacture thin film with high mechanical strength, fast integration rate, and good mouthfeel effects. Films are designed by using both synthetic and biopolymers [84]. Thin film has a fast dissociation rate as compared to other existing forms of dosage [85]. These are less fragile and convenient for storage and carrying purposes compared to other rapidly dissociating drugs requiring specialized packaging. These films can be carried individually on their own, as their single dose doesn't need any specific packaging or containers [86]. Poor stability of therapeutic drugs that are available in liquid forms should be improved also they need to be measured precisely and shaken every time before use [79]. Standard ophthalmic drug delivery systems including eye solutions and drops are more often used but these drugs do not show high bioavailability of drug and have less duration of their action [87]. Use of thin films in ophthalmology leads towards enhanced drug bioavailability to eyes. Transdermal films are preferred over transdermal patches as these films have enhanced permeation of water vapor and do not produce stickiness on the administered site resulting in reduced skin irritation [88]. Figure 7.6 represents the use of biopolymers and nanomaterials in drug delivery and drug designing.

7.5 Future Perspectives

Despite having plenty of published information based on natural nanomaterials drug delivery system, there is still a need to focus on some critical barriers. Most studies are in their initial stages and several strategies should be designed to fulfil various requirements. Mechanism of drug release and degradation of several types of nanomaterials can be altered and modified. We can overcome the current challenges by replacing native nanomaterials with genetically recombinant ones [89]. These modified and genetically engineered nanoparticles have significant role in drug delivery system as they enhance the specificity, therapeutic efficiency, functionality (cancer targeting), and lower the toxicity. The main challenges of protein-based nanomaterials include metabolic liability and less bioavailability. Proteins have difficulty crossing the epithelial barrier and face gastrointestinal degradation when taken orally, limiting their oral administration [90]. Drug delivery system based on oral administration is a commonly used method, but there are several barriers in this approach including challenges related to

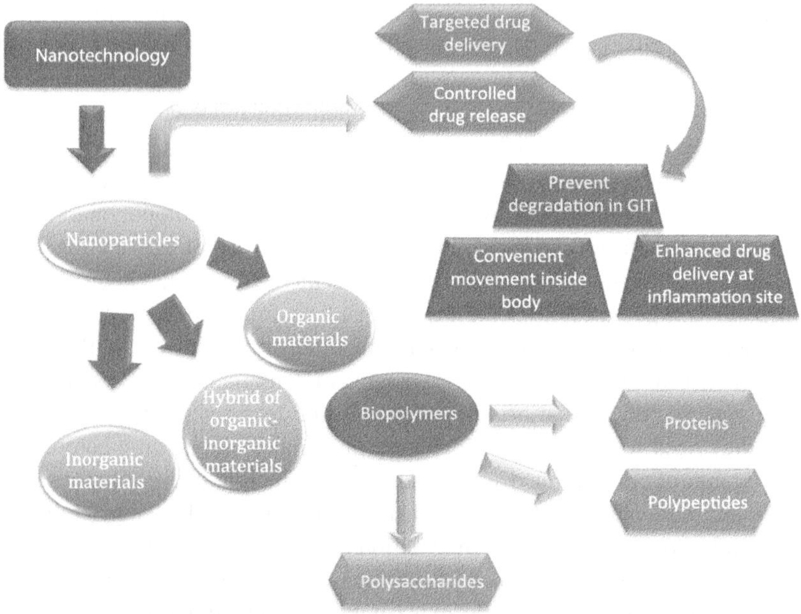

Figure 7.6 Biopolymers, blends, composites, gels and thin films in drug delivery and drug design.

pill-swallowing, especially in pediatric and elderly populations, and nanocomposites toxicology. In addition, clinical benefits, effective costing, and bad taste of orally administered drugs should be carefully monitored [60].

7.6 Conclusion

This review discusses current advancements in the manufacturing of naturally derived polymer-based nanoparticles and their use as gene/drug delivery carriers. These biopolymers exhibit various desired properties when compared to synthetic polymers in DDS, including biodegradability, ecological distribution, and biocompatibility. Polysaccharides are also promising candidates for the synthesis of nanomaterials because of having many advantageous characteristics such as biocompatibility, biodegradability, and convenience to get ionically and covalently modified. Nanoparticles can be manufactured using polysaccharides by ion-complex, covalent, or ionic crosslinking and grafting of several hydrophobic fragments to the polymer's backbone prior to their self-assembly. The nature of the material, such as the length of the polymer chain and charge, decide the technique being used to design nanoparticles. In addition to protein and polysaccharide-based nanoparticles, biopolymer and synthetic polymer conjugates have also been extensively studied. These biopolymer-based nanoparticles and biopolymer-synthetic polymer conjugates are the promising candidates for accurate drug/gene delivery in various fields of therapeutics.

References

1 Song, Z., Han, Z., Lv, S. et al. (2017). Synthetic polypeptides: from polymer design to supramolecular assembly and biomedical application. *Chemical Society Reviews* 46 (21): 6570–6599.

2 Tortorella, S., Buratti, V.V., Maturi, M. et al. (2020). Surface-modified nanocellulose for application in biomedical engineering and nanomedicine: a review. *International Journal of Nanomedicine* 15: 9909.

3 Malekzad, H., Mirshekari, H., Zangabad, P.S. et al. (2018). Plant protein-based hydrophobic fine and ultrafine carrier particles in drug delivery systems. *Critical Reviews in Biotechnology* 38 (1): 47–67.

4 Tavares, T.D., Antunes, J.C., Ferreira, F., and Felgueiras, H.P. (2020). Biofunctionalization of natural fiber-reinforced biocomposites for biomedical applications. *Biomolecules* 10 (1): 148.

5 Kataoka, K., Harada, A., and Nagasaki, Y. (2012). Block copolymer micelles for drug delivery: design, characterization, and biological significance. *Advanced Drug Delivery Reviews* 64: 37–48.

6 Wang, X., Yang, L., Chen, Z.G., Shin, D.M. (2008). Application of nanotechnology in cancer therapy and imaging. *CA: A Cancer Journal for Clinicians* 58 (2): 97–110.

7 Shi, J., Votruba, A.R., Farokhzad, O.C., and Langer, R. (2010). Nanotechnology in drug delivery and tissue engineering: from discovery to applications. *Nano Letters* 10 (9): 3223–3230.

8 Basarkar, A. and Singh, J. (2009). Poly (lactide-co-glycolide)-polymethacrylate nanoparticles for intramuscular delivery of plasmid encoding interleukin-10 to prevent autoimmune diabetes in mice. *Pharmaceutical Research* 26 (1): 72–81.

9 Furno, F., Morely, K.S., Wong, B. et al. (2004). Silver nanoparticles and polymeric medical devices: a new approach to prevention of infection? *Journal of Antimicrobial Chemotherapy* 54 (6): 1019–1024.

10 Roy, K., Mao, H.Q., Huang, S.K., and Leong, K.W. (1999). Oral gene delivery with chitosan–DNA nanoparticles generates immunologic protection in a murine model of peanut allergy. *Nature Medicine* 5 (4): 387–391.

11 Wilson, D.S., Dalmasso, G., Wang, L. et al. (2010). Orally delivered thioketal nanoparticles loaded with TNF-α–siRNA target inflammation and inhibit gene expression in the intestines. *Nature Materials* 9 (11): 923–928.

12 Kalainila, P., Subha, V., Sachidanandan, E.R.R., and Renganathan, S. (2014). Synthesis and characterization of silver nanoparticle from Erythrina indica. *Asian Journal of Pharmaceutical and Clinical Research* 7 (2): 39–43.

13 Yu, D.-G., Lin, W.-C., and Yang, M.-C. (2007). Surface modification of poly (L-lactic acid) membrane via layer-by-layer assembly of silver nanoparticle-embedded polyelectrolyte multilayer. *Bioconjugate Chemistry* 18 (5): 1521–1529.

14 Panyam, J. and Labhasetwar, V. (2003). Biodegradable nanoparticles for drug and gene delivery to cells and tissue. *Advanced Drug Delivery Reviews* 55 (3): 329–347.

15 Mironov, V., Kasyanov, V., and Markwald, R.R. (2008). Nanotechnology in vascular tissue engineering: from nanoscaffolding towards rapid vessel biofabrication. *Trends in Biotechnology* 26 (6): 338–344.

16 Heiati, H., Phillips, N.C., and Tawashi, R. (1996). Evidence for phospholipid bilayer formation in solid lipid nanoparticles formulated with phospholipid and triglyceride. *Pharmaceutical Research* 13 (9): 1406–1410.

17 Yang, Y.Y., Wang, Y., Powell, R., and Chan, P. (2006). Polymeric core-shell nanoparticles for therapeutics. *Clinical and Experimental Pharmacology and Physiology* 33 (5–6): 557–562.

18 Swamy, M.K. and Sinniah, U.R. (2016). Patchouli (*Pogostemon cablin* Benth.): botany, agrotechnology and biotechnological aspects. *Industrial Crops and Products* 87: 161–176.

19 Mohanty, S.K., Swamy, M.K., Sinniah, U.R. and Anuradha, M. (2017). *Leptadenia reticulata* (Retz.) Wight & Arn.(Jivanti): botanical, agronomical, phytochemical, pharmacological, and biotechnological aspects. *Molecules* 22 (6): 1019.

20 Rodrigues, T., Reker, D., Schneider, P., and Schneider, G. (2016). Counting on natural products for drug design. *Nature Chemistry* 8 (6): 531–541.

21 Siddiqui, A.A., Iram, F., Siddiqui, S., Sahu, K. (2014). Role of natural products in drug discovery process. *International Journal of Drug Development and Research* 6 (2): 172–204.

22 Beutler, J.A. (2009). Natural products as a foundation for drug discovery. *Current Protocols in Pharmacology* 46 (1): 9.11. 1–9.11. 21.

23 Thilakarathna, S.H. and Rupasinghe, H. (2013). Flavonoid bioavailability and attempts for bioavailability enhancement. *Nutrients* 5 (9): 3367–3387.

24 Liu, Z., Tabakman, S., Welsher, K., and Dai, H. (2009). Carbon nanotubes in biology and medicine: in vitro and in vivo detection, imaging, and drug delivery. *Nano Research* 2 (2): 85–120.

25 Razzacki, S.Z., Thwar, P.K., Yang, M. et al. (2004). Integrated microsystems for controlled drug delivery. *Advanced Drug Delivery Reviews* 56 (2): 185–198.

26 Patra, J.K. and Baek, K.-H. (2014). Green nanobiotechnology: factors affecting synthesis and characterization techniques. *Journal of Nanomaterials* 2014.

27 Joseph, R.R. and Venkatraman, S.S. (2017). Drug delivery to the eye: what benefits do nanocarriers offer? *Nanomedicine* 12 (6): 683–702.

28 Mirza, A.Z. and Siddiqui, F.A. (2014). Nanomedicine and drug delivery: a mini review. *International Nano Letters* 4 (1): 94.

29 Rudramurthy, G.R., Swamy, M.K., Sinniah, U.R., Ghasemzadeh, A. (2016). Nanoparticles: alternatives against drug-resistant pathogenic microbes. *Molecules* 21 (7): 836.

30 Lam, P.-L., Wong, W.-Y., Bian, Z. et al. (2017). Recent advances in green nanoparticulate systems for drug delivery: efficient delivery and safety concern. *Nanomedicine* 12 (4): 357–385.

31 Haba, Y., Kojima, C., Harada, A. et al. (2007). Preparation of poly (ethylene glycol)-modified poly (amido amine) dendrimers encapsulating gold nanoparticles and their heat-generating ability. *Langmuir* 23 (10): 5243–5246.

32 Lu, H. et al. (2016). Recent progress on nanostructures for drug delivery applications. *Journal of Nanomaterials* 2016: 5762431.

33 Blanco, E., Shen, H., and Ferrari, M. (2015). Principles of nanoparticle design for overcoming biological barriers to drug delivery. *Nature Biotechnology* 33 (9): 941–951.

34 Blanco, E., Shen, H., and Ferrari, M. (2021). Principles of nanoparticle design for overcoming biological barriers to drug delivery. In: *Nanomaterials and Neoplasms* (ed. V. Voliani), 585–622. Boca Raton: Jenny Stanford Publishing.

35 Kumari, A., Kumar, V., and Yadav, S. (2012). Nanotechnology: a tool to enhance therapeutic values of natural plant products. *Trends in Medical Research* 7 (2): 34–42.

36 Adiseshaiah, P.P., Hall, J.B., and McNeil, S.E. (2010). Nanomaterial standards for efficacy and toxicity assessment. *Wiley Interdisciplinary Reviews: Nanomedicine and Nanobiotechnology* 2 (1): 99–112.

37 Gaumet, M., Vargas, A., Gurny, R., and Delie, F. (2008). Nanoparticles for drug delivery: the need for precision in reporting particle size parameters. *European Journal of Pharmaceutics and Biopharmaceutics* 69 (1): 1–9.

38 Hobbs, S.K., Monsky, W.L., Yuan, F. et al. (1998). Regulation of transport pathways in tumor vessels: role of tumor type and microenvironment. *Proceedings of the National Academy of Sciences* 95 (8): 4607–4612.

39 Donaldson, K., Stone, V., Tran, C.L. et al. (2004). Occupational and environmental medicine. *Nanotoxicology* 61: 727–728.

40 Nakaoka, R., Tabata, Y., Yamaoka, T., and Ikada, Y. (1997). Prolongation of the serum half-life period of superoxide dismutase by poly (ethylene glycol) modification. *Journal of Controlled Release* 46 (3): 253–261.

41 Petros, R.A. and DeSimone, J.M. (2010). Strategies in the design of nanoparticles for therapeutic applications. *Nature Reviews Drug Discovery* 9 (8): 615–627.

42 Mohanraj, V. and Chen, Y. (2006). Nanoparticles-a review. *Tropical Journal of Pharmaceutical Research* 5 (1): 561–573.

43 Mansouri, S., Lavigne, P., Corsi, K. et al. (2004). Chitosan-DNA nanoparticles as non-viral vectors in gene therapy: strategies to improve transfection efficacy. *European Journal of Pharmaceutics and Biopharmaceutics* 57 (1): 1–8.

44 Li, S.-D. and Huang, L. (2006). Gene therapy progress and prospects: non-viral gene therapy by systemic delivery. *Gene Therapy* 13 (18): 1313–1319.

45 Dang, J.M. and Leong, K.W. (2006). Natural polymers for gene delivery and tissue engineering. *Advanced Drug Delivery Reviews* 58 (4): 487–499.

46 Weller, R.B. (2009). Nitric oxide–containing nanoparticles as an antimicrobial agent and enhancer of wound healing. *Journal of Investigative Dermatology* 129 (10): 2335–2337.

47 Carpenter, A.W. and Schoenfisch, M.H. (2012). Nitric oxide release: part II. Therapeutic applications. *Chemical Society Reviews* 41 (10): 3742–3752.

48 Eroy-Reveles, A.A. and Mascharak, P.K. (2009). Nitric oxide-donating materials and their potential in pharmacological applications for site-specific nitric oxide delivery. *Future Medicinal Chemistry* 1 (8): 1497–1507.

49 Friedman, A.J., Han, G., Navati, M.S. et al. (2008). Sustained release nitric oxide releasing nanoparticles: characterization of a novel delivery platform based on nitrite containing hydrogel/glass composites. *Nitric Oxide* 19 (1): 12–20.

50 Peng, L., Cheng, X., Zhuo, R. et al. (2009). Novel gene-activated matrix with embedded chitosan/plasmid DNA nanoparticles encoding PDGF for periodontal tissue engineering. *Journal of Biomedical Materials Research Part A* 90 (2): 564–576.

51 Anirudhan, T.S., Nima, J., and Divya, P.L. (2015). Synthesis, characterization, and in vitro cytotoxicity analysis of a novel cellulose-based drug carrier for the controlled delivery of 5-fluorouracil, an anticancer drug. *Applied Surface Science* 355: 64–73.

52 Liechty, W.B., Caldorera-Moore, M., Phillips, M.A. et al. (2011). Advanced molecular design of biopolymers for transmucosal and intracellular delivery of chemotherapeutic agents and biological therapeutics. *Journal of Controlled Release* 155 (2): 119–127.

53 Uebersax, L., Merkle, H.P., and Meinel, L. (2009). Biopolymer-based growth factor delivery for tissue repair: from natural concepts to engineered systems. *Tissue Engineering Part B: Reviews* 15 (3): 263–289.

54 Salerno, A. and Pascual, C.D. (2015). Bio-based polymers, supercritical fluids, and tissue engineering. *Process Biochemistry* 50 (5): 826–838.

55 Vuignier, K., Scappler, J., Veuthey, J.-L. et al. (2010). Drug–protein binding: a critical review of analytical tools. *Analytical and Bioanalytical Chemistry* 398 (1): 53–66.

56 Elzoghby, A.O., Samy, W.M., and Elgindy, N.A. (2012). Albumin-based nanoparticles as potential controlled release drug delivery systems. *Journal of Controlled Release* 157 (2): 168–182.

57 Bunschoten, A., Buckle, T., Kuil, J. et al. (2012). Targeted non-covalent self-assembled nanoparticles based on human serum albumin. *Biomaterials* 33 (3): 867–875.

58 Daamen, W.F., Veerkamp, J.H., van Hest, J.C.M., and van Kuppevelt, T.H. (2007). Elastin as a biomaterial for tissue engineering. *Biomaterials* 28 (30): 4378–4398.

59 Lammel, A.S., Hu, X., Park, S.-H. et al. (2010). Controlling silk fibroin particle features for drug delivery. *Biomaterials* 31 (16): 4583–4591.

60 Jacob, J., Haponiuk, J.T., Thomas, S, Gopi, S. (2018). Biopolymer based nanomaterials in drug delivery systems: a review. *Materials Today Chemistry* 9: 43–55.

61 Zhang, Y.-Q., Wang, Y.-J., Wang, H.-Y., et al. (2011). Highly efficient processing of silk fibroin nanoparticle-l-asparaginase bioconjugates and their characterization as a drug delivery system. *Soft Matter* 7 (20): 9728–9736.

62 Numata, K., Yamazaki, S., and Naga, N. (2012). Biocompatible and biodegradable dual-drug release system based on silk hydrogel containing silk nanoparticles. *Biomacromolecules* 13 (5): 1383–1389.

63 Maeda, M., Tani, S., Sano, A., Fujioka, K. (1999). Microstructure and release characteristics of the minipellet, a collagen-based drug delivery system for controlled release of protein drugs. *Journal of Controlled Release* 62 (3): 313–324.

64 Marty, J.J., Oppenheim, R.C., and Speiser, P. (1978). Nanoparticles-a new colloidal drug delivery system. *Pharmaceutica acta Helvetiae* 53 (1): 17–23.

65 Xie, M., Li, Y., Zhao, Z., et al. (2016). Development of silk fibroin-derived nanofibrous drug delivery system in supercritical CO2. *Materials Letters* 167: 175–178.

66 Coelho, J.F. (2014). Drug delivery systems: advanced technologies potentially applicable in personalised treatment, educational measures. *EPMA Journal* 5: A15.

67 Foox, M. and Zilberman, M. (2015). Drug delivery from gelatin-based systems. *Expert Opinion on Drug Delivery* 12 (9): 1547–1563.

68 Horne, D.S. (2006). Casein micelle structure: models and muddles. *Current Opinion in Colloid & Interface Science* 11 (2–3): 148–153.

69 Shapira, A., Davidson, I., Avni, N. et al. (2012). β-Casein nanoparticle-based oral drug delivery system for potential treatment of gastric carcinoma: stability, target-activated release, and cytotoxicity. *European Journal of Pharmaceutics and Biopharmaceutics* 80 (2): 298–305.

70 Tortorella, S., Maturi, M., Buratti, V.V. et al. (2021). Zein as a versatile biopolymer: different shapes for different biomedical applications. *RSC Advances* 11 (62): 39004–39026.

71 Sun, B., Zhang, M., Shen, J. et al. (2019). Applications of cellulose-based materials in sustained drug delivery systems. *Current Medicinal Chemistry* 26 (14): 2485–2501.

72 Abo-Elseoud, W.S., Hassan, M.L., Sabaa, M.W. et al. (2018). Chitosan nanoparticles/cellulose nanocrystals nanocomposites as a carrier system for the controlled release of repaglinide. *International Journal of Biological Macromolecules* 111: 604–613.

73 Wakaskar, R.R. (2017). Passive and active targeting in tumor microenvironment. *International Journal of Drug Development and Research* 9 (2): 37–41.

74 Mandal, A., Bisht, R., Rupenthal, I.D., Mitra, A.K. (2017). Polymeric micelles for ocular drug delivery: from structural frameworks to recent preclinical studies. *Journal of Controlled Release* 248: 96–116.

75 Kulthe, S.S., Choudhari, Y.M., Inamdar, N.N., and Mourya, V. (2012). Polymeric micelles: authoritative aspects for drug delivery. *Designed Monomers and Polymers* 15 (5): 465–521.

76 Nichifor, M., Lopes, A., Carpov, A., and Melo, E. (1999). Aggregation in water of dextran hydrophobically modified with bile acids. *Macromolecules* 32 (21): 7078–7085.

77 Maniruzzaman, M., Boateng, J.S., Snowden, M.J., and Douroumis, D. (2012). A review of hot-melt extrusion: process technology to pharmaceutical products. *International Scholarly Research Notices* 2012: 436763.

78 Patel, V.F., Liu, F., and Brown, M.B. (2011). Advances in oral transmucosal drug delivery. *Journal of Controlled Release* 153 (2): 106–116.

79 Borges, A.F., Silva, C., Coelho, J.F.J., Simões, S. (2015). Oral films: current status and future perspectives: I—galenical development and quality attributes. *Journal of Controlled Release* 206: 1–19.

80 Sharma, D., Kaur, D., Verma, S. et al. (2015). Fast dissolving oral films technology: a recent trend for an innovative oral drug delivery system. *International Journal of Drug Delivery* 7 (2): 60–75.

81 Castro, P.M., Fonte, P., Sousa, F. et al. (2015). Oral films as breakthrough tools for oral delivery of proteins/peptides. *Journal of Controlled Release* 211: 63–73.

82 Achouri, D., Alhanout, K., Piccerelle, P., and Andrieu, V. (2013). Recent advances in ocular drug delivery. *Drug Development and Industrial Pharmacy* 39 (11): 1599–1617.

83 Renukuntla, J., Vadlapudi, A.D., Patel, A. et al. (2013). Approaches for enhancing oral bioavailability of peptides and proteins. *International Journal of Pharmaceutics* 447 (1–2): 75–93.

84 Juluru, N.S. (2013). Fast dissolving oral films: a review. *International Journal of Advances in Pharmacy, Biology and Chemistry* 2 (1): 108–112.

85 Prabhu, S.C. et al. (2014). A review on fast dissolving sublingual films for systemic drug delivery. *International Journal of Pharmacy and Pharmaceutical Sciences* 3 (2): 501–511.

86 Russo, E., Selmin, F., Baldassari, S. et al. (2016). A focus on mucoadhesive polymers and their application in buccal dosage forms. *Journal of Drug Delivery Science and Technology* 32: 113–125.

87 Sultana, Y., Jain, R., Aqil, M., and Ali, A. (2006). Review of ocular drug delivery. *Current Drug Delivery* 3 (2): 207–217.

88 Guo, R., Du, X., Zhang, R., et al. (2011). Bioadhesive film formed from a novel organic–inorganic hybrid gel for transdermal drug delivery system. *European Journal of Pharmaceutics and Biopharmaceutics* 79 (3): 574–583.

89 Wenk, E., Merkle, H.P., and Meinel, L. (2011). Silk fibroin as a vehicle for drug delivery applications. *Journal of Controlled Release* 150 (2): 128–141.

90 Bruno, B.J., Miller, G.D., and Lim, C.S. (2013). Basics and recent advances in peptide and protein drug delivery. *Therapeutic Delivery* 4 (11): 1443–1467.

8

Biopolymers and Their Composites for Biotechnological Applications

Md Enamul Hoque and Md Ushama Shafoyat*

Department of Biomedical Engineering, Military Institute of Science and Technology, Dhaka, Bangladesh.
** Corresponding author*

8.1 Introduction

In recent years, the term "nanocomposite" has become widely used to describe nanocomposites that contain a naturally occurring polymers (biopolymers) along with an inorganic moiety and have at least one nanometer-scale dimension [1]. Bionanocomposites are a distinct type of materials that are created from nanocomposites that combine naturally existing polymers (biopolymers) with an inorganic nano moiety (BNCs). The term "biocomposite", often referred to as "nano biocomposites" (NCs), "green composites," or alternatively, "biohybrids", was first used in 2004 [2]. The development of unique features and the regulation of structure at the nanoscale depend heavily on a material's nanostructure. This makes nanotechnology a very promising topic for the twenty-first century. It will be necessary to completely reimagine the creative uses in the disciplines of semiconductors, inorganic and also natural materials, energy storage, and biotechnology [3]. In contrast, a composite is made from two or more materials that have different qualities. In composites, the different components are held together while maintaining their physical or compound qualities by specific physical or, on the other hand, concoction collaboration. The majority of the time, the qualities of composite materials are superior to those of separate, pure pieces in many ways [4]. A general definition of

Applications of Biopolymers in Science, Biotechnology, and Engineering, First Edition.
Edited by Arunima Reghunadhan, Akhina H., and Sabu Thomas.

nanotechnology is the development, processing, characterization, and use of material devices and systems having dimensions between 0.1 and 100 nm. It is possible to create nanocomposites with enhanced physical, thermal, and other distinctive features [5]. Due to its nanoscale dispersion with a size of less than 1,000 nm, bionanocomposites are significant [6]. Biopolymers are covalently linked long chains of monomeric units that make up polymeric biomolecules. Biopolymers, as opposed to manufactured polymers, have a sophisticated molecular assembly that results in a clearly defined three-dimensional structure. The term "bio" indicates that the polymer will spontaneously break down due to biological processes. They are able to mimic the characteristics of artificial polymeric materials. Biopolymers are very promising materials for very sensitive and selective gas and vapor sensors because of their abundance, biocompatibility, and distinctive characteristics. Materials made of biopolymers have good performance, are renewable, and have a small carbon footprint. These unique qualities of biopolymers make them a good option for the treatment of industrial wastewater. Potentially sustainable alternatives to petroleum-derived products for wastewater treatment include biopolymers and the materials they are made from. The development of innovative biopolymer-based composites and their practical applications in tissue engineering, medical implants, drug delivery systems, and wound healing, are greatly facilitated by biopolymers obtained from natural sources. Their applications in biomedical science, food packaging, and wrapping are numerous and varied. They are frequently used to create colloidal dispersions like foams or emulsions [7]. Although numerous researchers have examined polymer–polymer interactions [8], they continue to be among the hardest subjects to comprehend. They are favored in the food industry because they are environmentally friendly, non-toxic, non-immunogenic, biocompatible, have good chemical reactivity, are reasonably inexpensive [9], stable, have dietary advantages, and biodegradability [10]; additionally they're generally acknowledged as safe (GRAS) designation. In addition, they have a "clean" label since natural biopolymers have taken the role of synthetic stabilizers.

8.2 Application of Biopolymers, Composites, Nanocomposites, and Gels in the Field of Biotechnology

8.2.1 Application of Biopolymers

Biopolymers can be naturally produced by plants, animals, microbes, and agricultural waste. Plant sources, such as rice, maize, wheat, sorghum, yams, cassava, potatoes, banana, tapioca, corn, cotton, and barley, monomeric components like oils, carbohydrates, and amino acids, can be chemically converted into biopolymers. The most popular animal sources are cattle, while the most popular marine sources are corals, sponges, fish, lobster, and shrimp. The most typical microbiological sources include fungi, yeasts, and algae (Figure 8.1). Crops, paper wastes, green wastes, and wood wastes are examples of biomass-based sources that are high in carbohydrates. Vegetable oils like sunflower, soybean, safflower, jojoba, rapeseed, castor, and meadowfoam oil all contain triglycerides. In particular, vegetable oils from food producers make great substitutes for natural polymer production. Polyhydroxyalkanoates (PHAs), are a type of biopolymer that is produced by plants and microorganisms as secondary metabolites. PHAs are produced and aggregated intracellularly as translucent granules, and

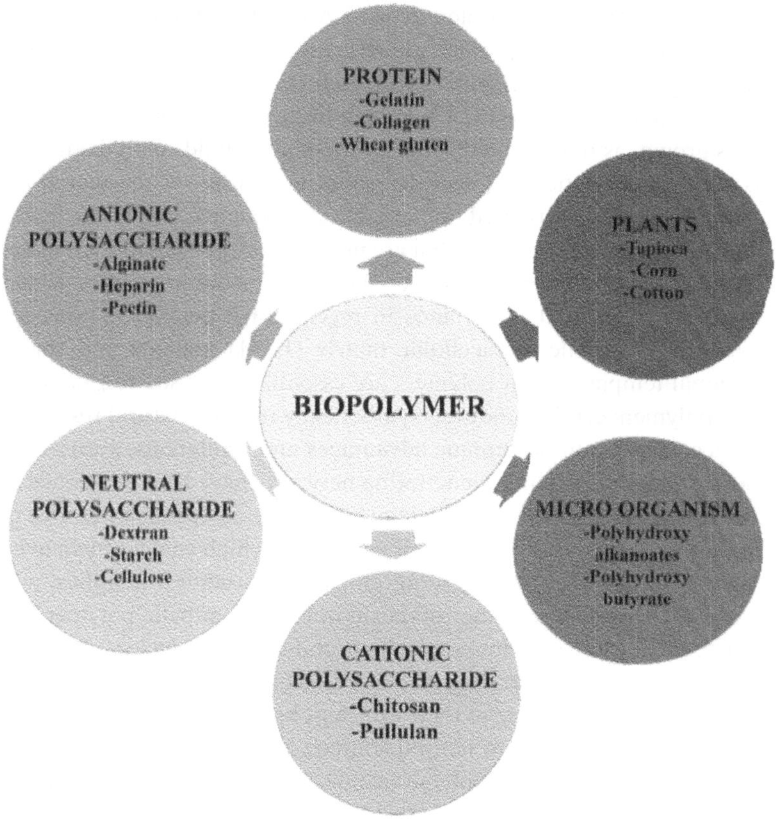

Figure 8.1 The visual representation of many sustainable natural biopolymers arranged by their source.

they are preserved as inclusion bodies in bacteria. Although these biopolymers can be melted and formed in the same way as chemical and synthetic thermoplastics, they are created naturally and broken down by microbial metabolisms [11].

8.2.1.1 Biopolymers for Medical Applications

The development of science and technology has significantly increased life expectancy. In this regard, innovative technologies and inventive strategies have been developed, leading to lower rates of morbidity and mortality. Treatment of disorders requires the use of bioactive molecules that are delivered by means of drugs, and research in this area has advanced significantly. In this regard, synthetic, semi-synthetic, and natural polymers are frequently used to develop drug delivery systems [12]. Contact inhibition, cell proliferation, tissue guidance, suturing, gluing, adhesion, covering, occlusion, isolation, and controlled drug administration [13, 14] are only a few examples of the potential medical uses for biopolymers. A number of biomaterial fabrication methods and techniques have been created in recent years to address and investigate the crucial functional, architectural, and compositional properties of natural tissues [15]. Understanding of the potential of biomaterials has grown, as has an interest in multimodal scaffolds with distinctive forms and physical–chemical characteristics, as a result of the search for better and more tissue-oriented implantable components. Due to the incorporation of various topographies that are not typically present in each material, these scaffolds have multifunctional or multimodal properties, which increases their potential importance in regenerative medical approaches [16]. Making synthetic extracellular matrix (ECM) habitats and three-dimensional templates with polymers are essential for tissue regeneration [17]. Biopolymers can be produced synthetically or from natural resources [18]. Since each group has unique advantages and limitations, a variety of composite materials and interpenetrating networks have been developed to achieve the necessary goals. Synthetic polymers are a dependable source of cutting-edge materials because of their versatility, which enables them to be altered to a wide range of degrading efficiency, structural qualities, and mechanical characteristics. The composition of the synthetic polymer can be adjusted to minimize immune reactions and integrate the key elements. Natural polymers have been suggested as a feasible alternative to frequently used synthetic materials due to their excellent biodegradability and lack of cytotoxicity. They come from natural sources, mimic soft tissues, and are made through chain-growth polymerization processes of activated monomers, which are frequently made inside cells during metabolic activities.

This category includes gels made of collagen, gelatin, dextran, agarose/alginate, hyaluronic acid, cellulose, and fibrin (Table 8.1). Natural polymers are frequently used in regenerative medicine, despite being cleansed to avoid a foreign body reaction after implantation. Biopolymers are widely used in a variety of well-known biomedical applications, such as soft-tissue replacement vascular grafts, breast implants, intraocular lenses, artificial hearts, contact lenses, plasmapheresis units, sutures, adhesives, and blood substitutes, as well as dialyzers, liver, pancreas, bladder, kidney, bone cement, catheters, external and internal ear repairs, coatings for pharmaceutical tablets and capsules, and cardiac assist devices [19–22]. PHAs can replace petrochemical polymers in biomedical applications since they are biodegradable. Particularly because of its versatility, poly (3-hydroxy octanoate) is potentially a good option as a biopolymer for drug delivery formulation and aspects of possible tissue engineering [23].

8.2.1.2 Biomedical Applications of Protein-Based Biopolymers

Biopolymers based on proteins are essential in medical applications. Biopolymers are non-toxic, biodegradable, and biocompatible materials. Their use in implanted devices is growing quickly and has a lot of potential [36]. The creation of nanoparticle proteins enables the development of precisely designed nanoparticles since the secondary structures of proteins determine their molecular sizes [37]. Table 8.2 summarizes the features of

Table 8.1 Examples of some biopolymers with their medical application.

Biopolymer	Medical application	References
Collagen	Tissue culture plate surface coating for cell culture, simple gels	[24]
Alginate	Medical regeneration, tissue engineering	[25]
Hyaluronic acid	Joint lubrication and treatment for injured joints, healing of skin and corneal wounds	[26]
Fibrin	Wound healing, blood clotting, surgical glue, sealant, and tumor development hemostatic agent	[27]
Silk fibroin	Medical regeneration, healing of injuries, tissue bioengineering	[28]
Agarose	Regeneration of skeletal tissues, fibroblast, and kidney encapsulation	[29]
Carrageenan	Regeneration of skeletal tissues, the delivery of cells	[30]

(Continued)

Table 8.1 (Continued)

Biopolymer	Medical application	References
Fibronectin	Wound recovery, cardiac rehabilitation, bone regrowth	[31]
PHA	Systems for delivering drugs, regeneration of one tissue	[32]
Elastin	Reconstruction of soft tissues, cell encapsulation, and orthopedics	[33]
Keratin	Engineering of corneal tissue, regenerating skin	[34]
Starch	Regeneration of cartilage and bones, therapy for spinal cord injuries	[35]

Table 8.2 Characteristics of several biodegradable nanoparticle kinds.

Nanoparticles	Properties	References
Chitosan	Nontoxic, safe for use in blood, tumor-fighting, antioxidant, antimicrobial, affordable, and biodegradable	[43]
Superparamagnetic iron oxide nanoparticles	Paramagnetic and superparamagnetic	[44]
Poly-L-lysine	Targeted distribution, high loading capacity, and biodegradability	[45]
Poly-D-L-lactide-co-glycolide	Non-toxic, biocompatible byproducts	[46]
Liposomes	Biocompatible, including hydrophobic substance	[47]
Alginate	Hydrophilic and biocompatible	[48]
Gold	Hyperthermia, biocompatible	[49]
Micelles	Able to transport a medication that is water soluble	[50]

various kinds of biodegradable nanoparticles. Some possible applications include medicinal chemical delivery and tissue engineering [38]. The goal of regenerative tissue engineering is to enhance the function of wounded or damaged tissues by utilizing the body's natural capacity for healing. For the purpose of creating an environment that is conducive to cell proliferation and tissue repair, synthetic polymers, and biopolymers are utilized as scaffolding. The building blocks for films, Pickering emulsions, hydrogels, nanofibers, connected porous scaffolds, and 3D-printed scaffolds are collagen, keratin, gelatin, sericin, and fibroin [14, 39–41]. Collagen-based nanospheres, nanoparticles, sponges, electrospun fiber, and hydrogels are used to deliver therapeutic molecules (gene delivery, medication delivery, and protein delivery) for tissue engineering [42].

Gelatin is a substance that has FDA approval and is non-toxic and biodegradable. To extract it from porcine or bovine sources, acid (type A gelatin) or essential (type B gelatin) hydrolysis is used. Many sources, including jellyfish, fins, bones, and sea urchins, have been used to extract gelatin [51]. Gelatin is capable of thickening, emulsification, foaming, and gel formation [52]. Chitosan is also frequently used in medicinal applications, because of its biodegradable and biocompatible qualities. It can be used in both biological and industrial applications and is useful for making gels, films, particles, membranes, and scaffolding. Moreover, because of its characteristics, chitosan is crucial for cell adhesion and proliferation. When produced with a porous structure it is frequently utilized as a matrix for tissue engineering. The principal applications of chitosan in implants include the regeneration of skin, liver, nerve, ligament, cartilage, and tendon. There have been reports of several chitosan-based matrices used in bone applications. Perhaps because of its mechanical brittleness [53], chitosan must be combined with other substances in bone tissue engineering applications [54]. Chitosan-created nanoparticles have intriguing characteristics such as low toxicity, good absorbability, permeability, and moisture retention, as well as being quickly degradable. They have poor long-term stability, though, and their rate of breakdown is greatly impacted by environmental factors including temperature and pH [55].

8.2.1.3 Biopolymers for Food Applications

With the use of various technical, scientific, and biotechnology tools, an increasing number of alternative and modified foods have been developed during the past 50 years. As a result, the majority of commercially produced food on the market today has undergone some type of modification or improvement in order to improve its appearance, flavor, or financial viability. The development of biotechnologies and applied engineering in the food business has been amazing. Many of the developments, from genetic engineering to improved preservatives and smart materials for creative materials, food quality monitoring, and packaging, are geared toward addressing the worldwide food problem in a population that is constantly growing. Recent studies reveal that new biopolymer products have a substantial impact on food research, even though medicinal science has gained the most from innovations in the field. These materials are currently being investigated for novel and distinctive food products that include traits like intelligent detection, superior nutritional attributes, bioresponsive capabilities, and a variety of biodegradable possibilities. Their use is advancing an environmentally responsible approach to food engineering and revolutionizing food design, production, and packaging. Since early on in our history humans have relied on biomaterials like silk, wool, cellulose, and leather as furniture, clothing, and food. Yet these polymers can be altered to fit particular requirements in the modern world. While humans live in both natural and artificial habitats, biopolymers derived from modern industrial economies and natural economies have a substantial positive impact on both ecosystems. Synthetic polymers are being used more frequently, which has raised issues with the environment and public health. As environmental concerns increase, the demand for materials manufactured from natural polymers is rising [56].

Food biopolymers are divided into two categories: proteins and carbohydrates. In both plant and animal tissues, proteins are amino acid polymers. Proteins in food are influenced structurally by shear, heat, and pH. In nature, polysaccharides, branching or linear polymers of sugars connected by glycosidic bonds, represent a substantial source of energy. Polysaccharides are hydrated chemicals that are present in both the endo- and exocore matrices of marine organisms and plants. Due to their potential to interact with other food components to improve their physicochemical characteristics and durability, biopolymers are currently a common component of specialized structures with qualities utilized in the processing of food. Polysaccharides are created from monosaccharides with similar or different residues. They provide a platform for visually appealing food packaging. Food is shielded against pathogenic and spoilage bacteria by active packaging, which is mostly made of polysaccharide biopolymers. Phase changes in food components occur during preparation and storage at various temperatures and pressures (liquid-gel or liquid-solid). The stability and quality of food are impacted by these modifications. This is due to the relationship between changes in the physical properties of foods and phase transitions in their constituent parts [57]. The food phase transition has been addressed in order to improve products and processes through process control. The design and attributes of food products that replace fat have been the subject of extensive investigation in recent years. In these food systems, biopolymers, especially hydrocolloids, are frequently used to mimic the sensory and rheological properties of lipids. It has been suggested that biopolymers are a suitable medium for the creation and stability of silver nanoparticles [58–60]. The polymer-assisted synthesis method encourages nanoparticle dispersion inside a polymer matrix, which affects the final stiffness and homogeneity of the nanocomposite film and preserves important antibacterial properties [61–63]. Improved food protection and longevity are two of the main benefits of nanoparticles, which can be attained by improving the physical and functional properties of materials used to make films [64].

8.2.2 Application of Biocomposites

8.2.2.1 Carbon Fiber-Reinforced Polymer Composites

Carbon fibers have characteristics of low density and high mechanical properties, specific tensile strength, and modulus. Polymers are added to carbon fibers providing a cohesive matrix in the desired geometric arrangement to form carbon fiber-reinforced polymer (CFRP). The polymer matrix transfers the load to the fibers and protects them from chemical and mechanical damages, thus yielding CFRPs with high strength-to-weight ratio, stiffness, and durability [65]. These features lend their widespread use for applications demanding high-performance, lightweight characteristics such as the aerospace, automotive, and defense industries.

In addition to industrial uses, CFRPs are a component of medical implants where mechanical strength and light weight are desired. They include cranioplasty, maxillofacial surgery, and scaffolds for tissue engineering in addition to orthopedic and dental implants. To improve stress transfer, CFRPs can obtain densities that are comparable to that of human bone depending on the polymer used and the manufacturing process. This is a desirable property for biomedical applications like orthopedic implants or prostheses. In comparison to conventional metal alloys, CFRPs provide superior fatigue and wear resistance [66]. Moreover, CFRPs are radiolucent and seem almost transparent when imaged with fluoroscopy, X-rays, or computed tomography (CT), which is significant in the clinic. For instance, bone tumor lesions, bone excision, or fractures are commonly treated with orthopedic stabilization with metal plate implants, followed by postoperative irradiation. Metal hardware obstructs the delivery of CT-guided radiotherapy for osteolytic lesions, resulting in imprecise target delineation and reduced dosage. Moreover, metal artifacts have an impact on follow-up imaging after treatment. When radiotherapy is required for spinal tumor lesions or head and neck cancers, respectively, persons who have metal spinal or dental implants also face these difficulties. As a result, CFRPs' radiolucency is a desirable property for usage as medical implants. Also, difficulties with metal detector security gates and worries about metal allergies are eliminated. CFRPs could be used as the enclosing reservoir for long-term durability in implantable drug delivery systems, which are of great interest for the prevention, management, and treatment of chronic diseases. Despite their advantages, CFRP-based medical implants have not yet been widely used in clinical settings due to a variety of reasons, including manufacturing issues.

8.2.2.2 Glass Fiber-Reinforced Polymer Composites

Because of their exceptional qualities, including low weight, high specific strength, and low cost, as well as its reasonably acceptable mechanical properties and non-abrasive nature, the glass fiber-reinforced polymer (GFRP) used as an implantable biomaterial. Afterward, synthetic glass fiber-reinforced polymer's mechanical characteristics in terms of tensile, flexural, hardness, and impact strength were assessed and contrasted with bone strength. Moreover, glass fiber sheet-reinforced plastic outperforms femur bone in terms of strength. The low porosity, stain resistance, and durable finish of GRP make it a popular choice for medical applications. GRP is used to construct everything from instrument enclosures to X-ray beds (where X-ray transparency is crucial).

8.2.2.3 Aramid Fiber-Reinforced Polymer Composites

A particularly popular aramid is Kevlar. Since concentrated sulfuric acid is the only useful solvent, it is particularly challenging to handle. Weight for weight, Kevlar is five times as strong as steel and is constructed of lightweight atoms while still being extremely strong and flexible. The alignment of the polymer chains and the hydrogen bonds that hold the flat molecules together are what gives it its strength. These molecular sheets may accumulate along the fiber axis (Figure 8.2). These fibers, which are already robust, are employed in ARPCs as fillers with phenolic or epoxy resin as the matrix. They are particularly helpful in situations requiring the absorption and release of energy, and they can withstand abrasion. These composite materials have found extensive use in the fields of sports equipment, medical device, aviation, helicopter rotor blades, and boats like kayaks and dinghies.

8.2.2.4 Natural Fiber-Reinforced Polymer Composites

Natural fiber-reinforced polymer-based composites are increasingly used in modern society due to their numerous benefits for biomedical engineering

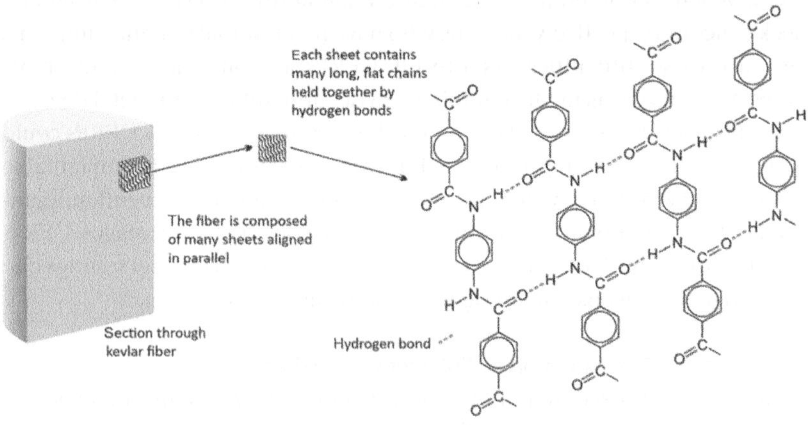

Figure 8.2 The structure of Kevlar.

applications. Polymer-based composite materials are utilized extensively in the creation of devices, prosthetics, construction, vehicles, aerospace, and many other fields due to their many benefits. Natural fibers such as jute, kenaf, pineapple, sugarcane, hemp, oil palm, flax, and leaves, etc., are cheap, environmentally benign, renewable, and totally and partially biodegradable and can be employed to generate novel high-performance polymer materials. These composites are more desirable than other composites due to their good mechanical characteristics, such as tensile qualities, flexural stress–strain behavior, fracture toughness, and fracture strength. Natural fibers can be utilized as a substitute for synthetic fibers as a reinforcing agent due to their accessibility and renewability.

8.2.3 Application of Nanocomposites

Products based on nanotechnology are now useful in every branch of engineering. Today, nanotechnology-based products are helpful in every area of engineering. Many clinical products make use of nanoparticles, and nanomedicine offers new healthcare options [67]. Nanomaterials can be composites of different materials or just have one main element. Because of their adaptable and functionally customizable architectures, nanomaterials and nanocomposites are thought to be effective tools for biomedical applications. Due to their qualities tailored to certain purposes, several of the materials can be employed in unique applications. For instance, due to their unique optical characteristics, high quantum yields, chemical and photostabilities, and potent absorption spectra, quantum dots (QDs) can be specifically chosen for medical imaging applications. Due to their excellent absorption qualities, iron oxide nanoparticles (NPs) and gold NPs can also be used in imaging applications. For medicine or therapeutic formulations, liposomes, hydrogels, and androgenic nanoparticles make excellent possibilities. Due to their great specificity, molecularly imprinted polymers have been considered for biosensor applications. Figure 8.3 summarizes the most popular nanomaterials and their composites as a classification by applications.

Materials used in biomedical applications should not be hazardous and should have a high level of biocompatibility, which is defined as the capacity to elicit the required biological response in a natural environment. There is currently potential for polymer nanocomposites (PNCs) in the biomedical sector due to the range of applications and significantly variable functional criteria for each of these applications. As a result, a PNC implant needs to adhere to a number of functional and design criteria, including those relating to biocompatibility, biodegradability, mechanical qualities, and, in some circumstances, aesthetics. Three basic components make up nanocomposites made by mixing nanosized fillers with thermoplastic or thermoset

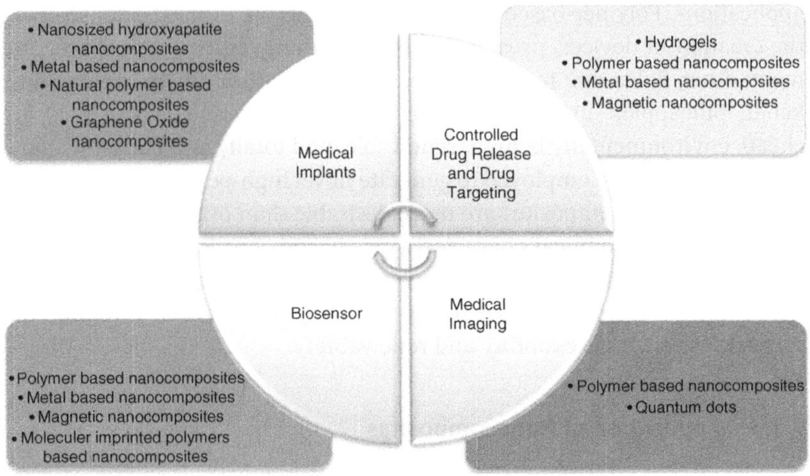

Figure 8.3 Nanocomposites in biomedical engineering applications.

polymers. The matrices primarily consist of nanoscale additives, fillers, and compatibilizers that in some cases improve or create interfacial contacts between the polymer phase and the filler. [68]. The forming of nanocomposite products and the physical performance of these materials are directly influenced by the interactions and compatibility of the constituent parts of PNCs. Due to their special characteristics, researchers created revolutionary composite nanosystems and smart polymers that react to changes in triggered illness situations. These materials are simple to use, have a high surface-to-volume ratio, and are environmentally benign. Nanomedicine based on composite materials aims to diagnose and cure conditions as accurately as possible without causing any adverse effects and to noninvasively evaluate the effectiveness of treatments [69].

8.2.3.1 Medical Implants Using Composite Nanomaterials

Composite materials made using nanotechnology have found use in the creation of medical implants. Better biological reactions and efficient mechanical contact between tissue and implant can be achieved by increasing the material's surface area and modifying the roughness of the nanoscale scale surface [70]. In tissue engineering, extracellular matrices or tissues can be built using nanotechnology and nanomaterials to improve cell responses to implants. There is a lot of interest in coatings with biomaterials for metallic implants as a result of the development of composite nanomaterials with novel biocompatible and antibacterial properties. The substance that makes up the majority of bones, and is known as bioceramics, is hydroxyapatite (HAp) [71]. Stainless steel, titanium, and cobalt–chrome alloys are the main materials used to create metal-based biomaterials. Moreover, these materials are hefty, and surgical procedures are required to remove these implants due to their insufficient biocompatibility qualities. Because of these restrictions, the ideal implant should have similar mechanical properties to native bone and be biodegradable and biocompatible [72]. Some drawbacks of medical implants include the development of biofilm on the implant surface, unfavorable host reactions, and sluggish tissue regeneration [73]. Moreover, the most typical illnesses brought on by urinary catheters are urinary tract infections. Antibacterial surface designs, such as bacterial repelling and bacteria-killing surfaces, are some methods to avoid the creation of biofilms. Bacteria-killing surfaces are created based on the disruption of bacterial cells upon contact, which results in cell death, as opposed to bacteria-repelling surfaces, which are focused on the prevention of biofilm development. Steric, electrostatic, contact-killing, and biocide-release effect-based antibacterial surface development methodologies can be seen in Figure 8.4 [74]. By promoting nearby cell migration, proliferation, and gene level regulation, these coatings hold great promise for long-term development and improvement of the host response. Adjusting the hydrophobicity and/or stiffness of the surface can achieve all of these goals, as well as shielding the implant from microbial attack and biofilm development [75]. So, it is necessary to modify medical implants to eliminate these drawbacks.

8.2.3.2 Controlled Drug Release and Drug Targeting with Composite Nanomaterials

Traditional nanocarriers cannot deliver medications to the intended spot in the proper concentration when subjected to internal or external stimulation.

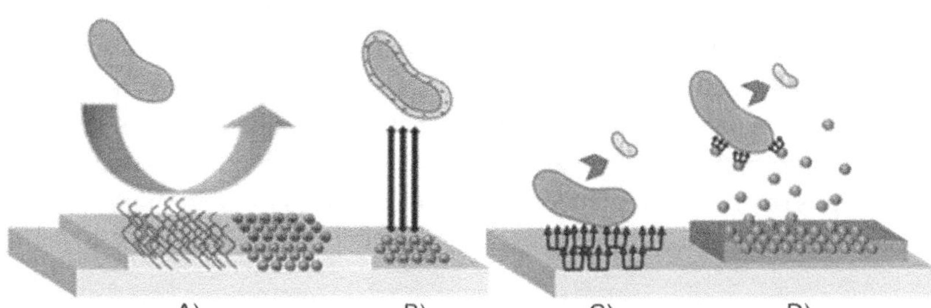

Figure 8.4 General antibacterial surface development strategies (A) steric, (B) electrostatic, (C) contact killing, and (D) biocide release effects. *Source:* Copyright 2020. Adapted from MDPI [74].

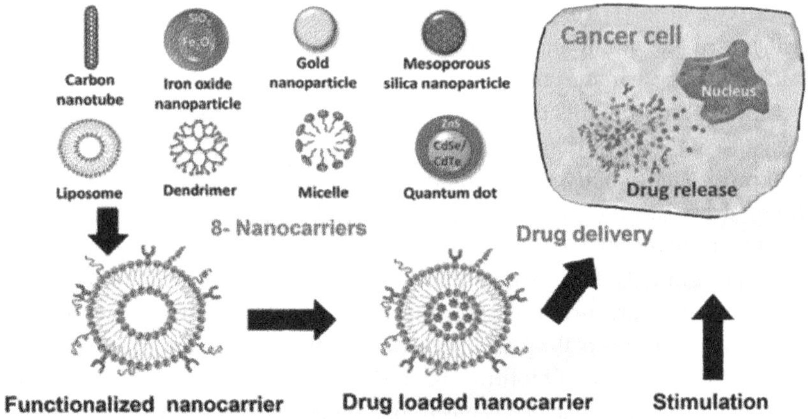

Figure 8.5 Application-specific functionalized nanosized carriers for drug delivery.

As a result, all common nanocarriers are regarded as intelligent. They must be altered or functionalized in order to serve as a smart nanocarrier. Figure 8.5 shows functionalized nanosized carriers for use in drug delivery systems. For various medication loading and delivery via stimulation, carbon nanotubes, metal oxide NPs, liposomes, mesoporous silica particles, QDs, polymer NPs, and their nanocomposites can be employed.

8.2.3.3 Composite Nanomaterials on Biosensors

Nanomaterials are particularly utilized in biosensors as active layers and mediators for biorecognition. The conductivity and arrangement of electron fluxes can be changed. The size, shape, or degree of aggregation can alter the properties of the generated nanosized materials. Figure 8.6 illustrates how nanomaterials affect electrochemical biosensors [76]. As the covering material for the sensor electrodes, graphene can be chosen due to its exceptional qualities, including high thermal conductivity, good mechanical capabilities, and high electronic transport. The idea of using it as a filler in polymer matrices to make cutting-edge multifunctional composites has aroused interest from both academia and industry. Flexible graphene-based nanocomposites have been created for new applications that call for high levels of mechanical, thermal, electrical, optical, and chemical efficiency [77].

8.2.3.4 Medical Imaging and Composite NPs

In the field of medical imaging, NPs are mostly used for magnetic resonance imaging (MRI) procedures. Superparamagnetic iron nanoparticles (SPIOs) and ultrasmall superparamagnetic iron nanoparticles are now used in

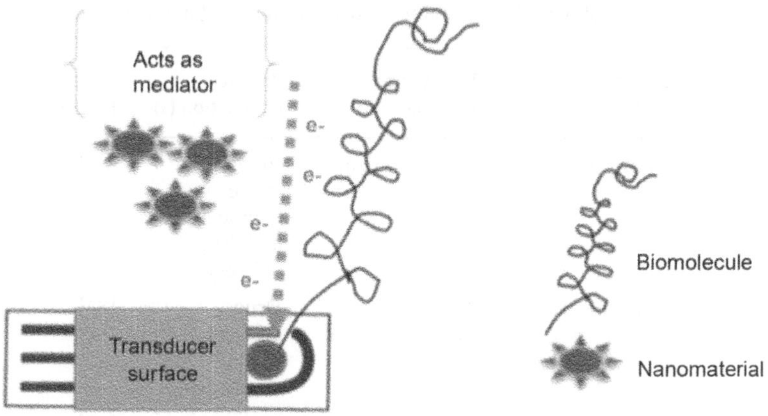

Figure 8.6 A nanomaterial's mediating effect on an electrochemical biosensor electrode [76]. *Source:* Copyright 2019. Adapted from MDPI.

clinical MRI (USPIOs). Moreover, a variety of NPs can be merged with imaging techniques based on X-rays, nuclear medicine, ultrasound, optical, and computer tomography [78]. Composite NPs offer improved picture resolution because of their size, lower risk, and photonic properties. Composite nanosized particle-based optical imaging techniques are supported by *in vivo* and *in vitro* stability, resistance to fluorine luminescence, high quantum yield and absorbance, resistance to metabolic degradation, and increased nontoxicity [79]. Successful molecular imaging requires the creation of intelligent, fully biodegradable NP/nanocomposite-based contrast agents that can be detected and are scaled properly for imaging targeted tissues. Determining the margins of brain tumors is a critical problem in clinical investigations. Imaging methods frequently have problems with insufficient sensitivity, specificity, and resolution.

8.2.4 Application of Hydrogels

8.2.4.1 Cell Delivery for Tissue Regeneration

Due to their extraordinary biocompatibility, simple production techniques, and variety of materials that can be utilized [80], recently biopolymer-based hydrogels have been used to transfer cells for the regeneration of various tissues, including skeletal muscle tissue, osteochondral tissue, and wound healing (Table 8.3). These hydrogels include cells for particular body locations that can be released from the hydrogel to promote cell growth and tissue regeneration. In some hydrogels, stem cells are encapsulated so they can differentiate into cells for particular tissues. Recently, injectable

Table 8.3 Overview of biopolymer materials used in hydrogel applications.

Material	Biopolymer category	Fabrication method	Recent application (s)	References
Collagen	Polypeptide	Demineralization	Implants, tissue engineering, and drug delivery	[83]
Gelatin	Polypeptide	Boiling natural products	Drug delivery, tissue engineering	[84]
Elastin	Polypeptide	Polymerization of tropoelastin monomers	Tissue engineering	[85]
Alginate	Polysaccharide	Extraction from cell walls of algae	Cell delivery	[86]
Cellulose	Polysaccharide	Extrusion from dissolved natural materials like wood pulp	Biomedical devices, tissue engineering, drug delivery	[87]
Glycosamino-glycans	Polysaccharide	Synthesis of UDP derived activated sugars	Tissue engineering	[88]
DNA	Nucleic acid	Light-directed combinational chemical synthesis	Bio sensing, cancer treatments, tissue engineering, and drug delivery	[89]

hydrogels have been revealed to have enormous potential in use as cell scaffolds for tissue engineering applications [81]. Without the need for large intrusive operations like surgery, tissues like skeletal muscles, heart tissues, and injured areas can be treated and mended using hydrogels [82].

8.2.4.2 Skeletal Muscle Regeneration

Self-healing hydrogels that operate to release skeletal muscle cells that aid in regeneration are frequently employed to aid in the regeneration of skeletal muscles. When employed in strenuous physical activities, these muscles can function as normal muscles since they are strong and flexible. An illustration of a biocompatible self-healing skeletal hydrogel is one consisting of N-carboxyethyl chitosan and dextran-graft-aniline tetramer-graft-4-formylbenzoic acid. The C2C12 cells in this hydrogel were released from the gel at a linear rate. This hydrogel containing the C2C12 cells helped with muscle regeneration as they were still able to proliferate and promote skeletal muscle regeneration. The hyaluronic acid–chitosan (HA–CS) hydrogel system, which also enclosed C2C12 cells, was a significant example of cell transport for skeletal muscle regeneration. Gelatin methacryloyl (GelMA) hydrogels are suitable as cell culture matrices that replicate native ECM due to their outstanding biocompatibility. Yet, one of the key prerequisites for creating tissue constructs that resemble living tissues is to produce structured assemblies of different types of cells that closely match the intricate architectures of the targeting tissues *in vitro* [90]. C2C12 cells were also enclosed in a hydrogel made of GelMA and poly(3,4-ethylenedioxythiophene) nanoparticles (PEDOT NPs) to promote the growth of muscle fibers [91]. This hydrogel was able to assist in the proliferation of muscle tissue and was demonstrated to be more effective than hydrogels made only with gelatin methacryloyl. These hydrogels with C2C12 encapsulations show how useful hydrogels may be in skeletal muscle regeneration, providing a more effective way to recover muscle following volumetric muscle loss.

8.2.4.3 Osteochondral Tissue Regeneration

Hydrogels are frequently self-healing materials used to promote cell growth and regeneration in osteochondral tissue regeneration. Because the cartilage between bones is highly avascular but protects the bones from grinding against each other, causing bone deformation, it is crucial to use hydrogels for osteochondral tissue [92]. An agarose/silk fibroin blended hydrogel that demonstrated good vitality for the seeded chondrocytes is an example of a biopolymer hydrogel. This agarose/silk fibroin hydrogel was tested *in vitro*, and the results showed that the chondrocytes proliferated, the silk fibers made the hydrogel more stable for load bearing, and the collagen production was increased [93]. Microfiber-reinforced silk hydrogels that were seeded with chondrocytes are another example of hydrogels utilized for osteochondral tissue regeneration. According to the results of testing this hydrogel, it was possible to create structures of cartilage that had an equilibrium modulus close to that of cartilage produced by the body [94]. Proliferation was also seen in a hydrogel consisting of fibrin gels with calcium ions and suspended bovine chondrocytes. This showed the promise of this hydrogel matrix as a potential treatment for damaged cartilage through osteochondral regeneration [95]. These hydrogels offer a potential route for osteochondral regeneration in the future without significant intrusive treatment.

8.2.4.4 Wound Repair and Regeneration

Differentiated cells or stem cells can frequently be used to mend wound sites in order to speed up the regeneration process. Because of their ability to differentiate, mesenchymal cells are stem cells that are frequently used in tissue regeneration. The drawback of utilizing mesenchymal stem cells is that they cannot survive in the wound site for long periods of time. A mixture of carboxymethyl chitosan and carboxymethyl cellulose dialdehyde to produce a gelatin cell carrier is an illustration of a biocompatible self-healing biopolymer hydrogel carrier for mesenchymal cells [96]. For the survival of mesenchymal cells and to allow stem cell paracrine signaling for effective wound repair and increased cell proliferation, this particular hydrogel has sheer-thinning and self-healing properties. Chitin nanofiber (CNF) hydrogel, which contained bone marrow mesenchymal stem cells (BMSC) for wound healing, is another hydrogel that is employed. By preserving the cells to keep them viable for wound repair, this hydrogel demonstrated the potential to stimulate bone marrow stem cell differentiation into a distinct and useful cell. The individual was able to repair the wound far more quickly with the BMSC hydrogel than with the other treatment option, demonstrating the enormous potential that hydrogels hold going forward. Because of their promising results, it is anticipated that in the future, treatments like these will be significantly less expensive to develop and administer. As a result, deeper wounds should heal more quickly and with less scarring and damage to the surrounding tissue.

8.2.4.5 Drug and Biomolecule Delivery

Recently, biopolymer-based hydrogels have also been utilized as controlled release techniques to deliver medications to the desired locations while defending against hostile bodily conditions. These hydrogels frequently include a medicine that can be released when particular triggers, including pH, temperature, or chemical ligands, are present. The diffusion coefficient of the drug across the hydrogel network determines how well the drug diffuses through the hydrogel. Hydrogels are frequently created to be biocompatible and biodegradable [97]. The release of insulin particles to decrease blood sugar levels and cancer therapy are two prominent examples of hydrogel pharmaceuticals and biomolecule deliveries [98].

8.2.4.6 Cancer Therapy

With the use of hydrogels, cancer medications can be delivered locally and under controlled release to the targeted areas without unfavorable off-target effects. Medication delivery that is thermoresponsive in order to help with the loading and release of cancer medications can be seen in hydrogel systems that have been observed to function well as they change states at various temperatures [94]. Chitosan/Disulfiram (DSF)-loaded hydrogel, which released the anticancer medication DSF, is an example of a thermoresponsive hydrogel. With a sustained administration, this hydrogel system for cancer therapy showed more cellular absorption than the cellular uptake of the DSF medication alone [95]. A chitosan-based hydrogel that included paclitaxel, an anti-cancer medication, and was neutralized with beta-glycerophosphate is another illustration of a hydrogel system. By releasing only 32% of the medication that was placed into the hydrogel, this hydrogel's targeted delivery capability revealed prolonged growth suppression of the tumor [96].

8.2.4.7 Insulin Drug Delivery

The present insulin delivery techniques have recently been found to be ineffective. In order to maintain blood glucose levels, patients must repeatedly inject themselves with insulin and constantly monitor their glucose levels. An improved technique for the regulated release of insulin has been developed using injectable SF (silk-fibroin) protein hydrogel (iSFH). This iSFH hydrogel has been found to control insulin release and maintain blood glucose levels. When injected into locations with a pH close to our gastric value, the alginate-gum tragacanth, another insulin delivery device, was able to keep almost all of the insulin contained. It was discovered that this hydrogel could release the insulin continuously in a pH environment similar to that of the digestive system, proving its capacity to adapt to various pH settings for localized insulin delivery [97]. Salecan hydrogels, a distinct type of insulin delivery device, were able to contain insulin and offer a regulated release because of the pH of the surrounding environment. For this hydrogel, it was discovered that insulin was not secreted at acidic pH levels, but was instead released at neutral pH levels, suggesting that insulin had reached the body's intended destinations [98]. These biopolymer-based hydrogel systems offer diabetics more effective ways to manage blood glucose levels without having to administer insulin on a regular basis or worry about whether all of it gets to the right place.

8.3 Conclusions and Future Prospects

Over the years, the commercial interest in biopolymers has continuously grown. Future biopolymer manufacturers are in desperate need of novel materials. Yet, given that it is specifically made available for sustainable development, the material's cost-effectiveness needs to be increased.

Biobased polymers are now more comparable to conventional polymers than before. Thanks to progress in biotechnology research and development as well as increased public awareness, biobased polymers are frequently used in a wide range of applications, from consumer items to high-tech ones. Food packaging is crucial in keeping food safe from outside contamination and preserving its quality, integrity, and safety over the course of its shelf life. Due to their simplicity of production, adaptability, affordability, utility, light weight, flexibility, and low cost, materials based on synthetic polymers are widely employed as packaging materials in the food business. These synthetic polymers, however, are not biodegradable, and the majority of plastic waste and litter severely damage the environment. To address these environmental issues, biodegradable polymer compounds must be created and used. Biopolymers or biopolymers derived from renewable resources include carboxymethyl cellulose, hemicellulose, pectins, starch, xanthan gum, pullulan, etc. Traditional petroleum-based food packaging materials can potentially be replaced by alginate, guar gum, gum karaya, agar, and gellan, among other substances. Medical applications for biopolymers like chitosan and PLA silk are being increasingly investigated. The special qualities of biopolymers, such biocompatibility and biodegradability, have important advantages and raise the possibility of their usage in implantable medical applications. As synthetic materials do not match the needs of biological systems, these novel materials are crucial in medicine. Recent studies have shown that combining biopolymers with synthetic materials can transform medicine.

The usage of non-renewable resources is raising environmental concerns, which is driving up demand for biodegradable polymers. Some of the most often produced materials for packaging and other uses include polystyrene and other polymers. These products pollute the land and water, and some studies have linked them to issues with human and animal health. Scientists and engineers are enthusiastically developing new biodegradable polymers to address these issues. They are extensively employed in many different industries and purposes, including packaging, farming, and healthcare. Scientists and engineers have investigated and tested the numerous kinds of materials used in the creation of biodegradable polymers to determine their effectiveness, safety, and environmental impact. Plastic packaging that is incorporated into society and supports a healthy, sustainable lifestyle is increasingly made of biodegradable polymers.

In order to replace present polymers, biobased polymers are currently undergoing study and development. Researchers are seeking for novel materials that can replace polymers generated from petroleum. Renewable raw ingredients make up biobased polymers. These polymers now account for a negligible (less than 1%) share of the plastics market. Biobased polymers are created using a bacterial fermentation method that produces monomers from organic waste, fatty acids, lignocellulosic biomass, and renewable resources from agricultural plants. Many proteins (including collagen) and nucleic acids naturally contain biobased polymers. Making polymers more biodegradable while keeping their strength and durability competitive with those of regular plastics is the most recent innovation anticipated in the plastics industry.

References

1. Darder, M., Aranda, P., and Ruiz-Hitzky, E. (2007). Bionanocomposites: a new concept of ecological, bioinspired, and functional hybrid materials. *Advanced Materials* 19: 1309–1319. doi: 10.1002/adma.200602328.
2. Kalsoom Khan, A., Saba, A.U., Nawazish, S. et al. (2017). Carrageenan based bionanocomposites as drug delivery tool with special emphasis on the influence of ferromagnetic nanoparticles. *Oxidative Medicine and Cellular Longevity* 2017: e8158315. doi: 10.1155/2017/8158315.
3. Biswas, M.C., Jony, B., Nandy, P.K. et al. (2022). Recent advancement of biopolymers and their potential biomedical applications. *Journal of Polymers and the Environment* 30: 51–74. doi: 10.1007/s10924-021-02199-y.
4. Mondal, S. (2015). Polymer nano-composite membranes. *Journal of Membrane Science & Technology* 05. doi: 10.4172/2155-9589.1000134.
5. Zhao, R., Torley, P., and Halley, P.J. (2008). Emerging biodegradable materials: starch- and protein-based bio-nanocomposites. *Journal of Materials Science* 43: 3058–3071. doi: 10.1007/s10853-007-2434-8.
6. Nuge, T., Liu, Z., Liu, X. et al. (2021). Recent advances in scaffolding from natural-based polymers for volumetric muscle injury. *Molecules* 26: 699. doi: 10.3390/molecules26030699.
7. Mohan, S., Oluwafemi, O.S., Kalarikkal, N. et al. (2016). Biopolymers-application in nanoscience and nanotechnology. In: *Recent Advances in Biopolymers* (ed. F.K. Perveen), 47–72. Croatia: InTech. http://dx.doi.org/10.5772/62225.
8. Evans, M., Ratclife, I., and Williams, P.A. (2013). Emulsion stabilisation using polysaccharide-protein complexes. *Current Opinion in Colloid & Interface Science* 18: 272–282. doi: 10.1016/j.cocis.2013.04.004.
9. Song, W.L., Wang, P., Cao, L. et al. (2012). Polymer/boron nitride nanocomposite materials for superior thermal transport performance. *Angewandte Chemie, International Edition* 51 (26): 6498–6501. doi: 10.1002/anie.201201689.

10 Yin, B., Zhang, R., and Yao, P. (2015). Influence of pea protein aggregates on the structure and stability of pea protein/soybean polysaccharide complex emulsions. *Molecules* 20 (3): 5165–5183. doi: 10.3390/molecules20035165.

11 Mojaveryazdi, F.S., Muhamad, I.I., Rezania, S., and Benham, H. (2014). Importance of glucose and *Pseudomonas* in producing degradable plastics. *Jurnal Teknologi* 69. doi: 10.11113/jt.v69.3194.

12 Gutierrez Cisneros, C., Bloemen, V., and Mignon, A. (2021). Synthetic, natural, and semisynthetic polymer carriers for controlled nitric oxide release in dermal applications: a review. *Polymers* 13: 760. doi: 10.3390/polym13050760.

13 Chen, Y., Hung, S.-T., Chou, E., and Wu, H.-S. (n.d.). Review of polyhydroxyalkanoates materials and other biopolymers for medical applications. *Mini-Reviews in Organic Chemistry* 15: 105–121. doi: 10.2174/1570193X14666170721153036.

14 Park, S.-B., Lih, E., Park, K.-S. et al. (2017). Biopolymer-based functional composites for medical applications. *Progress in Polymer Science*, Topical Volume on Biomaterials 68: 77–105. doi: 10.1016/j.progpolymsci.2016.12.003.

15 Yahya, E.B., Amirul, A.A., Khalil, H.P.S.A. et al. (2021). Insights into the role of biopolymer aerogel scaffolds in tissue engineering and regenerative medicine. *Polymers* 13: 1612. doi: 10.3390/polym13101612.

16 Song, W., Lima, A.C., and Mano, J.F. (2010). Bioinspired methodology to fabricate hydrogel spheres for multi-applications using superhydrophobic substrates. *Soft Matter* 6: 5868–5871. doi: 10.1039/C0SM00901F.

17 Hook, A.L., Anderson, D.G., Langer, R. et al. (2010). High throughput methods applied in biomaterial development and discovery. *Biomaterials* 31: 187–198. doi: 10.1016/j.biomaterials.2009.09.037.

18 Hassan, N., Verdinelli, V., Ruso, J.M., and Messina, P.V. (2012). Assessing structure and dynamics of fibrinogen films on silicon nanofibers: towards hemocompatibility devices. *Soft Matter* 8: 6582–6592. doi: 10.1039/C2SM25489A.

19 Pattanashetti, N.A., Heggannavar, G.B., and Kariduraganavar, M.Y. (2017). Smart biopolymers and their biomedical applications. Procedia Manufacturing, International Conference on Sustainable and Intelligent Manufacturing, RESIM 2016, 14–17 December 2016, Leiria, Portugal 12, 263–279. doi: 10.1016/j.promfg.2017.08.030.

20 Reddy, N., Reddy, R., and Jiang, Q. (2015). Crosslinking biopolymers for biomedical applications. *Trends in Biotechnology* 33: 362–369. doi: 10.1016/j.tibtech.2015.03.008.

21 Wróblewska-Krepsztul, J., Rydzkowski, T., Michalska-Pożoga, I., and Thakur, V.K. (2019). Biopolymers for biomedical and pharmaceutical applications: recent advances and overview of alginate electrospinning. *Nanomaterials* 9: 404. doi: 10.3390/nano9030404.

22 Gagliardi, A., Giuliano, E., Venkateswararao, E. et al. (2021). Biodegradable polymeric nanoparticles for drug delivery to solid tumors. *Frontiers in Pharmacology* 12. doi: 10.3389/fphar.2021.601626.

23 Sabbagh, F. and Muhamad, I.I. (2017). Production of poly-hydroxyalkanoate as secondary metabolite with main focus on sustainable energy. *Renewable and Sustainable Energy Reviews* 72: 95–104. doi: 10.1016/j.rser.2016.11.012.

24 Tripathi, D., Rastogi, K., Tyagi, P. et al. (2021). Comparative analysis of collagen and chitosan-based dressing for haemostatic and wound healing application. *AAPS PharmSciTech* 22: 76. doi: 10.1208/s12249-021-01944-9.

25 Álvarez-Castillo, E., Aguilar, J.M., Bengoechea, C. et al. (2021). Rheology and water absorption properties of alginate–soy protein composites. *Polymers* 13: 1807. doi: 10.3390/polym13111807.

26 Dulińska-Litewka, J., Dykas, K., Felkle, D. et al. (2022). Hyaluronic acid-silver nanocomposites and their biomedical applications: a review. *Materials* 15: 234. doi: 10.3390/ma15010234.

27 Nelson, D.W. and Gilbert, R.J. (2021). Extracellular matrix-mimetic hydrogels for treating neural tissue injury: a focus on fibrin, hyaluronic acid, and elastin-like polypeptide hydrogels. *Advanced Healthcare Materials* 10: 2101329. doi: 10.1002/adhm.202101329.

28 Grabska-Zielińska, S. and Sionkowska, A. (2021). How to improve physico-chemical properties of silk fibroin materials for biomedical applications?—Blending and cross-linking of silk fibroin—a review. *Materials* 14: 1510. doi: 10.3390/ma14061510.

29 Mahmood, A., Patel, D., Hickson, B. et al. (2022). Recent progress in biopolymer-based hydrogel materials for biomedical applications. *International Journal of Molecular Sciences* 23: 1415. doi: 10.3390/ijms23031415.

30 Khaliq, T., Sohail, M., Minhas, M.U. et al. (2022). Self-crosslinked chitosan/κ-carrageenan-based biomimetic membranes to combat diabetic burn wound infections. *International Journal of Biological Macromolecules* 197: 157–168. doi: 10.1016/j.ijbiomac.2021.12.100.

31 Klavert, J. and van der Eerden, B.C.J. (2021). Fibronectin in fracture healing: biological mechanisms and regenerative avenues. *Frontiers in Bioengineering and Biotechnology* 9. doi: 10.3389/fbioe.2021.663357.

32 Goswami, M., Rekhi, P., Debnath, M., and Ramakrishna, S. (2021). Microbial polyhydroxyalkanoates granules: an approach targeting biopolymer for medical applications and developing bone scaffolds. *Molecules* 26: 860. doi: 10.3390/molecules26040860.

33 Sarangthem, V., Singh, T.D., and Dinda, A.K. (2021). Emerging role of elastin-like polypeptides in regenerative medicine. *Advances in Wound Care* 10: 257–269. doi: 10.1089/wound.2019.1085.

34 Sellappan, L.K., Anandhavelu, S., Doble, M. et al. (2022). Biopolymer film fabrication for skin mimetic tissue regenerative wound dressing applications. *International Journal of Polymeric Materials and Polymeric Biomaterials* 71: 196–207. doi: 10.1080/00914037.2020.1817019.

35 Varma, K. and Gopi, S. (2021). Chapter 7 – Biopolymers and their role in medicinal and pharmaceutical applications. In: *Biopolymers and Their Industrial Applications* (ed. S. Thomas, S. Gopi, and A. Amalraj), 175–191. Elsevier. doi: 10.1016/B978-0-12-819240-5.00007-9.

36 Rebelo, R., Fernandes, M., and Fangueiro, R. (2017). Biopolymers in medical implants: a brief review. Procedia Engineering, 3rd International Conference on Natural Fibers: Advanced Materials for a Greener World, ICNF 2017, 21-23 June 2017, Braga, Portugal 200, 236–243. doi: 10.1016/j.proeng.2017.07.034.

37 Nitta, S.K. and Numata, K. (2013). Biopolymer-based nanoparticles for drug/gene delivery and tissue engineering. *International Journal of Molecular Sciences* 14: 1629–1654. doi: 10.3390/ijms14011629.

38 Nagarajan, S., Radhakrishnan, S., Kalkura, S.N. et al. (2019). Overview of protein-based biopolymers for biomedical application. *Macromolecular Chemistry and Physics* 220: 1900126. doi: 10.1002/macp.201900126.

39 Li, Y., Rodrigues, J., and Tomás, H. (2012). Injectable and biodegradable hydrogels: gelation, biodegradation, and biomedical applications. *Chemical Society Reviews* 41: 2193–2221. doi: 10.1039/C1CS15203C.

40 Kharkar, P.M., Kiick, K.L., and Kloxin, A.M. (2013). Designing degradable hydrogels for orthogonal control of cell microenvironments. *Chemical Society Reviews* 42: 7335–7372. doi: 10.1039/C3CS60040H.

41 Sridhar, R., Lakshminarayanan, R., Madhaiyan, K. et al. (2015). Electrosprayed nanoparticles and electrospun nanofibers based on natural materials: applications in tissue regeneration, drug delivery and pharmaceuticals. *Chemical Society Reviews* 44: 790–814. doi: 10.1039/C4CS00226A.

42 Jammalamadaka, U. and Tappa, K. (2018). Recent advances in biomaterials for 3D printing and tissue engineering. *Journal of Functional Biomaterials* 9: 22. doi: 10.3390/jfb9010022.

43 Thandapani, G., Prasad, P.S., Sudha, P.N., and Anil, S. (2017). Size optimization and in vitro biocompatibility studies of chitosan nanoparticles. *International Journal of Biological Macromolecules* 104: 1794–1806. doi: 10.1016/j.ijbiomac.2017.08.057.

44 Ahmed, H. and Hashim, A. (2021). Structure, optical, electronic, and chemical characteristics of novel (PVA-CoO) structure doped with silicon carbide. *Silicon* 13: 4331–4344. doi: 10.1007/s12633-020-00723-8.

45 Kodama, Y., Kuramoto, H., Mieda, Y. et al. (2017). Application of biodegradable dendrigraft poly-l-lysine to a small interfering RNA delivery system. *Journal of Drug Targeting* 25: 49–57. doi: 10.1080/1061186X.2016.1184670.

46 Song, R., Murphy, M., Li, C. et al. (2018). Current development of biodegradable polymeric materials for biomedical applications. *DDDT* 12: 3117–3145. doi: 10.2147/DDDT.S165440.

47 Ekici, S., Ilgin, P., Butun, S., and Sahiner, N. (2011). Hyaluronic acid hydrogel particles with tunable charges as potential drug delivery devices. *Carbohydrate Polymers* 84: 1306–1313. doi: 10.1016/j.carbpol.2011.01.028.

48 Ilkhanizadeh, S., Khalafy, J., and Dekamin, M.G. (2019). Sodium alginate: a biopolymeric catalyst for the synthesis of novel and known polysubstituted pyrano[3,2-c]chromenes. *International Journal of Biological Macromolecules* 140: 605–613. doi: 10.1016/j.ijbiomac.2019.08.154.

49 Huang, P., Lin, J., Li, W. et al. (2013). Biodegradable gold nanovesicles with an ultrastrong plasmonic coupling effect for photoacoustic imaging and photothermal therapy. *Angewandte Chemie* 125: 14208–14214. doi: 10.1002/ange.201308986.

50 Batrakova, E.V., Bronich, T.K., Vetro, J.A., and Kabanov, A.V. (2006). Polymer micelles as drug carriers. In: *Nanoparticulates as Drug Carriers*, 57–93. London: Imperial College Press. doi: 10.1142/9781860949074_0005.

51 Gómez-Guillén, M.C., Giménez, B., López-Caballero, M.E., and Montero, M.P. (2011). Functional and bioactive properties of collagen and gelatin from alternative sources: a review. *Food Hydrocolloids* 25: 1813–1827. doi: 10.1016/j.foodhyd.2011.02.007.

52 Gaspar-Pintiliescu, A., Stefan, L.M., Anton, E.D. et al. (2019). Physicochemical and biological properties of gelatin extracted from marine snail *Rapana venosa*. *Marine Drugs* 17: 589. doi: 10.3390/md17100589.

53 Abdul Kalil, H.P.S., Saurabh, C.K., Adnan, A.S., et al. (2016). A review on chitosan-cellulose blends and nanocellulose reinforced chitosan biocomposites: properties and their applications. *Carbohydrate Polymers* 150: 216–226. doi: 10.1016/j.carbpol.2016.05.028.

54 Sharifianjazi, F., Khaksar, S., Esmaeilkhanian, A. et al. (2022). Advancements in fabrication and application of chitosan composites in implants and dentistry: a review. *Biomolecules* 12: 155. doi: 10.3390/biom12020155.

55 Su, S. and Kang, P.M. (2020). Systemic review of biodegradable nanomaterials in nanomedicine. *Nanomaterials* 10: 656. doi: 10.3390/nano10040656.

56 Laine, C., Harlin, A., Hartman, J. et al. (2013). Hydroxyalkylated xylans – their synthesis and application in coatings for packaging and paper. *Industrial Crops and Products* 44: 692–704. doi: 10.1016/j.indcrop.2012.08.033.

57 Moraga, G., Talens, P., Moraga, M.J., and Martínez-Navarrete, N. (2011). Implication of water activity and glass transition on the mechanical and optical properties of freeze-dried apple and banana slices. *Journal of Food Engineering* 106: 212–219. doi: 10.1016/j.jfoodeng.2011.05.009.

58 Bankura, K.P., Maity, D., Mollick, M.M.R. et al. (2012). Synthesis, characterization, and antimicrobial activity of dextran stabilized silver nanoparticles in aqueous medium. *Carbohydrate Polymers* 89: 1159–1165. doi: 10.1016/j.carbpol.2012.03.089.

59 Kanmani, P. and Lim, S.T. (2013). Synthesis and characterization of pullulan-mediated silver nanoparticles and its antimicrobial activities. *Carbohydrate Polymers* 97: 421–428. doi: 10.1016/j.carbpol.2013.04.048.

60 Liu, Y., Chen, S., Zhong, L., and Wu, G. (2009). Preparation of high-stable silver nanoparticle dispersion by using sodium alginate as a stabilizer under gamma radiation. *Radiation Physics and Chemistry* 78: 251–255. doi: 10.1016/j.radphyschem.2009.01.003.

61 Islam, Md.S. and Yeum, J.H. (2013). Electrospun pullulan/poly(vinyl alcohol)/silver hybrid nanofibers: preparation and property characterization for antibacterial activity. *Colloids and Surfaces A: Physicochemical and Engineering Aspects* 436: 279–286. doi: 10.1016/j.colsurfa.2013.07.001.

62 Venkatesan, J., Singh, S.K., Anil, S. et al. (2018). Preparation, characterization, and biological applications of biosynthesized silver nanoparticles with chitosan-fucoidan coating. *Molecules* 23: 1429. doi: 10.3390/molecules23061429.

63 Youssef, A.M., Abou-Yousef, H., El-Sayed, S.M., and Kamel, S. (2015). Mechanical and antibacterial properties of novel high-performance chitosan/nanocomposite films. *International Journal of Biological Macromolecules* 76: 25–32. doi: 10.1016/j.ijbiomac.2015.02.016.

64 Kraśniewska, K., Galus, S., and Gniewosz, M. (2020). Biopolymers-based materials containing silver nanoparticles as active packaging for food applications–a review. *International Journal of Molecular Sciences* 21: 698. doi: 10.3390/ijms21030698.

65 Al-Mahmoud, F. (2018). 3.22 CFRP reinforcement rods. In: *Comprehensive Composite Materials II* (ed. P.W.R. Beaumont and C.H. Zweben), 578–591. Oxford: Elsevier. doi: 10.1016/B978-0-12-803581-8.03972-2.

66 Al Zahmi, S., Alhammadi, S., ElHassan, A., and Ahmed, W. (2022). Carbon fiber/PLA recycled composite. *Polymers* 14: 2194. doi: 10.3390/polym14112194.

67 Leong, H.S., Butler, K.S., Brinker, C.J. et al. (2019). On the issue of transparency and reproducibility in nanomedicine. *Nature Nanotechnology* 14: 629–635. doi: 10.1038/s41565-019-0496-9.

68 Saba, N., Tahir, P.M., and Jawaid, M. (2014). A review on potentiality of nano filler/natural fiber filled polymer hybrid composites. *Polymers* 6: 2247–2273. doi: 10.3390/polym6082247.

69 Singh, S.K., Kulkarni, P.P., and Dash, D. (2013). Biomedical applications of nanomaterials: an overview. In: *Bio-Nanotechnology: A Revolution in Food, Biomedical and Health Sciences* (ed. D. Bagchi, M. Bagchi, H. Moriyama, and F. Shahidi), 1–32. Chichester: John Wiley & Sons, Ltd. doi: 10.1002/9781118451915.ch1.

70 Ramos, A.P., Cruz, M.A.E., Tovani, C.B., and Ciancaglini, P. (2017). Biomedical applications of nanotechnology. *Biophysical Review* 9: 79–89. doi: 10.1007/s12551-016-0246-2.

71 Zhu, Y., Wang, H., Zhu, J. et al. (2015). Nanoindentation and thermal study of polyvinylalcohol/graphene oxide nanocomposite film through organic/inorganic assembly. *Applied Surface Science* 349: 27–34. doi: 10.1016/j.apsusc.2015.04.204.

72 Ding, P., Liu, Y., He, X. et al. (2019). In vitro and in vivo biocompatibility of Mg–Zn–Ca alloy operative clip. *Bioactive Materials* 4: 236–244. doi: 10.1016/j.bioactmat.2019.07.002.

73 McGuffie, M.J., Hong, J., Bahng, J.H. et al. (2016). Zinc oxide nanoparticle suspensions and layer-by-layer coatings inhibit staphylococcal growth. *Nanomedicine: Nanotechnology, Biology and Medicine* 12: 33–42. doi: 10.1016/j.nano.2015.10.002.

74 Olmo, J.A.-D., Ruiz-Rubio, L., Pérez-Alvarez, L. et al. (2020). Antibacterial coatings for improving the performance of biomaterials. *Coatings* 10: 139. doi: 10.3390/coatings10020139.

75 Erkoc, P. and Ulucan-Karnak, F. (2021). Nanotechnology-based antimicrobial and antiviral surface coating strategies. *Prosthesis* 3: 25–52. doi: 10.3390/prosthesis3010005.

76 Batool, R., Rhouati, A., Nawaz, M.H. et al. (2019). A review of the construction of nano-hybrids for electrochemical biosensing of glucose. *Biosensors* 9: 46. doi: 10.3390/bios9010046.

77 Silva, M., Alves, N.M., and Paiva, M.C. (2018). Graphene-polymer nanocomposites for biomedical applications. *Polymers for Advanced Technologies* 29: 687–700. doi: 10.1002/pat.4164.

78 Deng, L., Liu, Z., and Li, L. (2019). Chapter 5 – Hybrid nanocomposites for imaging-guided synergistic theranostics. In: *Nanomaterials for Drug Delivery and Therapy* (ed. A.M. Grumezescu), 117–147. William Andrew Publishing. doi: 10.1016/B978-0-12-816505-8.00017-5.

79 Marpu, S.B. and Benton, E.N. (2018). Shining light on chitosan: a review on the usage of chitosan for photonics and nanomaterials research. *International Journal of Molecular Sciences* 19: 1795. doi: 10.3390/ijms19061795.

80 Slaughter, B.V., Khurshid, S.S., Fisher, O.Z. et al. (2009). Hydrogels in regenerative medicine. *Advanced Materials* 21: 3307–3329. doi: 10.1002/adma.200802106.

81 Mantha, S., Pillai, S., Khayambashi, P. et al. (2019). Smart hydrogels in tissue engineering and regenerative medicine. *Materials* 12: 3323. doi: 10.3390/ma12203323.

82 Rammal, H., GhavamiNejad, A., Erdem, A. et al. (2021). Advances in biomedical applications of self-healing hydrogels. *Materials Chemistry Frontiers* 5: 4368–4400. doi: 10.1039/D0QM01099E.

83 Mirazul Islam, M., Cėpla, V., He, C. et al. (2015). Functional fabrication of recombinant human collagen–phosphorylcholine hydrogels for regenerative medicine applications. *Acta Biomaterialia* 12: 70–80. doi: 10.1016/j.actbio.2014.10.035.

84 Şelaru, A., Drăguşin, D.-M., Olăreţ, E. et al. (2019). Fabrication and biocompatibility evaluation of nanodiamonds-gelatin electrospun materials designed for prospective tissue regeneration applications. *Materials (Basel)* 12: 2933. doi: 10.3390/ma12182933.

85 Annabi, N., Mithieux, S.M., Boughton, E.A. et al. (2009). Synthesis of highly porous crosslinked elastin hydrogels and their interaction with fibroblasts in vitro. *Biomaterials* 30: 4550–4557. doi: 10.1016/j.biomaterials.2009.05.014.

86 Popa, E.G., Gomes, M.E., and Reis, R.L. (2011). Cell delivery systems using alginate–carrageenan hydrogel beads and fibers for regenerative medicine applications. *Biomacromolecules* 12: 3952–3961. doi: 10.1021/bm200965x.

87 Chaiyasat, A., Jearanai, S., Moonmangmee, S. et al. (2018). Novel green hydrogel material using bacterial cellulose. *Oriental Journal of Chemistry* 34: 1735–1740. http://dx.doi.org/10.13005/ojc/340404.

88 Narayanan, N., Jia, Z., Kim, K.H. et al. (2021). Biomimetic glycosaminoglycan-based scaffolds improve skeletal muscle regeneration in a Murine volumetric muscle loss model. *Bioactive Materials* 6: 1201–1213. doi: 10.1016/j.bioactmat.2020.10.012.

89 Mo, F., Jiang, K., Zhao, D. et al. (2021). DNA hydrogel-based gene editing and drug delivery systems. *Advanced Drug Delivery Reviews* 168: 79–98. doi: 10.1016/j.addr.2020.07.018.

90 Yue, K., Trujillo-de Santiago, G., Alvarez, M.M. et al. (2015). Synthesis, properties, and biomedical applications of gelatin methacryloyl (GelMA) hydrogels. *Biomaterials* 73: 254–271. doi: 10.1016/j.biomaterials.2015.08.045.

91 Wang, Y., Wang, Q., Luo, S. et al. (2021). 3D bioprinting of conductive hydrogel for enhanced myogenic differentiation. *Regenerative Biomaterials* 8: rbab035. doi: 10.1093/rb/rbab035.

92 Wang, J., Zhang, F., Tsang, W.P. et al. (2017). Fabrication of injectable high strength hydrogel based on 4-arm star PEG for cartilage tissue engineering. *Biomaterials* 120: 11–21. doi: 10.1016/j.biomaterials.2016.12.015.

93 Singh, Y.P., Bhardwaj, N., and Mandal, B.B. (2016). Potential of agarose/silk fibroin blended hydrogel for in vitro cartilage tissue engineering. *ACS Applied Materials & Interfaces* 8: 21236–21249. doi: 10.1021/acsami.6b08285.

94 Yodmuang, S., McNamara, S.L., Nover, A.B. et al. (2015). Silk microfiber-reinforced silk hydrogel composites for functional cartilage tissue repair. *Acta Biomaterialia* 11: 27–36. doi: 10.1016/j.actbio.2014.09.032.

95 Eyrich, D., Brandl, F., Appel, B. et al. (2007). Long-term stable fibrin gels for cartilage engineering. *Biomaterials* 28: 55–65. doi: 10.1016/j.biomaterials.2006.08.027.

96 Thomas, J., Sharma, A., Panwar, V. et al. (2019). Polysaccharide-based hybrid self-healing hydrogel supports the paracrine response of mesenchymal stem cells. *ACS Applied Bio Materials* 2: 2013–2027. doi: 10.1021/acsabm.9b00074.

97 Cikrikci, S., Mert, B., and Oztop, M.H. (2018). Development of pH sensitive alginate/gum tragacanth-based hydrogels for oral insulin delivery. *Journal of Agricultural and Food Chemistry* 66: 11784–11796. doi: 10.1021/acs.jafc.8b02525.

98 Ruel-Gariépy, E., Shive, M., Bichara, A. et al. (2004). A thermosensitive chitosan-based hydrogel for the local delivery of paclitaxel. *European Journal of Pharmaceutics and Biopharmaceutics*, Chitosan 57: 53–63. doi: 10.1016/S0939-6411(03)00095-X.

9

Biobased Polymers, Their Composites and Blends in Electronics

*Divyanshu Thakur and Maneesh Kumar Poddar**

Department of Chemical Engineering, National Institute of Technology Karnataka, Surathkal, Mangalore, India
** Corresponding author*

9.1 Introduction

Current environmental and economic concerns have attracted the research communities towards the production of biobased polymers due to their excellent biodegradability and environmental sustainability [1–3]. Biobased polymers are obtained from renewable resources of agricultural feedstocks (corn, potatoes, sugarcane, etc.), plants (starch, pectin, inulin, etc.), algae (alginates, ulvans), crustaceans (chitosan) and other sources from microorganisms (polyhydroxyalkanoates, bacterial cellulose), agricultural and food wastes (polylactic acid). Generally, the terms, biopolymers and biobased polymers are often used interchangeably with a little distinction that biopolymers are readily available in nature and can be created mainly by the biological action of living organisms. Silk, collagen, gelatin, bacterial cellulose, protein, etc., are several examples of biopolymers. However, biobased polymers are mainly produced from renewable but biological raw materials by the chemical or biological route. Polylactic acid (PLA), chitosan, polyhydroxyalkanoates (PHA), etc., are the common examples of biobased polymers [4–6]. A general layout and classification of biobased polymers are shown in Figure 9.1.

Advancement in technology has made a huge impact on every facet of our society and made humans very much dependent on the use of electrical and electronic appliances (TVs, computers, mobiles phones, electronic circuit

Applications of Biopolymers in Science, Biotechnology, and Engineering, First Edition.
Edited by Arunima Reghunadhan, Akhina H., and Sabu Thomas.

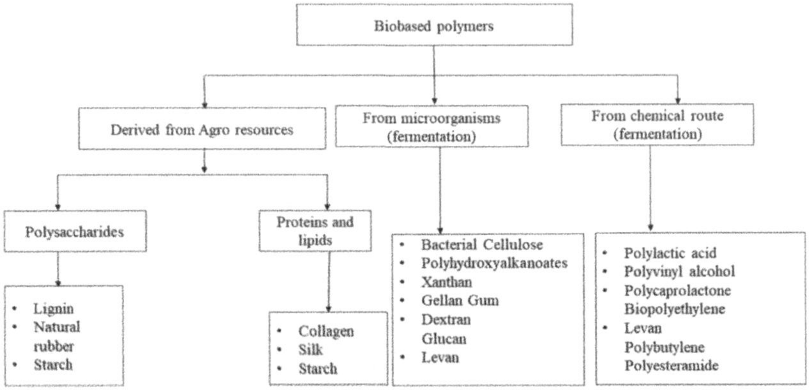

Figure 9.1 A general classification of biobased polymers based on their production routes.

boards, etc.), to fulfil their everyday needs and to experience a comfortable and smarter lifestyle. These electronics appliances fabricated using petroleum-derived polymers are mostly non-biodegradable and after their limited use are discarded into the environment in the form of e-wastes [7–9]. Moreover, these electronic wastes also contain hazardous and toxic chemicals that further lead to major environmental and ecological challenges.

Recently, electronic devices synthesized using biobased polymers either in pristine form or in combination with filler materials, or blending with other polymers, have shown tremendous potential for controlling the large e-waste along with the reduction in hazardous and toxic chemicals induced from petroleum-derived polymers [10, 11]. Various electrically insulating biobased polymers such as cellulose, chitosan, starch, etc., are being aggressively used as a potential substrate for many electronics devices [12]. However, the incorporation of electrically conductive nanofillers such as graphene, single and multiwalled carbon nanotubes (SWCNTs, MWCNTs), carbon nanofibers, metal oxides (AgO, AuO, ZnO, etc.), into pristine biobased polymers have widened their application in the fabrication of modern and advanced electronics appliances like solar cells, electromagnetic interference (EMI) shielding, transistors, sensors, flexible electronics, and other electronics devices [13, 14]. This chapter summarizes the numerous biobased polymers, their composites with different filler materials along the synthesis methods for application in electronics. Moreover, the latest advances and significant achievements in electronics appliances made from biobased polymers are also discussed. A summary of selected biobased polymers, their composites and blends with nanomaterials and other polymers are summarized in Table 9.1.

Table 9.1 A summary of properties of biobased polymers and their nanocomposites in electronics application.

Polymer and composite types	Synthesis method and experimental conditions	Major findings and conclusion	Reference
Cellulose	Nanocrystalline cellulose (CNC) and nanofibrillated cellulose (NFC) prepared using solvent (water) casting, stirring time: 12 h, drying temperature: 45 °C, drying time: 5 min	• Glycerol used as a plasticizer • Cellulose used as substrate in fabrication of organic photovoltaics (OPVs) device • Power conversion efficiency (PCE) in CNCs found higher than NFC substrates	[15]
Cellulose	Solvent (water) casting followed by atomic layer deposition (ALD), electrochemical method used in deposition of MnO_2 on CNT-Papers, stirring time: 20 m, temperature: 100 °C	• Cellulose fibers used as a substrate • Cellulose fibers with an open pore channel showed better electrochemical performance	[82]
Chitosan	Solvent (acetic acid) casting followed by NaOH treatment, chitosan molecular weights: 70 kDa and 300 kDa, chitosan 2% (w/v)	• Chitosan made flexible electronic skin showed low swelling ratio • Chitosan with 300 KDa showed Young's modulus of ~151 kPa	[83]
PLA	Electrospinning with dispersion of PLA into mixture of chloroform and acetone, chloroform, and acetone: 3:1 (volume ratio), applied voltage: 15 kV, feed rate: 5–10 μL/min	• Biosensor made using PLA was discussed • Biotin was successfully fixed on the PLA substrate	[84]

(Continued)

Table 9.1 (Continued)

Polymer and composite types	Synthesis method and experimental conditions	Major findings and conclusion	Reference
BC/MWCNTs	Ex–situ polymerization, incubation time: 14 days, incubation temperature: 30 °C, pH: 7	• Electrical conductivity: 1.4×10^{-1} Scm^{-1} • Surfactant was recovered at the end of the process	[85]
BC/RGO	Vacuum-assisted self-assembly technique, BC suspension in the alkali medium, cooling temp: −20 °C, suspension time: 6 h	• Highly flexible and mechanically strong BC/RGO nanocomposites, • Electrical conductivity: 1.1×10^{-6} Scm^{-1}, • Young's modulus: 1.7 GPa; tensile strength: 242 MPa; elongation at break: 5.4%	[86]
Chitosan/graphene/ glucose oxidase (GOD) biosensor	Three steps: (i) graphene prepared in mixture of sulfuric and nitric acid and potassium chlorate, (ii) graphene–chitosan prepared on the surface of glassy carbon electrode (GCE) and (iii) GOD coated on graphene–chitosan film, graphene preparation time: 96 h, drying temperature: 4 °C	• Biosensor used for glucose detection • Detection limit: 0.02 mM with sensitivity of 37.93 μAmM^{-1} cm^{-2} • High conductivity and better enzyme absorption	[61]
Sodium alginate/CuO	Solvent (water) casting, CuO: alginate weight ratio:7.5:1 or 1:1, pH:8.4	• Greatly improve lithium storage capacity of CuO • High structural ability	[20]

Silk/CNTs	Two stages: (i) preparation of silk nanofibers, (ii) preparation of silk/CNTs nanofibers using solvent (water) casting, silk fibers boiled in 0.06% (w/v) Na_2CO_3 for 30 m, silk CNT: 0.15 wt%, sonication: 300 W.	• High dimensional stability • Young's modulus: 498.7 MPa • Excellent chemical durability and electrical heating properties	[71]
PLA/MWCNTs/ polyaniline	Solvent (dichloromethane) casting, PLA (0.06 g mL^{-1}), PEG (0.04 g mL^{-1}), CNTs (0.006 g mL^{-1}), aniline (0.1 mL^{-1}), time: 30 min	• High flexible stability of 93.4% after 180°-bending for 700 cycles • Specific capacitance of 510.3 F g^{-1}	[76]
Chitosan/ polypyrrole (PPy)	Solvent (acetic acid) casting followed by electrochemical polymerization, acetic acid (1 vol%), three electrodes system	• High electrical conductivity • Shielding effectiveness > 98%	[79]
Polycarbonate (PC)/ poly(lactic acid) (PLA)	High temperature blending, PLA (32 wt%) and PC (62 wt%) maximum temperature: 265 °C	• High heat resistance, excellent flexibility • High flexural and tensile strength	[80]

9.2 Biobased Polymers and Their Application in Electronics

Biobased polymers can be produced directly from biomass available in nature using microorganisms or via the polymerization of biobased monomers (lactic acid, butylene succinate, etc.), with controlled polymerization. Pristine biopolymers have been continuously used as a potential substrate for the fabrication of modern electronic devices like organic light-emitting diodes (OLEDs), organic field-effect transistors (OFETs), sensors, etc. Among various biopolymers, polysaccharides are complex carbohydrates and are made of repeating units of glucose molecules bonded together. Based on the chemical composition, polysaccharides can be further divided into two parts: the first one is homopolysaccharides which are composed of a single type of monosaccharide like cellulose and chitin molecules, and the second one is heteropolysaccharides that consists of more than one type of monosaccharides like algin, glycoproteins, etc. Polysaccharides are acquired from sustainable sources like green plants and microorganisms such as parasites and microbes. With their ever-evolving physicochemical and structural properties, low-cost and excellent biodegradability, polysaccharides are extensively used as potential substrates in the fabrication of various modern and advanced electronics such as solar cells, supercapacitors for energy storage applications, flexible electronics [15, 16]. A summary of various important polysaccharides and their applications in the electronics industries are discussed in the subsequent sections.

9.2.1 Cellulose

Cellulose, a type of polysaccharide, is a naturally available biopolymer found in most cell walls of plants along with lignin and hemicellulose. Cellulose is composed of linear chains of glucose molecules linked repetitively via beta 1–4 glycosidic bonds. Due to its low cost, light weight and easy availability, it is frequently used as a biodegradable substrate in the fabrication of many electronic devices such as printed circuit boards (PCB), light-emitting diodes (LED), nanogenerators, etc. [15, 17, 18]. Bacterial cellulose (BC), which possesses a similar chemical structure to cellulose and is produced by bacteria is the purest form of cellulose. Because of its high crystallinity, excellent tensile strength, and unique fibrous structure, both its pristine and its composites forms are widely applied in advanced electronics. Cellulose nanofibers (CNF) and cellulose nanocrystals (CNC), which are the two main categories of nanocellulose, after incorporation of conductive nanofillers like graphene carbon nanotubes (CNT), into the cellulose matrix have enhanced electrical and mechanical properties for potential applications in sensors, flexible devices, actuators, etc. [19, 20]. BC has abilities to withstand high sonic velocity at a low dynamic loss and is used in the fabrication of high-quality sound devices like hi-fidelity (Hi-Fi) loudspeakers, headphones, and speakers [21]. Moreover, the three-dimensional (3D) fibrous structure of BC makes it a compatible biopolymer for the development of flexible high-performance nanodevice structures. Molybdenum disulfide (MoS_2), a phototransistor built on a highly transparent, ecofriendly, and flexible nanofibrillated cellulose (NFC) fiber substrate for application in printed circuit electronics [22]. The highly flexible and broad spectral range phototransistor has an outstanding photo responsivity with a magnitude of ~1.5 kA/W. The functioning of electronics appliances at low operating voltage is an important approach to save energy and operating costs. Considering these factors, ecofriendly and highly flexible organic field-effect transistors (OFET) ionic conductive cellulose nanopaper at a low operating voltage ($< 2V$) for high capacitance electronics have been developed. The devices exhibited numerous interesting features such as high optical transparency ($> 80\%$), ultra-smooth surface, and high ionic conductivity high dielectric constant, and thermal resistance for their application in low-voltage OFET devices in electronics.

9.2.2 Chitosan

Chitosan is the second most abundant biopolymer available on Earth after cellulose. It is extracted from chitin, which is found in the exoskeletons of crustaceans and insects as well as in fungi and yeast cell walls. Chitosan is biodegradable, non-toxic, non-allergic and therefore is used mostly in food packaging applications. Because of its electrical insulating and lightweight characteristics, it is used as a substrate in the fabrication of dielectric materials. The highly reactive hydroxyl and amine (-OH and -NH_2) groups present in chitosan allow it to form a strong composite material with externally added filler materials (micro- or nanoscale) and widen its application to electronics in the fabrication of supercapacitors, disposable transistors, biosensors, fuel cells, etc. [23, 24]. Moreover, low crystallinity reported in chitosan further increases its mechanical strength and elastic modulus (150 GPa) and allows its application in transparent and flexible electronics, logic circuits, strain sensor, and diodes. Other properties of chitosan, such as slow methanol penetrability, amphiphilicity, high thermal stability, and water barrier properties, further extend its use in solid polymer electrolytes in electrolyte-based fuel cells applications [25].

9.2.3 Alginate

Alginate is another natural polysaccharide and anionic biopolymer that can be derived from brown algae. The presence of functionalities such as carboxyl and hydroxyl groups allow its easy blending and composite formation with other polymers or filler materials [26]. The presence of β-1,4-linked D-mannuronic acid (M Block) and L-guluronic acid (G Block) in popular sodium alginate can form a gel under mild conditions and can be readily utilized in flexible and bioelectronics device applications like the development of strain sensors, dielectric materials, actuator, artificial skin and muscles [27, 28]. The presence of active sites on the dense functional groups makes the alginate a suitable candidate for exceptional proton conductivity with a magnitude of $\sim 5.5\times 10^{-3}$ S cm^{-1} and remarkable electric double layer (EDL) capacitance of ~ 2.0 μF/cm^2 for its potential application in the fabrication of bioelectronic materials [29]. The high gel-forming ability of alginate helps its effective utilization in the fabrication of biosensing devices for application in the medicine and healthcare industries [30].

9.3 Polylactic Acid

Polylactic acid (PLA) is a recyclable polyester and is obtained by condensation polymerization of the lactic acid monomer. The production of lactic acid is done through the fermentation of renewable materials of plant starch like sugarcane, cassava, corn, etc. Due to its abundance in nature and ability to biodegrade, it is extensively used as a substrate surface for many electronics devices, supercapacitors, 3D printing, EMI shielding, actuators. PLA also shows enough biocompatibility and less toxicity to be used as a substrate in bioelectronics areas where it is employed in applications like pacemakers, recording brains, heart waves, nano diagnostics devices, etc. [31]. PLA, in combination with conductive fillers, can be utilized as a biodegradable substrate in the fabrication of printed circuit boards (PCB) [32]. Moreover, PLA can be used as a substrate in the preparation of organic and inorganic transistors [33]. An excellent work by Ullah and Flint [34] proposed an electro textile-based patch antenna on PLA substrate with varying thicknesses of 1.7 and 3 mm. The results exhibited a gain of 2.7 and 3.45 dBi with an efficiency of 35 and 43% for 1.7- and 3-mm thickness substrates.

9.4 Others Biobased Polymers and Their Applications in Electronics

Polymers that are derived from 20 natural amino acids are categorized as protein-based biopolymers [35]. These polymers offer numerous advantages over other biopolymers and synthetic polymers in terms of high loading capacity, ease of production and their non-immunomodulatory characteristics [36]. Silk protein, which is an animal-derived fibrous biopolymer, is used as a supporting substrate in electronics devices due to its excellent optical transparency and mechanically robust characteristics [37, 38]. The high optical transparency allows silk to be used in anti-glare display panels and potential substrate for OLEDs. Moreover, the high physical flexibility and impressive thermal stability of silk enable its use as a strong and durable platform in numerous modern transistors and photonic devices [39, 40]. The incorporation of nanofillers into the silk matrix further enhance its physical and chemical properties and thus appears to be an excellent candidate for biosensing and bioimaging applications extending its uses.

Collagen is another protein-based biopolymer derived from the skin, tendon, and bones of vertebrates. Collagen exhibits high dielectric constant and excellent biodegradability, which enables its application in high-performance electronics to build a high hierarchical structure for microwave absorbance and the polarization of electromagnetic waves to antenna systems. Gelatin, a colorless, odorless and non-immunogenic biopolymer that is obtained after denaturation of collagen, has opened up new avenues in electronics due to its excellent biocompatibility and free radical degradation mechanism [41]. Gelatin can be used in the development of edible actuators in combination with other biomaterials in the synthesis of edible robots [42]. The edible robotics show potential applications in medical and food industries to administer drugs in humans and animals to carry out specific biological actions in their bodies. Gelatin also exhibits multifaceted adsorption behavior and compatibility with various inorganic nanofillers, which further help its binding during the synthesis of gelatin-based composites [43]. Moreover, the properties of high surface area and vigorous mechanical integrity make gelatine an attractive choice over the use of other biopolymers in the fabrication of numerous electronics devices. Figure 9.2 provides a visual representation of various biopolymers for their application in electronics.

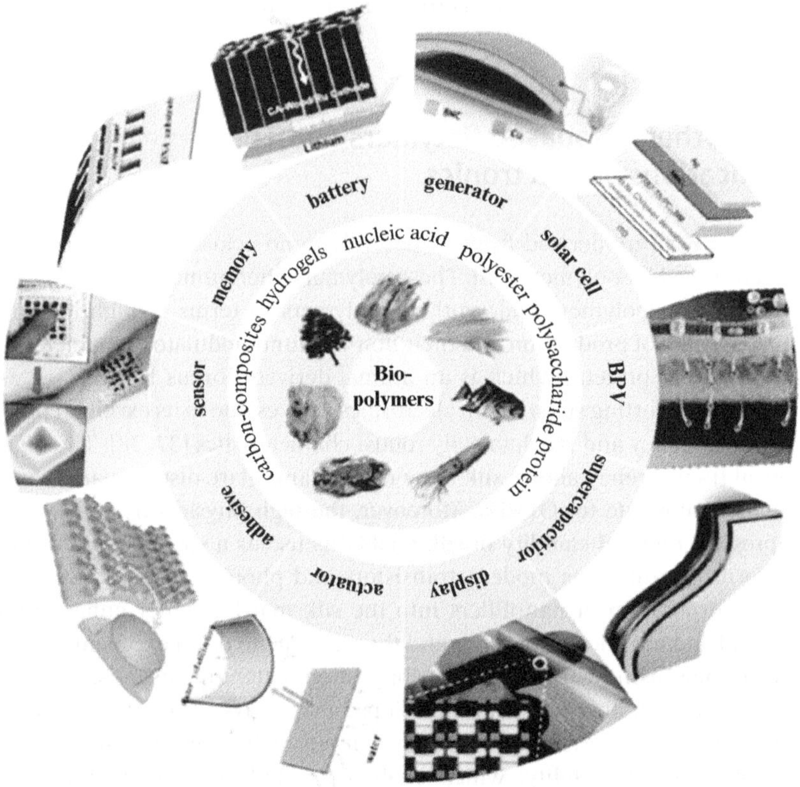

Figure 9.2 Schematic of numerous biobased/biopolymers for their application in electronics. *Source:* Adapted from [44] / with permission of Elsevier.

9.5 Biobased Polymer Composites and Their Application in Electronics

Polymer composites with successful dispersion of organic or inorganic filler materials into the polymer matrix have shown remarkable improvement in the physicochemical properties of final composites. Advancements in polymer composites with the use of nanosize filler materials (<100 nm) have significantly improved the properties of the composite due to the distinct properties of the nanofillers: its high surface to volume ratio and high surface area as compared to micron size fillers. Since work on modern composites is mainly focused on composites with the incorporation of nanosize fillers/additives we will focus on biobased polymer nanocomposites with nanosize fillers materials.

Several synthesis methods, e.g., in-situ polymerization, solvent casting, hybrid latex castings, melt intercalation, electrospinning, etc., have been utilized in the synthesis of polymer nanocomposites [45–47]. Each synthesis method has its own advantages and disadvantages and affects the properties of the final composite either directly or indirectly. Greater detail of the synthesis methods of polymer composites can be found in the literature [45, 48, 49], standard textbooks [50, 51], and popular reviews [52]. A brief description of biobased composites comprising the biobased polymer as the main matrix and fillers as a dispersing agent for their application in modern and advanced electronic devices are discussed in the subsequent sections.

9.5.1 Polysaccharide-Based Nanocomposites

Polysaccharide-based nanocomposites are prepared by incorporating the nanoscale carboneous (graphene, CNTs, CNFs, etc.), metal and metal oxides (Ag, Au, AgO, TiO_2) nanofillers into the bulk matrix of polysaccharides (cellulose, chitosan, glycogen, etc.). Successful encapsulation of these nanomaterials into pristine polysaccharides imparts distinct properties to final nanocomposites and makes it a promising candidate for the fabrication of modern electrical and electronic appliances. Given below, we have discussed various polysaccharides-based nanocomposites and their important properties featuring in electronics applications.

9.5.1.1 Cellulose-Based Nanocomposites

Carbonaceous and metal oxide-based nanofillers are being used in combination with cellulose materials in electronic industries. For example, a mechanically flexible and high capacitance electrode made of nanocellulose/graphene nanocomposites is widely used in the fabrication of supercapacitors [53]. Xu et al. [54] have developed a bacterial cellulose/polypyrrole flexible supercapacitor electrode that exhibits impressive electrical conductivity of 3.9 Scm^{-1} and maximum discharge capacity 101.9 $mAhg^{-1}$. Flexible and conductive paper made from a combination of cellulose, multiwalled carbon nanotubes (MWCNTs) and 1-butyl-3-methylimidazolium chloride (used as electrolyte) is utilized in the fabrication of high energy supercapacitors with an operating voltage of ~2.3 V and a power density of 1.5 $kW.kg^{-1}$ [55]. Further cellulose/graphene oxide nanocomposite prepared after grafting of functionalized graphene with cellulose can be used in sensing organic solvents with variation in capacitance of the nanocomposites [56]. The dielectric constant of cellulose/graphene nanocomposite was 64 times higher than the pristine cellulose. However, Young's modulus of final nanocomposites was also improved to 3.18 GPa as compared to 2.37 GPa of pristine cellulose. BC, in combination with a hybrid nanocomposite of BC/ZnO/CuO, was reported in improving the EMI shielding capacity of electronic devices in protecting the electronic circuits against harmful electrical and magnetic wave flux [57].

9.5.1.2 Chitosan-Based Nanocomposites

Chitosan-based nanocomposites have shown a wide range of applications in green electronics in the development of biosensors, capacitors, EMI shielding, portable electronics, etc. [58, 59]. Biosensors fabricated using chitosan-based nanocomposites are highly stable and selective towards the sensing materials and yield high reproducibility. Shan et al. [60] have developed a biosensor using chitosan/graphene/AuNP nanocomposites via solvent casting method. The biosensor showed outstanding amperometric response to glucose molecules in the range of 2–10 mM at −0.2 V and 2 to 14 mM at 0.5 V. In another study, Kang et al. [61] introduced a biosensor prepared from glucose oxidase/chitosan/graphene nanocomposites for glucose detection. The study revealed very broad linearity in the range of 0.08–12 mM along with the high glucose sensitivity of 37.93 $\mu AmM^{-1}cm^{-2}$. Further, chitosan nanocomposites prepared with electrically conductive nanofillers (CNTs, graphene, CNFs, etc.) are used to protect electronic appliances against harmful EMI. In this, chitosan/MWCNTs nanocomposites synthesized using high energy ball milling method coupled with 3D printed ink technology showed outstanding EMI effectiveness of 25.07 dB [62]. High EMI effectiveness in this study is attributed to the formation of porous composite structure along with continuous conductive network structure formed in nanocomposites. Carboxymethyl chitosan–palladium (CMCS–Pd), a chemically modified chitosan after addition with palladium (Pd), showed a considerable rise in shielding effectiveness in between 40–60 dB [63].

9.5.1.3 Alginate-Based Nanocomposites

Alginate-based composites are frequently used in tissue engineering and food packaging application. Moreover, the composites have also shown their potential application in electronics devices, biosensors, capacitors, EMI shielding, etc. [64, 65]. Work carried out by Kang et al. [20] has shown the substantial increase in storage capacity of lithium batteries prepared using algae/copper oxide (CuO) nanocomposites. The study confirmed a very high discharge capacity of lithium battery electrodes of 1304.9 mAh g^{-1} when operated at 0.1C due to the reduction in electrochemical polarization. Zhao et al. [66] have designed pyramid shape pressor-driven sensors for their effective utilization in prosthetic skins, smartwatches, flexible touch displays, etc. Hybrid nanocomposites prepared in an aerogel form with the addition of polydimethylsiloxane (PDMS) into alginate/N-CNTs/Ag improved both the elastic and conductive properties of hybrid nanocomposites and widened their application in the preparation of flexible electronics. Alginate/graphene nanocomposites prepared using the vacuum filtration method have greatly improved the EMI shielding effectiveness of final composites materials [67]. In another study, Shen et al. [68] have developed alginate/RGO nanocomposites (film thickness of ~12μm) and showed almost 96% improvement in EMI shielding effectiveness with EMI SE of 25.7 dB.

9.6 Protein-Based Nanocomposites

Recently, protein-based nanocomposites, in which protein (silk, collagen and gelatin, elastin, resilin) are commonly used as polymer matrix and nanofillers are used as reinforcing phase, augments the intrinsic property of proteins and extend their potential application in electronics [69]. The subsequent section provides a brief description of protein-based nanocomposites and their multiple application in electronics.

9.6.1 Silk-Based Nanocomposites

Silk-based nanocomposites have shown their potential application in energy storage and wearable electronics. Silk-based nanocomposites have the potential to convert thermal, mechanical, and solar energy into useful electrical energy for its utilization in energy storage devices [70]. The electrode made of silk/$NiCO_2S_4$ nanocomposites using the hydrothermal vulcanization method has confirmed its effective utilization in the development of supercapacitors. The study revealed that the final nanocomposites exhibited a very high specific capacitance of 630 F g^{-1} with a good energy density of 52.34 Wh kg^{-1}. Encapsulation of carbon nanotubes into silk matrix makes the silk electrically conductive and widens its use in electronics devices where high electrical conductivity is of utmost importance [71]. Preparation of silk/Au nano-bioconjugate prepared with the combination of spider silk and gold nanoparticles opened a new way to produce biobased vapor sensors for detection of organic vapors, which is frequently required in chemical industries [36]. Recently, hybrid silk composites, with properties of being highly malleable and stretchable, have been used in solar devices for energy conversion [72]. The highly thermally stable hybrid sheet can be utilized in the fabrication of photovoltaic cells with minimum energy loss.

9.6.2 Collagen and Gelatin-Based Nanocomposites

Collagen-based nanocomposites are used in the fabrication of biosensors, EMI shielding, capacitors, and other electronic devices. Successful incorporation of magnetic nanoparticles into collagen matrix has displayed significant improvement in the magnetic dipole movement (~8.32 emu/g), thermal stability and EMI shielding of the final nanocomposites [73]. A combination of size-controlled silver nanoparticles (AgNPs) with collagen fiber (CF) matrix acted as a quality stabilizer against microwave absorption [74]. The studies showed a high reflection loss (RL) of greater than 10dB in the frequency range of 13–18 GHz and could be considered a potential candidate to be used in electronic communication, information processing and broadcasting applications. Recently, biodegradable fish gelatin (FG/Ag NWs) nanocomposites of alternating current electroluminescent (ACEL) for use in foldable and supported electronics along with flat panel displays have been reported [41]. The addition of AgNWs into the gelatin matrix remarkably improved the elastic modulus as well as the luminosity factor (56.0 cdm^{-2}) of the final nanocomposites.

9.7 PLA-Based Nanocomposites

PLA/graphene nanocomposites, after the addition of graphene nanosheets into the PLA matrix, have significantly improved its electrical conductivity in the order of 3.5×10^{-2} S cm^{-1} [75]. Moreover, the nanocomposite further showed an increase in EMI shielding effectiveness with the magnitude of 45dB and has the potential to promote protection against harmful EM radiation. Other than electrical conductivity, enhancement in thermal conductivity (0.577 W/mK) has also been reported, hence its effective utilization in the fabrication of electronics devices where both thermal and electrical conductivity are of utmost importance. PLA hybrid nanocomposites with the combined use of nanofillers and conductive polymer like polylactic acid/CNTs/polyaniline nanocomposites have shown an exceptional rise in specific capacitance and tensile strength of 510.3 Fg^{-1} and 18.7 MPa respectively and can be used in the development of free-standing electrodes as an intrinsic part of new generation capacitors [76]. PLA nanocomposites formed with encapsulation of metal and its oxide nanofillers into PLA matrix have shown considerable improvement in optical, electrical and mechanical properties of final nanocomposites and have proven to be a good candidate for the fabrication of flexible and optoelectronics devices [77].

9.8 Polymer Blends and Their Application in Electronics

Physiochemical properties of polymers are further enhanced via mixing with other polymers and are called polymer blends, sometimes referred to as copolymers. Polymer blends are essential when polymer composites prepared with the addition of filler materials adversely affect the final composite properties due to the non-compatibility and high agglomerating behavior of fillers with polymer matrix. Also, sometimes blending is important when the final polymer properties desired are in between the properties of blends polymers [69, 78]. A blend of chitosan and conductive organic polymer of polypyrrole obtained after electrochemical polymerization is used in attenuation of harmful EM waves with excellent EMI shielding effectiveness (> 98%) in the frequency range of 8–12 GHz due to the increase in electrical conductivity of the final blends [79]. The blends of polycarbonate and polylactic acid are employed in the preparation of glass fiber composite for enhancement in mechanical, heat resistant and dielectric properties of final blends in electronics and automation industries [80]. Perfectly blended two polymers can be blended with another polymer to further improve the properties of final blends. For example, the blends of biopolymers of PBAT (polybutyleneadipate-co-terephthalate) and PHB (polyhydroxybutyrate) prepared in a weight ratio of 1:4 using melt extrusion technique have reported a resonant frequency of 2.44 GHz, the forward antenna gain of 4.01 dBi and half-power beamwidth of 48.2^{0} that allow the blends to be used in wireless and other communication devices [81]. The copolymer nanocomposites collagen/polyaniline/IONPs with the addition of iron oxide nanoparticles (IONPs) prepared using in-situ polymerization have shown potential application in electronics for safeguarding against harmful electromagnetic radiation. The hybrid composite showed electrical conductivity of 2.84×10^{-2} and is much higher than the pristine collagen [73].

9.9 Conclusion and Future Perspectives

This chapter summarized the present status of biobased polymers and their composites for applications in green electronics and can provide a viable solution to control e-waste generation. Pristine biobased polymers, mainly cellulose, alginate, chitosan, PLA, and collagen, are largely used as substrates in the fabrication of electronic devices due to their lightweight, high optical transparency, and excellent biodegradability. Biobased nanocomposites formed after the addition of nanoparticles into pristine biobased polymer widen its application in advanced electronics in the production of biosensors, capacitors, conducting materials, EMI shielding, etc. In this, incorporation of carbonaceous nanofillers into biobased polymers is found to significantly increase the electrical conductivity and mechanical properties of final composites by several times as compared to pristine biopolymers. However, higher loading of these conductive fillers also causes a reduction in final properties due to their poor dispersion and improper adhesion with the polymer matrix. The critical issues of poor filler dispersion and their agglomeration can be overcome with advancement in synthesis methods. A blending of polymers also helps in transferring their physicochemical properties with another polymer and widens their use in the fabrication of numerous electronics devices. Regarding the future trends of biobased polymers and their composites in electronics, a significant amount of scientific research is continuously being added to the literature. Currently, it is estimated that the production of biobased polymer grows at 3%–4% annually. The continuous depletion of petroleum reservoirs and increase in crude oil cost has further opened a new window to find an alternative way to increase the production of biobased polymers. An increasing amount of research work in the development of biobased polymers and their nanocomposites is underway for the advanced electronics and automobile industries. This information is summarized in Table 9.1.

References

1 Gopi, S., Amalraj, A., Sukumaran, N.P. et al. (2018). Biopolymers and their composites for drug delivery: a brief review. *Macromolecular Symposia* 380 (1): 1800114. doi: 10.1002/masy.201800114.

2 Liu, H., Jian, R., Chen, H. et al. (2019). Application of biodegradable and biocompatible nanocomposites in electronics: current status and future directions. *Nanomaterials* 9. doi: 10.3390/nano9070950.

3 Vladu, M.I. (2012). Green electronics: biodegradable and biocompatible materials and devices for sustainable future. *Chemical Society Reviews* 43 (2): 588–610. doi: 10.1039/C3CS60235D.

4 Avérous, L. (2008). Polylactic acid: synthesis, properties, and applications. In: *Monomers, Polymers and Composites from Renewable Resources* (ed. M.N. Belgacem and A. Gandini), 433–450. Amsterdam: Elsevier. doi: 10.1016/B978-0-08-045316-3.00021-1.

5 Bobade, S.K., Paluvai, N.R., Mohanty, S., and Nayak, S.K. (2016). Bio-Based thermosetting resins for future generation: a review. *Polymer-plastics Technology and Engineering* 55 (17): 1863–1896. doi: 10.1080/03602559.2016.1185624.

6 Zargar, V., Asghari, M., and Dashti, A. (2015). A review on chitin and chitosan polymers: structure, chemistry, solubility, derivatives, and applications. *ChemBioEng Reviews* 2: 204–226. doi: 10.1002/cben.201400025.

7 Arjmand, M. and Sundararaj, U. (2015). Electromagnetic interference shielding of nitrogen-doped and undoped carbon nanotube/polyvinylidene fluoride nanocomposites: a comparative study. *Composites Science and Technology* 118: 257–263. doi: 10.1016/j.compscitech.2015.09.012.

8 Ogunseitan, O.A., Schoenung, J.M., Saphores, J.-D.M., and Shapiro, A.A. (2009). The electronics revolution: from e-wonderland to e-wasteland. *Science* 326: 670–671. doi: 10.1126/science.1176929.

9 Zou, Z.N., Zhu, C.P., Li, Y. et al. (2018). Rehealable, fully recyclable, and malleable electronic skin enabled by dynamic covalent thermoset nanocomposite. *Science Advances* 4: 1–8. doi: 10.1126/sciadv.aaq0508.

10 Chang, J.K., Chang, H.P., Guo, Q. et al. (2018). Biodegradable electronic systems in 3D, heterogeneously integrated formats. *Advanced Materials* 30: 1–10. doi: 10.1002/adma.201704955.

11 He, Y., Sun, J., Qian, C. et al. (2016). Solution-processed natural gelatin was used as a gate dielectric for the fabrication of oxide field-effect transistors. *Organic Electronics* 38: 357–361. doi: 10.1016/j.orgel.2016.09.017.

12 Mühl, S. and Beyer, B. (2014). Bio-organic electronics—overview and prospects for the future. *Electronics* 3: 444–461. doi: 10.3390/ELECTRONICS3030444.

13 Lah, N.A.C., Zubir, M.N.M., and Samykano, M.A. (2018). Engineered nanomaterial in electronics and electrical industries. In: *Handbook of Nanomaterials for Industrial Applications* (ed. C.M. Hussain), 324–364. Amsterdam: Elsevier. doi: 10.1016/B978-0-12-813351-4.00021-3.

14 Poddar, M.K. and Dikshit, P.K. (2021). Recent development in bacterial cellulose production and synthesis of cellulose based conductive polymer nanocomposites: a review paper. *Nanoselect* 1–24. doi: 10.1002/nano.202100044.

15 Costa, S.V., Pingel, P., Janietz, S., and Nogueira, A.F. (2016). Inverted organic solar cells using nanocellulose as substrate. *Journal of Applied Polymer Science* 133 (28). doi: 10.1002/APP.43679.

16 Mohammed, A.S.A., Naveed, M., and Jost, N. (2021). Polysaccharides; classification, chemical properties, and future perspective applications in fields of pharmacology and biological medicine (a review of current applications and upcoming potentialities). *The Journal of Polymers and the Environment* 29: 2359–2371. doi: 10.1007/s10924-021-02052-2.

17 Eichhorn, S.J., Dufresne, A., Aranguren, M. et al. (2010). Current international research into cellulose nanofibers and nanocomposites. *Journal of Materials Science* 45: 1–33. doi: 10.1007/s10853-009-3874-0.

18 Nakayama, Y., Mori, T., Tsuruta, R. et al. (2018). Surface crystallographic structures of cellulose nanofiber films and overlayers of pentacene. *Journal of Applied Physics* 57: 1–4. doi: 10.7567/jjap.57.03ee02.

19 Huang, Q., Xu, M., Sun, R., and Wang, X. (2016). Large scale preparation of graphene oxide/cellulose paper with improved mechanical performance and gas barrier properties by conventional papermaking method. *Industrial Crops and Products* 85: 198–203. doi: 10.1016/J.INDCROP.2016.03.006.

20 Kang, W., Zhao, C., and Shen, Q. (2012). Lithium storage capability of nanocrystalline CuO improved by its water-based interactions with sodium alginate. *International Journal of Electrochemical Science* 7: 8194–8204. doi: 10.1016/s1452-3981(23)17986-7.

21 Iguchi, M., Yamanaka, S., and Budhiono, A. (2000). Bacterial cellulose—a masterpiece of nature's arts. *Journal of Materials Science* 35: 261–270. doi: 10.1023/A:1004775229149.

22 Zhang, Q., Bao, W., Gong, A. et al. (2016). A highly sensitive, highly transparent, gel-gated MoS_2 phototransistor on biodegradable nanopaper. *Nanoscale* 8 (29): 14237–14242. doi: 10.1039/C6NR01534D.

23 Agrawal, P., Strijkers, G.J., and Nicolay, K. (2010). Chitosan-based systems for molecular imaging. *Advanced Drug Delivery Reviews* 62 (1): 42–58. doi: 10.1016/J.ADDR.2009.09.007.

24 Anandhavelu, S. and Thambidurai, S. (2013). Single step synthesis of chitin/chitosan-based graphene oxide–ZnO hybrid composites for better electrical conductivity and optical properties. *Electrochimica Acta* 90: 194–202. doi: 10.1016/J.ELECTACTA.2012.12.024.

25 Peter, S., Lyczko, N., Gopakumar, D. et al. (2020). Chitin and chitosan-based composites for energy and environmental applications: a review. *Waste Biomass Valorization* 12: 4777–4804. doi: 10.1007/s12649-020-01244-6.

26 Teng, K., An, Q., Chen, Y. et al. (2021). Recent development of alginate-based materials and their versatile functions in biomedicine, flexible electronics, and environmental uses. *ACS Biomaterials Science & Engineering* 7: 1302–1337. doi: 10.1021/acsbiomaterials.1c00116.

27 An, Q., Huang, T., and Shi, F. (2018). Covalent layer-by-layer films: chemistry, design, and multidisciplinary applications. *Chemical Society Reviews* 47: 5061–5098. doi: 10.1039/c8cs90072h.

28 Chen, M., Fu, X., Chen, Z. et al. (2021). Protein engineered functional materials for bioelectronics. *Advanced Functional Materials* 31: 2006744. doi: 10.1002/adfm.202006744.

29 Liu, Y., Wan, X., Zhu, L.Q. et al. (2014). Laterally coupled dual-gate oxide-based transistors on sodium alginate electrolytes. *IEEE Electron Device Letters* 35 (12): 1257–1259. doi: 10.1109/LED.2014.2362143.

30 Kaklamani, G., Kazaryan, D., Bowen, J. et al. (2018). On the electrical conductivity of alginate hydrogels. *Regenerative Biomaterials* 5 (5): 293. doi: 10.1093/RB/RBY019.

31 Chandrashekar, G. and Udupa, N. (1996). Biodegradable injectable implant systems for long term drug delivery using poly (lactic-co-glycolic) acid copolymers. *The Journal of Pharmacy and Pharmacology* 48 (7): 669–674. doi: 10.1111/J.2042-7158.1996.TB03948.X.

32 Géczy, A., Kovács, M., and Hajdu, I. (2012). Conductive layer deposition and peel tests on biodegradable printed circuit boards. *IEEE 18th International Symposium for Design and Technology in Electronic Packaging (SIITME)*, 139–142. doi: 10.1109/SIITME.2012.6384363.

33 Bettinger, C. and Bao, Z. (2010). Organic thin film transistors fabricated on resorbable biomaterial substrates. *Advanced Materials* 22: 651–655. doi: 10.1002/adma.200902322.

34 Ullah, S. and Flint, J.A. (2014). Electro-textile based wearable patch antenna on biodegradable poly lactic acid (PLA) plastic substrate for 2.45 Ghz, ISM band applications. *International Conference on Emerging Technologies (ICET)*, 158–163. doi: 10.1109/ICET.2014.7021036.

35 Numata, K. (2020). How to define and study structural proteins as biopolymer materials. *Polymer Journal* 52: 1043–1056. doi: 10.1038/s41428-020-0362-5.

36 Singh, A.A., Hede, S., and Saster, M. (2007). Spider silk as an active scaffold in the assembly of gold nanoparticles and application of the gold-silk bioconjugate in vapor sensing. *Nano Macro Small* 3 (3): 466–473. doi: 10.1002/SMLL.200600413.

37 Lawrence, B.D., Cronin-Golomb, M., Georgakoudi, I. et al. (2008). Bioactive silk protein biomaterial systems for optical devices. *Biomacromolecules* 9: 1214–1220. doi: 10.1021/bm701235f.

38 Sofia, S., McCarthy, M.B., Gronowicz, G., and Kaplan, D.L. (2001). Functionalized silk-based biomaterials for bone formation. *Journal of Biomedical Materials Research* 54: 139–148. doi: 10.1002/1097-4636.

39 Amsden, J.J., Perry, H., Boriskina, S.V. (2009). Spectral analysis of induced colour change on periodically nanopatterned silk films. *Optics Express* 17: 21271–21279. doi: 10.1364/oe.17.021271.

40 Kim, D.-H., Viventi, J., Amsden, J.J. et al. (2009). Dissolvable films of silk fibroin for ultrathin conformal bio-integrated electronics. *Nature Materials* 9 (6): 511–517. doi: 10.1038/nmat2745.

41 Zhang, X., Ye, T., Meng, X. et al. (2020). Sustainable and transparent fish gelatin films for flexible electroluminescent devices. *ACS Nano*. doi: 10.1021/ACSNANO.9B09880.

42 Shintake, J., Sonar, H., Piskarev, E. et al. (2017). Soft pneumatic gelatin actuator for edible robotics. *IEEE International Conference on Intelligent Robots and Systems, 2017-September*, 6221–6226. doi: 10.1109/IROS.2017.8206525.

43 Hoorick Van, J., Tytgat, L., Dobos, A. et al. (2019). Photo crosslinkable gelatin derivatives for biofabrication applications. *Acta Biomaterialia* 97: 46–73. doi: 10.1016/J.ACTBIO.2019.07.035.

44 Li, X., Ding, C., Li, X. et al. (2020). Electronic biopolymers: from molecular engineering to functional devices. *Chemical Engineering Journal* 397: 125499. doi: 10.1016/j.cej.2020.125499.

45 Alexandre, M. and Dubois, P. (2000). Polymer layered silicate nanocomposites: preparation, properties, and uses of a new class of materials. *Materials Science and Engineering* 28: 1–63.

46 Kumar, M., Shanmuga Priya, N., Kanagaraj, S., and Pugazhenthi, G. (2016). Melt rheological behavior of PMMA nanocomposites reinforced with modified nanoclay. *Nanocomposites* 2: 109–116. doi: 10.1080/20550324.2016.1221876.

47 Poddar, M.K., Sharma, S., and Moholkar, V.S. (2016). Investigations in two-step ultrasonic synthesis of PMMA/ZnO nanocomposites by in-situ emulsion polymerization. *Polymer* 99: 453–469. doi: 10.1016/j.polymer.2016.07.052.

48 Beyer, G. (2002). Nanocomposites: a new class of flame retardants for polymers. *Plastics, Additives and Compounding* 4 (10): 22–70.

49 Solomon, M.J., Almusallam, A.S., Seefeldt, K.F. et al. (2001). Rheology of polypropylene/clay hybrid materials. *Macromolecules* 34: 1864–1872.

50 Koo, J.H. (2020). *Polymer Nanocomposites: Processing, Characterization, and Applications*, 2e. London: McGraw Hill.

51 Thomas, S., Ponnamma, D., and Zachariah, A.K. (eds.) (2012). *Polymer Processing and Characterization*, 1e. Apple Academic Press. doi: 10.1201/b13105.

52 Pavlidou, S. and Papaspyrides, C.D. (2008). A review on polymer–layered silicate nanocomposites. *Progress in Polymer Science* 33: 1119–1198. doi: 10.1016/j.progpolymsci.2008.07.008.

53 Xing, J., Tao, P., Wu, Z. et al. (2019). Nanocellulose-graphene composites: a promising nanomaterial for flexible supercapacitors. *Carbohydrate Polymers* 207: 447–459. doi: 10.1016/J.CARBPOL.2018.12.010.

54 Pushparaj, V.L., Shaijumon, M.M., Kumar, A. et al. (2007). *Sciences of the USA 13574-13577 PNAS* 104 (34). doi: 10.1073/pnas.0706508104.

55 Xu, J., Zhu, L., Bai, Z. et al. (2013). Conductive polypyrrole–bacterial cellulose nanocomposite membranes as flexible supercapacitor electrode. *Organic Electronics* 14 (12): 3331–3338. doi: 10.1016/J.ORGEL.2013.09.042.

56 Kafy, A., Sadasivuni, K.K., Akther, A. et al. (2015). Cellulose/graphene nanocomposite as multifunctional electronic and solvent sensor material. *Materials Letters* 159: 20–23. doi: 10.1016/J.MATLET.2015.05.102.

57 Wasim, M., Mushtaq, M., Khan, S.U. et al. (2020 December). Development of bacterial cellulose nanocomposites: an overview of the synthesis of

Table 9.1 A summary of properties of biobased polymers and their nanocomposites in electronics application.

Polymer and composite types	Synthesis method and experimental conditions	Major findings and conclusion	Reference
Cellulose	Nanocrystalline cellulose (CNC) and nanofibrillated cellulose (NFC) prepared using solvent (water) casting, stirring time: 12 h, drying temperature: 45 °C, drying time: 5 min	• Glycerol used as a plasticizer • Cellulose used as substrate in fabrication of organic photovoltaics (OPVs) device • Power conversion efficiency (PCE) in CNCs found higher than NFC substrates	[15]
Cellulose	Solvent (water) casting followed by atomic layer deposition (ALD), electrochemical method used in deposition of MnO_2 on CNT-Papers, stirring time: 20 m, temperature: 100 °C	• Cellulose fibers used as a substrate • Cellulose fibers with an open pore channel showed better electrochemical performance	[82]
Chitosan	Solvent (acetic acid) casting followed by NaOH treatment, chitosan molecular weights: 70 kDa and 300 kDa, chitosan 2% (w/v)	• Chitosan made flexible electronic skin showed low swelling ratio • Chitosan with 300 KDa showed Young's modulus of ~151 kPa	[83]
PLA	Electrospinning with dispersion of PLA into mixture of chloroform and acetone, chloroform, and acetone: 3:1 (volume ratio), applied voltage: 15 kV, feed rate: 5–10 μL/min	• Biosensor made using PLA was discussed • Biotin was successfully fixed on the PLA sub-strate	[84]

(Continued)

Table 9.1 (Continued)

Polymer and composite types	Synthesis method and experimental conditions	Major findings and conclusion	Reference
BC/MWCNTs	Ex–situ polymerization, incubation time: 14 days, incubation temperature: 30 °C, pH: 7	• Electrical conductivity: 1.4×10^{-1} Scm^{-1} • Surfactant was recovered at the end of the process	[85]
BC/RGO	Vacuum-assisted self-assembly technique, BC suspension in the alkali medium, cooling temp: −20 °C, suspension time: 6 h	• Highly flexible and mechanically strong BC/RGO nanocomposites, • Electrical conductivity: 1.1×10^{-6} Scm^{-1}, • Young's modulus: 1.7 GPa; tensile strength: 242 MPa; elongation at break: 5.4%	[86]
Chitosan/graphene/ glucose oxidase (GOD) biosensor	Three steps: (i) graphene prepared in mixture of sulfuric and nitric acid and potassium chlorate, (ii) graphene–chitosan prepared on the surface of glassy carbon electrode (GCE) and (iii) GOD coated on graphene–chitosan film, graphene preparation time: 96 h, drying temperature: 4 °C	• Biosensor used for glucose detection • Detection limit: 0.02 mM with sensitivity of 37.93 μAmM^{-1} cm^{-2} • High conductivity and better enzyme absorption	[61]
Sodium alginate/CuO	Solvent (water) casting, CuO: alginate weight ratio:7.5:1 or 1:1, pH:8.4	• Greatly improve lithium storage capacity of CuO • High structural ability	[20]

bacterial cellulose nanocomposites with metallic and metallic-oxide nanoparticles by different methods and techniques for biomedical applications. *Journal of Industrial Textiles*. doi: 10.1177/1528083720977201.

58 Avcu, E., Baştan, F.E., Abdullah, H.Z. et al. (2019). Electrophoretic deposition of chitosan-based composite coatings for biomedical applications: a review. *Progress in Materials Science* 103: 69–108. doi: 10.1016/j.pmatsci.2019.01.001.

59 Li, M.Z., Jia, L.C., Zhang, X.P. et al. (2018). Robust carbon nanotube foam for efficient electromagnetic interference shielding and microwave absorption. *Journal of Colloid and Interface Science* 530: 113–119. doi: 10.1016/J.JCIS.2018.06.052.

60 Shan, C., Yang, H., Han, D. et al. (2010). Graphene/AuNPs/chitosan nanocomposites film for glucose biosensing. *Biosensors and Bioelectronics* 25 (5): 1070–1074. doi: 10.1016/J.BIOS.2009.09.024.

61 Kang, X., Wang, J., Wu, H. et al. (2009). Glucose oxidase-graphene-chitosan modified electrode for direct electrochemistry and glucose sensing. *Biosensors and Bioelectronics* 25 (4): 901–905. doi: 10.1016/j.bios.2009.09.004.

62 Pei, X., Zhao, M., Li, R. et al. (2021). Porous network carbon nanotubes/chitosan 3D printed composites based on ball milling for electromagnetic shielding. *Composites Part A: Applied Science and Manufacturing* 145: 106363. doi: 10.1016/J.COMPOSITESA.2021.106363.

63 Zhao, Y., Cai, Z., Zhou, Z., and Fu, X. (2011). Adsorption behavior of monomers and formation of conducting polymers on polyester fibers. *Journal of Applied Polymer Science* 119 (2): 662–669. doi: 10.1002/APP.32743.

64 Esmat, M., Farghali, A.A., Khedr, M.H., and El-Sherbiny, I.M. (2017). Alginate-based nanocomposites for efficient removal of heavy metal ions. *International Journal of Biological Macromolecules* 102: 272–283. doi: 10.1016/J.IJBIOMAC.2017.04.021.

65 Zia, F., Sobhani, H., Mohammadi, M. et al. (2017). Alginate-based hybrid nanocomposite materials. Algae based polymers, blends, and composites: chemistry. *Biotechnology and Materials Science* 603–648. doi: 10.1016/B978-0-12-812360-7.00017-3.

66 Zhao, S., Zhang, G., Gao, Y. et al. (2014). Strain-driven and ultrasensitive resistive sensor/switch based on conductive alginate/nitrogen-doped carbon-nanotube-supported ag hybrid aerogels with pyramid design. *ACS Applied Materials & Interfaces* 6 (24): 22823–22829. doi: 10.1021/AM5069936.

67 Jia, L.-C., Sun, W.-J., Zhou, C.-G. et al. (2018). Integrated strength and toughness in graphene/calcium alginate films for highly efficient electromagnetic interference shielding. *Journal of Materials Chemistry C* 6 (34): 9166–9174. doi: 10.1039/C8TC03151G.

68 Shen, B., Zhai, W., and Zheng, W. (2014). Ultrathin flexible graphene film: an excellent thermal conducting material with efficient EMI shielding. *Advanced Functional Materials* 24 (28): 4542–4548. doi: 10.1002/ADFM.201400079.

69 Markovic, G. and Visakh, P.M. (2017). Polymer blends: state of art. In: *Recent Developments in Polymer Macro, Micro and Nano Blends: Preparation and Characterisation* (ed. P.M. Visakh, G. Markovic, and D. Pasquini), 1–15. Amsterdam: Elsevier. doi: 10.1016/B978-0-08-100408-1.00001-7.

70 Strassburg, S., Zainuddin, S., and Scheibel, T. (2021). The power of silk technology for energy applications. *Advanced Energy Materials* 2100519. doi: 10.1002/AENM.202100519.

71 Wang, Q., Yan, S., Han, G. et al. (2020). Facile production of natural silk nanofibers for electronic device applications. *Composites Science and Technology* 187: 107950. doi: 10.1016/J.COMPSCITECH.2019.107950.

72 Ma, R. and Tsukruk, V.V. (2017). Seriography-guided reduction of graphene oxide biopapers for wearable sensory electronics. *Advanced Functional Materials* 27. doi: 10.1002/adfm.201604802.

73 Mekonnen, B.T., Ragothaman, M., and Palanisamy, T. (2017). Bifunctional hybrid composites from collagen biowastes for heterogeneous applications. *ACS Omega* 2 (8): 5260–5270. doi: 10.1021/ACSOMEGA.7B01011.

74 Guo, J., Wu, H., Liao, X., and Shi, B. (2011). Facile synthesis of size-controlled silver nanoparticles using plant tannin grafted collagen fiber as reductant and stabilizer for microwave absorption application in the whole ku band. *The Journal of Physical Chemistry C* 115 (48): 23688–23694. doi: 10.1021/JP207194A.

75 Wang, G., Zhao, G., Wang, S. et al. (2018). Injection-molded microcellular PLA/graphite nanocomposites with dramatically enhanced mechanical and electrical properties for ultra-efficient EMI shielding applications. *Journal of Materials Chemistry C* 6 (25): 6847–6859. doi: 10.1039/C8TC01326H.

76 Wang, Q., Wang, H., Du, P. et al. (2019). Porous polylactic acid/carbon nanotubes/polyaniline composite film as flexible free-standing electrode for supercapacitors. *Electrochimica Acta* 294: 312–324. doi: 10.1016/J.ELECTACTA.2018.10.108.

77 Fallahi, H., Azizi, H., Ghasemi, I., and Karrabi, M. (2017). Preparation and properties of electrically conductive, flexible, and transparent silver nanowire/poly (lactic acid) nanocomposites. *Organic Electronics* 44: 74–84. doi: 10.1016/J.ORGEL.2017.01.043.

78 Berezina, N. and Martelli, S.M. (2014). Chapter 1: bio-based polymers and materials. 1–28. doi: 10.1039/9781782620181-00001.

79 Abdi, M.M., Kassim, A.B., Mahmud, H.N.M.E. et al. (2009). Electrical and shielding properties of conductive polymer composite matrix with chitosan. *Solid State Science and Technology* 17: 12–21.

80 Yuryev, Y., Mohanty, A.K., and Misra, M. (2017). Novel biocomposites from biobased PC/PLA blend matrix system for durable applications. *Composites Part B: Engineering* 130: 158–166. doi: 10.1016/J.COMPOSITESB.2017.07.030.

81 Serres, G.K.F., Freire, R.R.C.S., Morais, S.M.A. et al. (2020). Eco-friendly metamaterial antenna for 2.4 Ghz WLAN applications. *14th European Conference on Antennas and Propagation, EuCAP*. doi: 10.23919/EUCAP48036.2020.9135785.

82 Gui, Z., Zhu, H., Gillette, E. et al. (2013). Natural cellulose fiber as substrate for supercapacitor. *ACS Nano* 7: 6037–6046. doi: 10.1021/nn401818t.

83 Lin, Y.H., Kang, P.L., Xin, W. et al. (2018). Preparation and evaluation of chitosan biocompatible electronic skin. *Computers in Industry* 100: 1–6. doi: 10.1016/j.compind.2018.03.040.

84 Li, D., Frey, M.W., and Baeumner, A.J. (2006). Electrospun polylactic acid nanofiber membranes as substrates for biosensor assemblies. *Journal of Membrane Science* 279: 354–363. doi: 10.1016/j.memsci.2005.12.036.

85 Yoon, S.H., Jin, H.J., Kook, M.C., and Pyun, Y.R. (2006). Electrically conductive bacterial cellulose by incorporation of carbon nanotubes. *Biomacromolecules* 7: 1280–1284.

86 Feng, Y., Zhang, X., Shen, Y. et al. (2012). A mechanically strong, flexible, and conductive film based on bacterial cellulose/graphene nanocomposite. *Carbohydrate Polymers* 87: 644–649. doi: 10.1016/j.carbpol.2011.08.039.

10

Polymers and Biopolymers in Sensing

*Greeshma U. Chandran, Hema S., Malavika Sajith, Rashid Sulthan, and Sreedha Sambhudevan**

Department of Chemistry, Amrita Vishwa Vidyapeetham, Amritapuri, Kerala, India
** Corresponding author*

10.1 Introduction

Modern electronics use polymers as essential components. They have countless promising applications in both industrial and non-industrial sectors, and may be employed in their pure, composite, as well as nanocomposite forms. Polymer nanocomposites have drawn ample theoretical recognition and found a lot of practical uses in a variety of technological domains because they offer materials with remarkable functions in addition to virtues in processing. There is proof that the conductive filler networks inside the polymer matrix regulate the electrical conductivity of polymer nanocomposite. Therefore, even a small change in the conductive networks could bring about noticeable modulations in the electrical impulse emission of a polymer nanocomposite. Polymer nanocomposites may be used to create new, sensitive sensors to measure physical and chemical characteristics by taking advantage of the conductive networks' ability to respond to stimuli. The mechanisms driving the production of various polymer nanocomposites are being investigated in the field of material science, and more sophisticated answers are being sought [1].

Polymer-based sensors with the proper modifications or functionality may be utilized as highly specialized sensors. Polymers can be used in

Applications of Biopolymers in Science, Biotechnology, and Engineering, First Edition.
Edited by Arunima Reghunadhan, Akhina H., and Sabu Thomas.

sensing directly or indirectly by immobilising certain receptors on them. The advantages of utilising polymers as sensors go beyond their simplicity in functionalization, biocompatibility, extended lives, lightness, and affordability in manufacture. Significant improvements have been prioritized for enhanced efficiency and progression of the sensors from the lab to commercial applications [2]. When assessing the progress made in polymer-based sensors, Majumdar and Adhikari talked about a number of themes that may prove to be of utmost significance, which include: using chemical modifications to enhance receptor molecule immobilization, looking for innovative processing methods and smart materials, creation of sensing components that may react to diverse inputs in various ways, and examining the biochemical processes taking place in non-aqueous media [2]. Nowadays, biopolymers are widely employed in a broad-spectrum, from low-tech to high-tech, as a result of improvements in biotechnological processes and growing consumer interest. The traditional petroleum-based composites are starting to be replaced through the development of biocomposites employing biopolymers. Additionally, the reinforcement of chemical sensors that use biopolymers may make it easier to detect chemicals.

10.2 Type of Sensors and Sensor Action

Sensors are sophisticated electronic instruments that may analyse an electrical impulse as an input to detect physiochemical properties both subjectively and quantitatively. Numerous modules that are part of the sensor's signal processing section carry out the following functions: target analysis, signal analysis and processing, sampling, data gathering and interpreting. The detecting element of the sensor alters its physical or chemical properties whenever it is in close proximity to the analyte, and these modifications are subsequently converted into perceptible electrical impulses [3].

Depending on how their transducer functions, sensors may be divided into a wide variety of categories, such as optical sensors, electrochemical sensors, mass sensitive sensors, and electrical sensors.

In optical sensors, an accurate diagnostic and readable signal is produced from changes in optical characteristics such as reflectance, absorbance, luminescence, fluorescence, and light scattering. The optical sensor detects the amount of light that is absorbed, dispersed, and released as a result of any chemical reactions that are occurring inside the analyte and that cause absorbance and fluorescence phenomena. The electrochemical sensor is capable of producing a useful signal from the electrochemical reaction between the analyte and the sensing electrode. The sensor is made up of numerous electrode systems, with the right electrode system being able to become oxidized or reduced. For instance, in a potentiometric sensor, the sensing electrode computes the potential in relation to a reference electrode, whereas in an amperometric sensor, the current is measured in either AC or DC mode, and in a conductometric sensor, the conductance or resistance is measured. Gases and liquids are the main applications for mass sensitive or receptive sensors. The fundamental tenet of this sensor is that a fluctuation in mass causes a change in a property. Mass change is brought on by the accumulation of analyte. For instance, in piezoelectric sensors, a mass of the analyte is physically adsorbed, causing a change in frequency to be detected by the detecting electrode or substance. While in surface radio wave sensors the sensing electrode picks up changes in the propagation velocity of a radio wave that is created and is impacted by the accumulation of a specific amount of the analyte. Electrical sensors pick up electrical signals caused by changes in electrical properties brought on by interactions between the detecting substance and the analyte. In chemiresistive sensors, for instance, the sensing electrode is used to monitor changes in resistance brought on by the analyte's contact with the electrode or substance [4].

10.3 Selection of Polymers for Sensing Applications

Chemical sensors are essential for monitoring the society that we live in since they provide data on production technologies, food and beverage quality control, and other purposes. Since the majority of such electronic-based entities are complex and expensive to manufacture, existing sensor technology may be enhanced by using inexpensive, portable equipment with basic, flexible substrates [5]. In recent years, nanomaterials have advanced, bringing significant increases in the efficiency of chemical sensors and biosensors [6]. As well as having a high availability, being biocompatible, and being biodegradable, cellulose possesses unique material features including a high Young's modulus, is dimensionally stable, and has a low thermal expansion coefficient. An adulterant modification of cellulose has recently been researched [7] to improve the material characteristics of cellulose for electrical and actuator solicitation. The chemical stability, electrical conductivity, and photosensitivity of cellulose-based sensors were significantly improved by the incorporation of metal oxide [8]. Using a hybrid nanocomposite of cellulose paper, glucose oxidase (GOx), immobilized cellulose, and stannic oxide, Mahadeva and Kim [9] created a glucose biosensor. Through covalent bonds between porous GOx and stannic oxide, cellulose-SnO_2 hybrid nanocomposite was joined to permeable glucose oxidase. Rudall and Kenchington [10] employed glossy paper to make a pliable paper-based detector for the measurement of ethanol using multiwalled carbon nanotubes as electrodes and In_2O_5Sn nanoparticulate powder as a detecting medium. Silver nanoparticles (AgNP) were recently used on paper-based devices to show colorimetric detection of copper by Ratnarathorn et al. [9]. Homocysteine (Hcy) and dithiothreitol (DTT) functionalized AgNPs were immobilized in the sampling area to create the paper-based instruments [11].

Numerous studies have been made on conducting polymers and silver nanocomposites that may have biological implications. Electrical conductivity assays showed that Ag-starch nanocomposites' ionic conductivity increased with temperature, which is an example of a thermic triggered conduction technique [12]. This feature results from the energy of charge carriers (Ag^+ ions) increasing with temperature. This possibility might result in the creation of innovative biosensors for biotechnological uses like DNA detection. In order to create an inert immobilising environment for optical experiments, aqueous polymeric gels are often utilized. Cognet et al. [13] constructed a biosensor employing SWCNTs wrapped in sodium dodecylbenzenesulfonate (SDBS) and implanted in agarose gel. This are useful for local pH gradient detection.

Chitosan (CS) is generated from altered seafood scrap, like crab and shrimp, which is a highly effective component for gas and electrochemical sensors. Due to its astounding film-forming abilities and ability to preserve its inherent features, chitosan has also been employed as a biosensor. A chitosan-based ammonia sensor was made via electrochemical deposition in a contemporary work by Chandrasakaran et al. [14]. Carbon nanotubes (CNTs) and graphene (G) are two of the most cutting-edge nanomaterials with distinct electrochemical characteristics [15]. The creation of CNT-Chitosan composites as manufacturing platforms for sensors for biotechnological, pharmacological, and environmental applications has received a lot of attention in recent years [16]. Using a MWCNT-Chitosan composite, Wu [17] enhanced a glassy carbon electrode (GCE) for detecting morphine (MO) and dopamine (DA) in human blood serum or urine. Due to its strong mechanical strength, water resistance, and electron transfer characteristic, CS enhanced the electrode stability and permitted the simultaneous qualitative and quantitative analysis of DA and MO under physiological settings. In order to quickly detect Sudan I in samples of chilli powder [11], a MWCNT-CS modified GCE was developed using the same procedure. A flow injection chemiluminescence sensor for the detection of sulfamethoxazole was successfully created by Huamin et al. [18] employing the chitosan–graphene oxide–MIP as a key component. The capacity of biopolymers to gather and aggregate analytes on sensor surfaces, their suitability for use on objects like thin adhesive films, and their chemical structure-dependent scrupulous synergy with the vapor make them useful materials for chemical vapor analysis.

10.4 Different Types of Polymer-Based Sensor

10.4.1 Conducting Polymer

Due to their unique combination of organic polymer characteristics and semiconductor electrical capabilities, conducting polymers (CPs) are new materials that have generated considerable interest. Shirakawa et al. reported the electrical conductivity of conjugated polymers for the first time in 1977. Due to CP's distinct electrical properties, they have since received a lot of research attention. The remarkable electronic characteristics of CPs, including as low energy optical transitions, ionic properties, mechanical pliability, high large specific surface area, and high electron affinity, are caused by charge mobility along their p-electron frame. Conjugation length, total chain length, and charge transmission to neighboring molecules are some of the variables that affect the polymer's conductivity. They are being used in electrocatalysis, rejuvenate cells, super capacitors, membrane separation, optoelectronics, extraction, solar cells, electrochemical biosensors [19], and batteries [20]. Ion-selective electrodes [21] and molecularly imprinted polymers (MIP) have both been created using CPs [22].

Scientists and engineers employ conducting polymers, a crucial class of polymers with exceptional electrical characteristics, as suitable materials for sensors. Common conducting polymers include polyaniline (PANI), polypyrrole (PPy), and poly(3,4-ethylenedioxythiophene) (PEDOT) and are widely used as sensing materials. The resistance of the polymer material alters as a result of the analyte's synergy with the conducting polymer. The detection method is based on a change in resistance brought on by dopant addition to the polymers or swelling of the matrix of the polymer. The focus of sensing material research has recently shifted from primordial conducting polymers to conducting polymer nanocomposites. In comparison to traditional PANI-based sensors, PANI-nanofiber-based sensor performed well [23]. However, the inadequacies of conducting polymer-based sensors that limited their global expansion are their low mechanical firmness and elevated susceptibility to ambient moisture, which continues to be a challenge for materials research.

10.4.2 Acrylic Polymer

The covalently attached sensing functions to the polymeric backbone, which might be customized to process products with enhanced characteristics, are what make acrylic polymers useful in the field of sensing. The most popular acrylic polymers utilized for this purpose include methacrylic acid/acrylamide copolymers, acrylamide derivatives, and their copolymers. When creating sensor-based materials, acrylic moieties can integrate a variety of functional groups [24]. Acrylic-based sensors with fluorescent/colorimetric characteristics for a diverse set of analytes were described by Garcia et al. [25]. The method involved creating thin film polymers using radical polymerization using a sensory unit as the hydrophobic component, methyl methacrylate (MMA) monomer as the hydrophilic base, and 1-vinyl-2-pyrrolidone as the hydrophilic base [2].

A chemiluminescence polymer film that alters color when subjected to amino acids is included in the kit. The system enables the measurement of the whole amino acid amount by analysing the color characteristics of a sensory film made from photographs taken with a smartphone. By offering an analytical process that is unaffected by subjective judgement, this device can help in the identification of chronic human illnesses [26]. A remarkable study [27] described the development of a spintronic humidity sensor elicited from scandium polyacrylics. On flat borosilicate surfaces, nanostructured scandium polyacrylate was positioned to study the sorption and desorption of humidity at room temperature. Polymerization that is more durable and ecologically friendly is made possible by the application of ultrasonic energy. In default of any subsidiary reducing agents or UV/gamma radiations, polyacrylic acid (PAA) is used as a capping agent in integrating of silver nanoparticles. The generated Ag-PA sol was utilized to visually assess pH. This study serves as an illustration of a workable, inexpensive pH sensor built using a colorimetric innovative polymer. A spintronic humidity sensor based on scandium polyacrylic was the subject of another significant study [28]. Nanoscandium-polyacrylate was applied on flat borosilicate surfaces to study the sorption and desorption of humidity at standard temperatures. It seems to be an effective device for monitoring humidity [29].

10.4.3 EVOH Polymer

The ethylene and vinyl alcohol di-block copolymer, known as EVOH, is widely utilized in food packaging and biomedical applications due to its impeding properties against gases and moisture. The framework cannot be formed because, when C_2H_4 and C_2H_4O monomers are copolymerized, the keto-enolic tautomeric equilibrium shifts to the aldehydic form. As a result, the most used synthetic approach uses hydrolysis after adding ethylene and vinyl acetate as monomers. The ethylene monomer, vinyl acetate monomer, and vinyl alcohol monomer structures are shown in Figures 10.1 and Figure 10.2

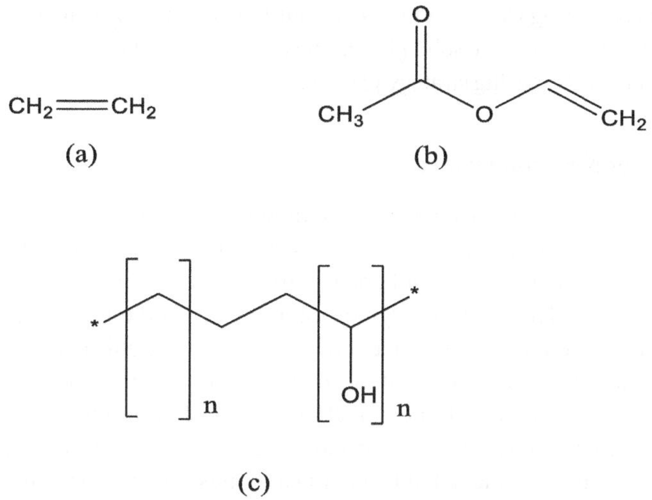

Figure 10.1 Monomer structures of (a) ethylene monomer (b) vinyl acetate monomer, and (c) vinyl alcohol [2].

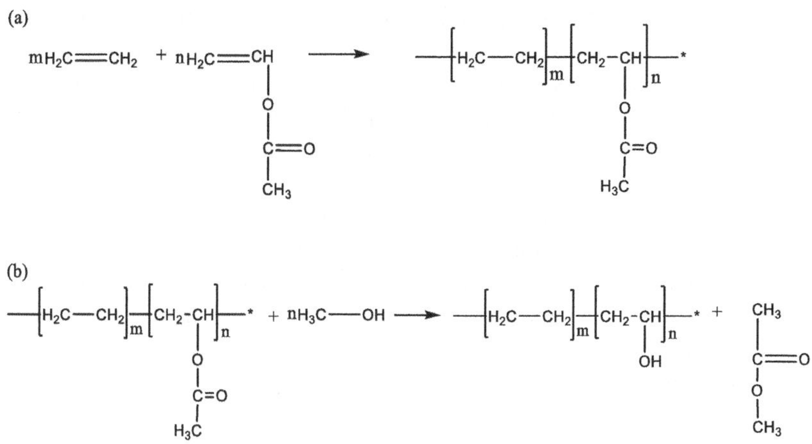

Figure 10.2 Illustration of the two-step procedure used to create EVOH [29].

illustrate the two-step procedure used to create EVOH [30]. On a wide scale, EVOH is available in several ethylene proportions, including 27%, 32%, 38%, and 44%. The different monomer percentages produce different barrier characteristics against gases and volatility in organic solvents. It is important to stress that the polymer's surface tactility and new functionalities affect its solubility. Owing to its excellent mechanical properties and ease of surface functionalization, EVOH is a strong candidate for the construction of sensors.

A sensor for encountering Cu^{2+} in a hydrated solution was built by Cui et al. [31] using EVOH as a polymeric solid phase. The dampening of light by Cu^{2+} is the foundation of this sensing strategy [27, 30].

10.4.4 Polymer Nanocomposites

A new class of advanced composite materials known as polymer nanocomposites has drawn significant interest from the scientific and business communities. These materials have nanofillers that are just a few nanometers in size, giving the nano-element and polymer matrix a very large interfacial specific area. As a result, polymer nanocomposites exhibit notable gains in strength, thermal stability, etc., without sacrificing their natural properties, including rigidity or optoelectronic conciseness. Weak intermolecular forces are often used to interact between nanofillers and the matrix. But, occasionally, the contact is mediated by chemical bonds. The uniform distribution of the nanofiller throughout the polymer matrix is the most important feature in the production of polymer nanocomposites. Generally speaking, it is challenging to evenly scatter nanofillers in a polymer matrix due to the nanoparticles' high surface energy, which promotes agglomeration [32]. Due to their special character traits, polymer nanocomposites are utilized in electronics and nanoelectrical equipments. These qualities include gauzy, high surface area and aspect ratio, flexible charge carriage, chemical particularity, ease of access, low maintenance cost, ease of fabrication, and scalable production [33].

Due to the differences in structure, reciprocity, and characteristics between the various components present in polymer nanocomposites, a number of key elements have an impact on the function of nanoparticles as reinforcement fillers in the polymer matrix. It is crucial to evenly distribute nanoparticles throughout the matrix such that each particle is covered in a polymer. An effective load transmission to the matrix of nanofillers is made possible by the enhanced dispersion, which leads to a more uniform distribution of stress. These crucial concerns have a significant impact on the apparent characteristics and consequently on the responsiveness of the polymer nanocomposites to physiological arousal in sensing execution. So far, several surface modification techniques for nanofillers and various methods of synthesis for nanocomposite materials have been put forth in an effort to tweak the scattering of the nanofiller inside the polymer matrix [1].

Successful polymer nanocomposites synthesis and manufacturing have sparked widespread interest in sensing applications due to their enhanced physiochemical characteristics. A broad range of carbon-based reinforced nanostructures are included in the polymer matrix of the polymer nanocomposites. Among chemiresistive gas sensors, the use of carbon-based polymer nanocomposite as a sensing component is a promising area for research [34]. The chemical conformations of the polymer composites, which are easily accomplished by changing structural geometry, guarantee the specificity of the polymer nanocomposite-based chemiresistive sensor. Due to the larger zone of contact, the polymer nanocomposite-based sensor displays excessive variation of electrical properties, which include capacitance and resistance, when exposed to analyte molecules. Technology has been used to create and design the characteristic qualities, turning the nanocomposites in to the elevated materials enabling sensor applications possessing distinctive attributes [1].

10.4.5 Polyvinylidene Fluoride

Ferroelectric polymers are a crucial class of functional materials that may flip between spontaneous and thermodynamically stable polarization states when exposed to a powerful enough external electric field. Ferroelectric polymers' outstanding electric characteristics are driving increased research into them. They merit investigation as potential active components in non-volatile memories, piezoelectric or pyroelectric sensors, electromechanical sensors or artificial muscles, high electric energy density capacitors, organic solar cells, nanogenerators for electric power generation, and self-powered electronics, as well as in high electric energy density capacitors, organic solar cells, and nanogenerators [35]. Polymers based on poly(vinylidene fluoride) (PVDF) are among the ferroelectric polymers that are generating a lot of interest [36].

The most widely used piezopolymer in electromechanical devices is PVDF, a semi-crystalline polymer created by polymerizing $H_2C{=}CF_2$ monomers [37]. This is because PVDF has the greatest piezoelectric coefficient among bulk polymers, 20–28 pC N^{-1}. The molecular structure of it is shown in Figure 10.3 [38]. Since the Japanese researcher Kawai discovered PVDF's potent piezoelectricity in 1969, several specialists have conducted studies on the material. In PVDF, there are two different transformation mechanisms: while 33-mode is the polarization parallel to the path of the tension, 31-mode is the polarization orthogonal to the plane of the stress applied to the object. When compared to the other modes, the 31-mode may produce a greater electric field with the same stress. Figure 10.4 [39] depicts the two ways. The magnitude of the piezoelectricity of PVDF is significantly lower than PZT, the most popular type of piezoelectric material. PVDF has become a good material for use in practical applications versatility, broad frequency range, cost-effectiveness, ease of manufacture, and other characteristics

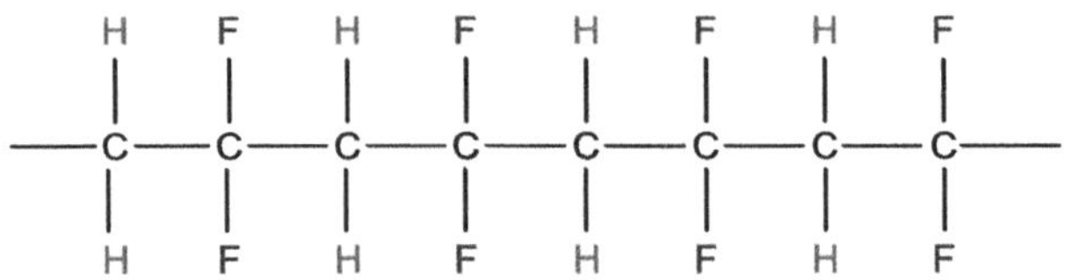

Figure 10.3 The molecular structure of polyvinylidene fluoride.

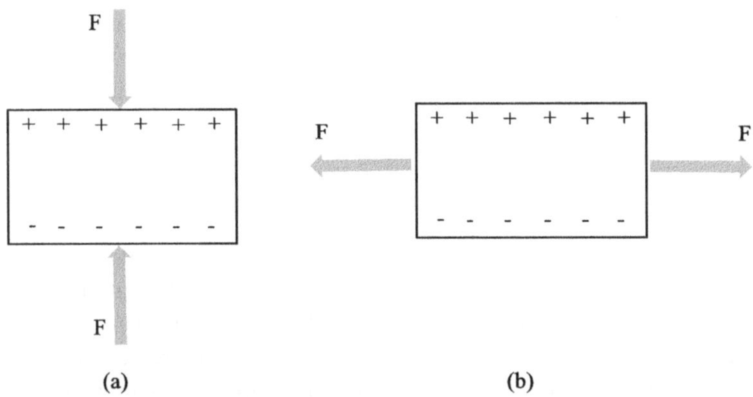

Figure 10.4 Two modes of PVDF: (a) 33- and (b) 31-modes.

including low permittivities, low thermal conductivities, and low acoustic resistances. Since PVDF sensors are ideal for sensing vibrations, their output signal may be boosted by a curved structure, and their resonance frequencies change as their geometric parameters change [39].

10.4.6 Flexible Polymers

The future study and creation of high-performance flexible sensors still presents a number of potential difficulties. A critical area of research, particularly for invasive applications, is the biocompatibility of the active substances and flexible substrates, comprising protracted malignancy assessments. These problems can be solved by creatively utilising device designs, components, erection techniques, surface engineering, and interface systematizing. Soft, flexible sensors may result from development of novel substances for active layers, substrates, and conductive layers; this new paradigm may broaden the use of existing technologies for various sensing tasks. Sensing applications frequently use flexible substrates made of polyimide (PI), polyetheretherketone (PEK), polyethersulfone (PES), polycarbonate, poly(ethylene napthalate) (PEN), and polyester resins like polyethylene terephthalate (PET). For real-world sensing enactment, a material's temperature resistance, chemical resistance, transparency, as well as flexibility should be taken into account [40].

A low elastic modulus enables the mechanical stability of flexible devices during deformation without resulting in physical damage, which is one of their most crucial capabilities. Candidates for printed materials in flexible devices should have a suitable elastic modulus as well as strong electrical dispatch that can be used over large regions at a reasonable cost. Table 10.1 compares the materials and associated attributes utilized in flexible circuit boards with that of printed electronic devices, some of which are essential factors for deciding device manufacturing techniques [41]. Sensor applications can use a variety of flexible substrates. The most popular flexible substrate is polyimide (PI), since it has a strong heat and chemical resistance. However, due to its inherent coloration, its use in transparent devices is constrained. Most recently, research has focused on incorporating elastic properties into the sensor frame virtually via connections and features like nanowires and electrodes that have been specifically developed for greater flexibility [42].

Table 10.1 Flexible substrates character traits [41].

Material	Characteristics	Temperature [°C]
Polyimide (PI)	Orange color, high CTE, good chemical resistance, expensive, high moisture absorption	275
Polyetheretherketone (PEEK)	Amber color, good chemical resistance, expensive, low moisture absorption	250
Polyethersulfone (PES)	Light yellow, good dimensional stability, poor solvent resistance, expensive, moderate moisture absorption	230
Polycarbonate (PC)	Clear, poor CTE, inexpensive, moderate moisture absorption	155
Poly(ethylene napthalate) (PEN)	Clear, moderate CTE, good chemical resistance, inexpensive, moderate moisture absorption	150
Polyester (PET)	Clear, moderate CTE, good chemical resistance, inexpensive, moderate moisture absorption	120

10.5 Biobased Polymers for Sensing

A novel class of materials known as "biobased polymeric materials" is produced utilising biological, chemical, and physical processes employing sustainable biomass, such as crops, trees, or other plants, as well as their by-products and inclusions as raw materials. Biobased materials contain a variety of unique qualities, including some that standard polymers lack, such as being renewable, ecologically benign, and biodegradable. Numerous biobased polymers had already reported 20%–30% yearly growth rates in recent years. Consequently, research on stimuli-responsive polymers has increased significantly. These materials have a wide variety of uses in several industries. When exposed to internal or external stimuli, they can alter their structures and other physicochemical properties. These polymers may also change one or more of their properties, such as wettability and color. Polylactic acid (PLA), microbial polyesters, and polysaccharose cellulose are the three primary microbial sources for synthetic biobased polymers. In terms of botanical origin, celluloses, hemicelluloses, and lignin are typical biobased polymers. Research on starch-based polymers, which are currently a popular topic, is also derived from plants [38].

Biobased sensors feature one constituent that comes from a biobased source, which might be the detecting element or the substrate. The majority of biobased material sensors used in food packaging rely on photometric analyte recognition. Natural substances like anthocyanins and curcumin, which are known as natural pH indicators, are used to color many staples, and flowers, giving them hues that cover almost the whole visible spectrum. These molecules undergo protonation and deprotonation, which causes a rearrangement of their displaced electronic frame and a change in the number of resonant electrons overall plus their detention, which changes the color of the molecules. Choi et al. [43] produced a pH sensor with potato starch and agar with anthocyanin extracted from purple sweet potatoes that displayed color fluctuations at pH 2.0–10.0. With a somewhat wider pH range of 2.0 to 12.0, red radish anthocyanin was employed in a gelatin–gellan gum matrix by Zhai et al. [44]. As decaying proteins create alkaline volatile nitrogen compounds, it is frequently adequate to analyse the change of pH, even if photometric pH-sensitive sensors are generally not handy for specific examination [45].

Ammonia, ethylamine, cadaverine, and putrescine have been detected colorimetrically and electrochemically using trifluoroacetyl azobenzene dyes created by Mohr et al [46]. Since the trifluoroacetyl group's carbonyl carbon is severely electron-deficient, it interacts easily with electron donors like amines or alcohols. It transforms into a hemiaminal group with primary, secondary, or tertiary amines, which reduces the number of unpaired electrons in the diazobenzene backbone and shortens their confinement length, causing an optical absorption shift to blue. The vapors of the analytes with a concentration of 1.0–0.1 vol.% may be detected using sensors printed on paper. Additionally, polydiacetylenes that were crosslinked in self-assembled vacuoles supported by cellulose nanocrystals in the chitosan matrix were used to exhibit very delicate ammonia sensors that function even at low temperatures. The sensing method is based on the polydiacetylene backbone's conformal shift in response to environmental factors like pH, mechanical stress, or temperature [47]. The films of the sensors were able to detect 100 ppm of ammonia [48].

10.5.1 Biocompatible Polymers

The "ability to interact with a biological system without generating an undesirable consequence" is the definition of biocompatibility given by the IUPAC. In general, a substance is deemed biocompatible if it causes a tolerable host reaction when in contact with the body or bodily fluids. Biocompatible substances frequently cause varied degrees of humoral and cellular immune reactions, although these reactions are either harmless or typical of the body. When contemplating wearable and implanted electronics, a material must not only be physically compatible but also chemically compatible with its environment. Before a material is categorized as biocompatible, it must undergo a thorough evaluation through *in vivo* implantation or *in vitro* culture trials.

Natural polymers, such as those derived from proteins and polysaccharides, are widely accessible, reasonably priced, and generally biocompatible. Furthermore, these naturally produced polymers may unavoidably provoke an undesirable immunogenic response. Gene transfection, tissue engineering, medication delivery, and other fields have all made extensive use of synthetic polymers that have been biocompatiblely developed. Some polymers that are often used in implanted electronics that have been given clinical use approval by the U.S. FDA are parylene, PLA, PLGA, and PEG. While there are several well-known and widely-used biocompatible polymers, new molecular design allows for the synthetic tuning of certain functions, opening the door to a wider range of wearable and implantable electronic potential in numerous body parts and physiological fluids [49]. Even though many organic and synthetic polymers are being used as parts of biodegradable electronic devices, there is still a pressing need to further enhance the electrical and mechanical characteristics of polymer materials. The mechanical characteristics of synthetic polymers may be tuned more easily, notably their flexibility, which may make it easier to fabricate stretchy biodegradable electronics and provide greater conformal contact between the implanted device and dynamic tissue surface [50].

10.5.2 Natural Polymers

Due to their low toxicity, biocompatibility with organisms and tissues, and degradability, natural polymer nanocomposites have recently received more attention than synthetic polymers. Nanostarch and nanocellulose are excellent materials for bionanocomposites in a variety of applications. The creation of beads, nanofibers, gels, sponges, microparticles, scaffolds, and NPs from chitosan and chitin is simple. These forms have several uses in tissue engineering, cancer medication delivery, nanobiotechnology, and wound dressing. NPs made of capectinate and mg-pectinate can be utilized to transfer genes. MB and Cu^{2+} ions may be removed from aqueous solution using pectin-modified Fe_3O_4 NPs, whereas pectin-based nanospheres containing Ca^{2+} and CO_3^{2-} ions can be employed for *in vitro* medications. HA-based devices are particularly suitable for ocular medication delivery and are crucial for wound healing. The addition of magnetic NPs to carrageenan hydrogels produced hydrogel nanocomposites that are crucial for biological applications because they respond to external magnetic stimuli. For a walking gadget with a quick reaction, a smart actuator made of a polypyrrole/agar nanocomposite is employed.

The major category of biodegradable polymers, natural polyesters or biopolyesters, are made from biological monomers such polycaprolacton and polylactic acid. Melt intercalation is utilized to create PHA-based nanobiocomposites, and solution casting is used to create ZnO-reinforced poly(3-hydroxybutyrate-co-3-hydroxyvalerate) ecological nanocomposites, which may be used as packaging for edible goods. Biodegradable packaging materials include PLA-PHB/CNC bionanocomposite sheets that have been plasticized. For the treatment of hereditary diseases, AIDS, cardiovascular illnesses, Alzheimer's disease, cancer, and multiple sclerosis, bionanocomposites can be employed as nonviral gene carriers. By encasing DNA, DNAthionineCNTs nanocomposites may be created. Thionine-based bionanocomposites are applied to the surface of carbon nanotubes to form biosensor applications with strong electrochemical responsiveness and high stability. For sensor applications, polyisoprenehigh structure carbon NPs bionanocomposites and polyisoprenecarbon black can be employed [51].

10.6 Molecularly Imprinted and Surface Imprinted Polymers for Sensor Applications

Molecular imprinting is the process of chemically synthesising transmitters in a polymer under the influence of a template. As enticing alternatives to natural transmitters, specialized recognition sites created by molecular imprinting offer outstanding qualities including high specificity endurance and low costs. The goal of contemporary sensor research is to create synthetic receptors that might mimic the interactions between antibodies and antigens in nature while also taking sensitivity and specificity into account. The method of electrochemical deposition involves making templates for recognition sites for polymer molecules. The crosslinking in the polymer stabilized and maintained the connections between scaffolding and functional monomers during polymerization. Figure 10.5 depicts a general imprinting procedure [52]. When a monomer interacts with a molecular template, a cavity often forms around it; when the template is eventually removed, the cavity remains for target molecular rebinding. The sensor's sensitivity measures at 2.1 W/% RH, and its repeatability is 89%, with recovery and reaction times of 155 and 25 s, respectively. The target analyte may be detected preferentially in the template-derived areas by the generated MIP as more than merely a result [2].

MIPs are better in terms of their protracted chemical and thermal durability because they selectively attach their targets in a reversible way using functional nanocavities. MIP synthesis is often quick, inexpensive, and simple, especially when bulk polymerization methods are employed. For MIP synthesis, a suitable solvent ("porogen") is dissolved in the chosen target molecule type (template) with optimal ratios of monomer units, a cross-linker, and an initiator. Following polymerization, the precursor molecules are removed to produce a MIP that can selectively rebind the target. The spotting of substances with low molecular weight is a key application area for bulk-polymerized MIPs, and a number of transducers, including spectroscopy, microbalances,

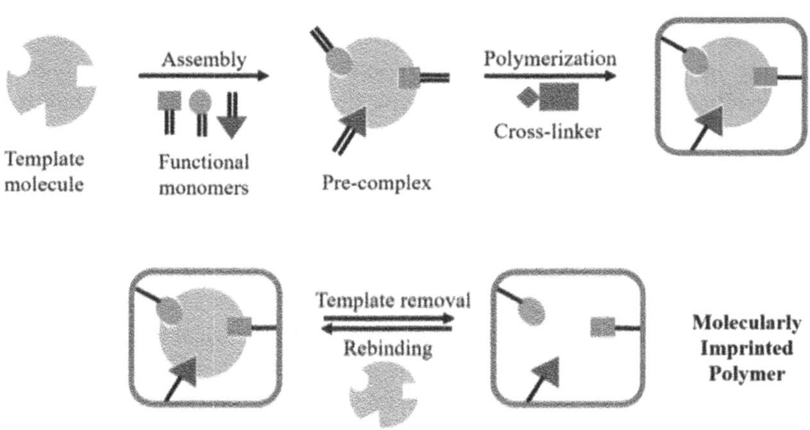

Figure 10.5 A general imprinting procedure [52].

and thermal detection, have been used effectively. Additionally, MIPs were created for macromolecular entities such viruses, DNA, and proteins [53].

10.7 Electroactive Polymers

In reaction to an applied electrical field, polymers known as electroactive polymers (EAPs) alter in form and/or dimension. Electro-responsive polymers (ERPs), which generally display electrically linked responses, are broken down into EAPs as a subset. Roentgen and Sacerdote [54] conducted the first investigation on EAPs while researching how an electric field might cause a dielectric polymer to distort. People who work on electrically driven composite materials are drawn to EAPs because some EAPs, such dielectric elastomers, can withstand significantly greater stresses than their comparable traditional transducer materials. EAPs are appropriate for a variety of sensory applications, including chemical sensing, tactile sensing, blood pressure and heartbeat monitoring, etc. Numerous advantageous characteristics, including ease of manufacturing, good mechanical elasticity, adaptable electromechanical coupling qualities, and tailorable geometries, are the reasons for this. Ionic and electrical EAPs are the two main types of EAPs. Ion or molecule movement is what causes the electrical activation of ionic EAPs. Conducting polymers (CPs) and ionic polymer–metal composites (IPMCs) are two examples of ionic EAPs [55]. Coulomb forces and applied electric fields are used to activate electronic EAPs. The electronic EAP category includes dielectric elastomers, electrostrictive polymers, liquid crystal polymers, and piezoelectric polymers [56].

Low voltages are typically used by ionic EAPs. Polypyrrole (PPy) and poly(3,4-ethylenedioxythiophene) (PEDOT), two materials that are frequently employed in the fabrication of ionic EAPs, are also frequently biocompatible. To function, ionic EAPs typically need an ion reservoir that can transport ions or molecules. The majority of ionic EAPs fall within CP, IPMC, and polymer gel categories. Recently, carbon nanotubes have been added to this category.

Electronic EAPs are devoid of any electrolyte medium, in contrast to ionic EAPs. The electromechanical connection does not need ion movement. Therefore, compared to ionic EAPs, electronic EAPs need less time to respond or relax, as little as 1 ms [57]. As a result, electrical EAPs can operate at greater frequencies for dynamic sensing than ionic EAPs. The primary subgroups of electronic EAPs are dielectric elastomers, electrostrictive polymers, liquid-crystal elastomers, and piezoelectric polymers.

10.8 Conclusion

Biobased polymers are appealing materials in a variety of industries due to their exceptional qualities. Smart materials and gadgets are now in-demand trends in both science and commercialized sector. Electronics that are biodegradable and biocompatible not only offer a solution for managing e-waste and safeguarding the environment, but also opens door for biomedical uses of disposable electronics. Although several organic and synthetic polymers have been used as components in biodegradable electronic devices, there is still a pressing need to boost the electrical and mechanical characteristics of polymer materials. The mechanical characteristics of synthetic polymers may be tuned more easily, notably their flexibility, which may make it easier to fabricate stretchy biodegradable electronics and provide greater conformal contact between implanted device and dynamic tissue surface. The development of synthetic design and polymer chemistry in recent years has greatly increased the possibilities for integration into multifunctional electronics. Biodegradable polymers are now being used extensively in society, and if they break down during reprocessing, they might easily end up as pollutants in a recycling stream. By using biobased or renewable materials and green synthesis techniques, an even wider spectrum of sustainability may be applied to descendants of electronics. Combinatorial libraries of biocompatible, biodegradable, and recyclable polymers might be developed as instruments for the creation of best future sustainable and sentient products.

References

1 Shukla, P. and Saxena, P. (2021). Polymer nanocomposites in sensor applications: a review on present trends and future scope. *Chinese Journal of Polymer Science* 39 (6): 665–691. doi: 10.1007/s10118-021-2553-8.

2 Alam, M.W., Bhat, S.I., Al Qahtani, H.S., et al. (2022). Recent progress, challenges, and trends in polymer-based sensors: a review. *Polymers* 14 (11), Art. no. 11. doi: 10.3390/polym14112164.

3 Hulanicki, A., Glab, S., and Ingman, F. (1991). Chemical sensors: definitions and classification. *Pure and Applied Chemistry* 63 (9): 1247–1250. doi: 10.1351/pac199163091247.

4 Aswal, D.K. and Gupta, S.K. (2007). *Science and Technology of Chemiresistor Gas Sensors*. New York: Nova Publishers.

5 Dufresne, A. and Vignon, M.R. (1998). Improvement of starch film performances using cellulose microfibrils. https://pascal-francis.inist.fr/vibad/index.php?action=getRecordDetail&idt=2216750 (accessed 28 July 2022).

6 Kim, J.-H., Mun, S., Ko, H.-U. et al. (2014). Disposable chemical sensors and biosensors made on cellulose paper. *Nanotechnology* 25 (9): 092001. doi: 10.1088/0957-4484/25/9/092001.

7 Marques, P.A.A.P., Trindade, T., and Neto, C.P. (2006). Titanium dioxide/cellulose nanocomposites prepared by a controlled hydrolysis method. *Composites Science and Technology* 66 (7): 1038–1044. doi: 10.1016/j.compscitech.2005.07.029.

8 Pinto, R.J.B., Marques, P.A.A.P., Barros-Timmons, A.M., et al. (2008). Novel SiO2/cellulose nanocomposites obtained by in situ synthesis and via polyelectrolytes assembly. *Composites Science and Technology* 68 (3): 1088–1093. doi: 10.1016/j.compscitech.2007.03.001.

9 Mahadeva, S.K. and Kim, J. (2011). Conductometric glucose biosensor made with cellulose and tin oxide hybrid nanocomposite. *Sensors and Actuators B: Chemical* 157 (1): 177–182. doi: 10.1016/j.snb.2011.03.046.

10 Rudall, K.M. and Kenchington, W. (1973). The chitin system. https://onlinelibrary.wiley.com/doi/abs/10.1111/j.1469-185X.1973.tb01570.x (accessed 28 July 2022).

11 Abhilash, M. and Thomas, D. (2017). 15 – Biopolymers for biocomposites and chemical sensor applications. In: *Biopolymer Composites in Electronics* (ed. K.K. Sadasivuni, D. Ponnamma, J. Kim et al.), 405–435. Amsterdam: Elsevier. doi: 10.1016/B978-0-12-809261-3.00015-2.

12 Ratnarathorn, N., Chailapakul, O., Henry, C.S., and Dungchai, W. (2012). Simple silver nanoparticle colorimetric sensing for copper by paper-based devices. *Talanta* 99: 552–557. doi: 10.1016/j.talanta.2012.06.033.

13 Cognet, L., Tsyboulski, D.A., Rocha, J.-D.R., et al. (2007). Stepwise quenching of exciton fluorescence in carbon nanotubes by single-molecule reactions. https://www.science.org/doi/abs/10.1126/science.1141316 (accessed 28 July 2022).

14 Chandrasakaran, D.S., Nainggolan, I., Derman, M.N.B., and Ikhsan, T. (2014). Chloroform gas sensor based on chitosan biopolymer. *Applied Mechanics and Materials* 679: 45–49. doi: 10.4028/www.scientific.net/AMM.679.45.

15 Salavagione, H.J., Díez-Pascual, A.M., Lázaro, E. et al. (2014). Chemical sensors based on polymer composites with carbon nanotubes and graphene: the role of the polymer. *Journal of Materials Chemistry A* 2 (35): 14289–14328. doi: 10.1039/C4TA02159B.

16 Kianipour, S. and Asghari, A. (2013). Room temperature ionic liquid/multiwalled carbon nanotube/chitosan-modified glassy carbon electrode as a sensor for simultaneous determination of ascorbic acid, uric acid, acetaminophen, and mefenamic acid. *IEEE Sensors Journal* 13 (7): 2690–2698. doi: 10.1109/JSEN.2013.2259588.

17 Wu, M., Tang, W., Guimarães, J. et al. (2013). Electrochemical detection of Sudan I using a multi-walled carbon nanotube/chitosan composite modified glassy carbon electrode. 2013. doi: 10.4236/ajac.2013.46A001.

18 Huamin, Q., Lulu, F., Li, X. et al. (2013). Determination sulfamethoxazole based chemiluminescence and chitosan/graphene oxide-molecularly imprinted polymers. *Carbohydrate Polymers* 92 (1): 394–399. doi: 10.1016/j.carbpol.2012.09.092.

19 Saranya, K., Rameez, Md., and Subramania, A. (2015). Developments in conducting polymer based counter electrodes for dye-sensitized solar cells – an overview. *European Polymer Journal* 66: 207–227. doi: 10.1016/j.eurpolymj.2015.01.049.

20 Yadav, A., Kumar, H., Sharma, R., and Kumari, R. (2021). Influence of polyaniline on the photocatalytic properties of metal nanocomposites: a review. *Colloid and Interface Science Communications* 40: 100339. doi: 10.1016/j.colcom.2020.100339.

21 Naseri, M., Fotouhi, L., and Ehsani, A. (2018). Recent progress in the development of conducting polymer-based nanocomposites for electrochemical biosensors applications: a mini-review. *The Chemical Record* 18 (6): 599–618. doi: 10.1002/tcr.201700101.

22 Ates, M. (2013). A review study of (bio)sensor systems based on conducting polymers. *Materials Science and Engineering C* 33 (4): 1853–1859. doi: 10.1016/j.msec.2013.01.035.

23 Tran, H.D., Li, D., and Kaner, R.B. (2009). One-dimensional conducting polymer nanostructures: bulk synthesis and applications. *Advanced Materials* 21 (14–15): 1487–1499. doi: 10.1002/adma.200802289.

24 Guembe-García, M., Santaolalla-García, V., Moradillo-Renuncio, N. et al. (2021). Monitoring of the evolution of human chronic wounds using a ninhydrin-based sensory polymer and a smartphone. *Sensors and Actuators B: Chemical* 335: 129688. doi: 10.1016/j.snb.2021.129688.

25 Chowdhury, P., Hazra, A., Mondal, M.K. et al. (2019). Facile synthesis of polyacrylate directed silver nanoparticles for pH sensing through naked eye. *Journal of Macromolecular Science, Part A* 56 (8): 773–780. doi: 10.1080/10601325.2019.1607376.

26 Leibl, N., Haupt, K., Gonzato, C., and Duma, L. (2021). Molecularly imprinted polymers for chemical sensing: a tutorial review. *Chemosensors* 9 (6), Art. no. 6. doi: 10.3390/chemosensors9060123.

27 Xiao, N., Deng, J., Cheng, J. et al. (2016). Carbon paste electrode modified with duplex molecularly imprinted polymer hybrid film for metronidazole detection. *Biosensors and Bioelectronics* 81: 54–60. doi: 10.1016/j.bios.2016.02.041.

28 Smolinska-Kempisty, K., Ahmad, O.S., Guerreiro, A. et al. (2017). New potentiometric sensor based on molecularly imprinted nanoparticles for cocaine detection. *Biosensors and Bioelectronics* 96: 49–54.

29 Maes, C., Luyten, W., Herremans, G. et al. (2018). Recent updates on the barrier properties of ethylene vinyl alcohol copolymer (EVOH): a review. *Polymer Reviews* 58 (2): 209–246. doi: 10.1080/15583724.2017.1394323.

30 Wu, Y. and Meyerhoff, M.E. (2008). Nitric oxide-releasing/generating polymers for the development of implantable chemical sensors with enhanced biocompatibility. *Talanta* 75 (3): 642–650. doi: 10.1016/j.talanta.2007.06.022.

31 Ko, H.-F., Sfeir, C., and Kumta, P.N. (2010). Novel synthesis strategies for natural polymer and composite biomaterials as potential scaffolds for tissue engineering. *Philosophical Transactions of the Royal Society A: Mathematical, Physical and Engineering Sciences* 368 (1917): 1981–1997. doi: 10.1098/rsta.2010.0009.

32 Rong, M.Z., Zhang, M.Q., and Ruan, W.H. (2006). Surface modification of nanoscale fillers for improving properties of polymer nanocomposites: a review. *Materials Science and Technology* 22 (7): 787–796. doi: 10.1179/174328406X101247.

33 Sree, U., Yamamoto, Y., Deore, B. et al. (2002). Characterisation of polypyrrole nano-films for membrane-based sensors. *Synthetic Metals* 131 (1): 161–165. doi: 10.1016/S0379-6779(02)00179-0.

34 Yoon, H., Xie, J., Abraham, J.K. et al. (2005). Passive wireless sensors using electrical transition of carbon nanotube junctions in polymer matrix. *Smart Materials and Structures* 15 (1): S14–S20. doi: 10.1088/0964-1726/15/1/004.

35 Fan, F.R., Tang, W., and Wang, Z.L. (2016). Flexible nanogenerators for energy harvesting and self-powered electronics. *Advanced Materials* 28 (22): 4283–4305. doi: 10.1002/adma.201504299.

36 Chen, X., Han, X., and Shen, Q.-D. (2017). PVDF-based ferroelectric polymers in modern flexible electronics. *Advanced Electronic Materials* 3 (5): 1600460. doi: 10.1002/aelm.201600460.

37 Liu, M., Nie, F.-Q., Wei, Z. et al. (2010). In situ electrochemical switching of wetting state of oil droplet on conducting polymer films. *Langmuir* 26 (6): 3993–3997.

38 Gao, S., Tang, G., Hua, D. et al. (2019). Stimuli-responsive bio-based polymeric systems and their applications. *Journal of Materials Chemistry B* 7 (5): 709–729. doi: 10.1039/C8TB02491J.

39 Xin, Y., Sun, H., Tian, H., et al. (2016). The use of polyvinylidene fluoride (PVDF) films as sensors for vibration measurement: a brief review. *Ferroelectrics* 502 (1): 28–42. doi: 10.1080/00150193.2016.1232582.

40 Han, S.-T., Peng, H., Sun, Q. et al. (2017). An overview of the development of flexible sensors. *Advanced Materials* 29 (33): 1700375. doi: 10.1002/adma.201700375.

41 Nathan, A., Ahnood, A., Cole, M.T. et al. (2012). Flexible electronics: the next ubiquitous platform. *Proceedings of the IEEE* 100 (Special Centennial Issue): 1486–1517. doi: 10.1109/JPROC.2012.2190168.

42 Rim, Y.S., Bae, S.-H., Chen, H. et al. (2016). Recent progress in materials and devices toward printable and flexible sensors. *Advanced Materials* 28 (22): 5.

43 Choi, I., Lee, J.Y., Lacroix, M., and Han, J. (2017). Intelligent pH indicator film composed of agar/potato starch and anthocyanin extracts from purple sweet potato. *Food Chemistry* 218: 122–128.

44 Wang, S., Suh, J.H., and Zheng, X. et al. (2017). Identification and quantification of potential anti-inflammatory hydroxycinnamic acid amides from wolfberry. *Journal of Agriculture and Food Chemistry* 65 (2): 364–372.

45 Bulushi, I.A., Poole, S., Deeth, H.C., and Dykes, G.A. (2009). Biogenic amines in fish: roles in intoxication, spoilage, and nitrosamine formation—a review. *Critical Reviews in Food Science and Nutrition* 49 (4): 369–377. doi: 10.1080/10408390802067514.

46 Mohr, G.J. (2004). Tailoring the sensitivity and spectral properties of a chromoreactand for the detection of amines and alcohols. *Analytica Chimica Acta* 508 (2): 233–237.

47 Nguyen, L.H., Naficy, S., McConchie, R. et al. (2019). Polydiacetylene-based sensors to detect food spoilage at low temperatures. *Journal of Materials Chemistry C* 7 (7): 1919–1926. doi: 10.1039/C8TC05534C.

48 Halonen, N., Pálvölgyi, P., Bassani, A., et al. (2020). Bio-based smart materials for food packaging and sensors – a review. *Frontiers in Materials* 7.

49 Chiong, J.A., Tran, H., Lin, Y. et al. (2021). Integrating emerging polymer chemistries for the advancement of recyclable, biodegradable, and biocompatible electronics. *Advanced Science* 8 (14): 2101233. doi: 10.1002/advs.202101233.

50 Cao, Y. and Uhrich, K.E. (2019). Biodegradable and biocompatible polymers for electronic applications: a review. *Journal of Bioactive and Compatible Polymers* 34 (1): 3–15. doi: 10.1177/0883911518818075.

51 Noreen, A., Sultana, S., Sultana, T., et al. (2020). Natural polymers as constituents of bionanocomposites. In: *Bionanocomposites* (ed. K. Mahmood Zia, F. Jabeen, M.N. Anjum, and S. Ikram), 55–85. Amsterdam: Elsevier. doi: 10.1016/B978-0-12-816751-9.00003-9.

52 Alberti, G., Zanoni, C., Losi, V. et al. (2021). Current trends in polymer based sensors. *Chemosensors* 9 (5), Art. no. 5. doi: 10.3390/chemosensors9050108.

53 Eersels, K., Lieberzeit, P., and Wanger, P. (2016). A review on synthetic receptors for bioparticle detection created by surface-imprinting

techniques—from principles to applications. *ACS Sensors* 1 (10): 1171–1187.

54 Sacerdote, M.P. (1899). On the electrical deformation of isotropic dielectric solids. *Journal of Physics* 3: 282–285.

55 Baughman, R.H. (1996). Improved piezoelectric 0–3 ceramic particle/polymer composites. *Synthetic Metals* 78: 339.

56 Wang, T., Farajollahi, M., Choi, Y.S., et al. (2016). Electroactive polymers for sensing. *Interface Focus* 6 (4): 20160026. doi: 10.1098/rsfs.2016.0026.

57 Jean-Mistral, C., Sylvestre, A., Basrour, S., and Chaillout, J.-J. (2010). Dielectric properties of polyacrylate thick films used in sensors and actuators. *Smart Materials and Structures* 19 (7): 075019. doi: 10.1088/0964-1726/19/7/075019.

11

Applications of Biopolymers in Construction and Civil Engineering

Eapen Philip[1] and Reshmy R.[2,]*

[1] Post Graduate and Research Department of Chemistry Bishop Moore College, Mavelikara, Kerala, India
[2] Department of Science and Humanities, Providence College of Engineering, Chengannur, Kerala, India
[*] You may contact the corresponding author at reshmypkumar@gmail.com

11.1 Introduction

The development of new manufacturing sectors that utilize sustainable resources is among the major concerns in the 21st century. Biopolymers are now widely used in the construction field. The traditional organic chemical, materials, and pharmaceutical firms rely on natural fuels, especially petrochemicals, to create the materials that are employed in daily items like textiles, colors, packaging, medicines, building supplies, and electronic materials [1]. However, because fossil sources are limited, the availability of petrochemicals will decline as the century goes on. The biopolymer-based feedstock is a genuine long-term alternative. Nature creates over 170,000 million tonnes of sustainable carbon each year as "biomass", but only about 3% of this is for edible and non-edible food use [2]. The foreseeable requirements of the chemical and associated sectors, which presently use fossil fuels, might be satisfied with less than 1% of the above.

The primary development technique and the materials used in building can be used to classify various types of soil buildings [3]. However, soil's durability and strength are constrained as a construction material. There have been several cases of earthquakes, flooding, and heavy rains damaging

Applications of Biopolymers in Science, Biotechnology, and Engineering, First Edition.
Edited by Arunima Reghunadhan, Akhina H., and Sabu Thomas.

earthen structures [4]. However, residents of Southwest Asia, the Middle East, and Africa continue to depend on habitations formed of soil. Regardless of the style of walls development, the utilization of binders has a significant impact on the durability of soil constructions. Many different kinds of binders created for building projects have been utilized in buildings, but producing binders also involves the emission of carbon dioxide. Cement manufacture has been identified as a source of around 5% of the world's greenhouse gas emissions necessitating cement usage limits [5]. The agro-resource polymers such as lignin, cellulose, starch, chitosan, gums, and proteins are examples of biopolymers. They are typically made of recyclable polymers and produced by organisms like algae, bacteria, and fungi, by utilizing carbon during development. In numerous disciplines, various biopolymers have been produced for a variety of purposes. The markets for plastics items made from biopolymers has been growing as a substitute for high CO_2 emitting products, especially in light of rising petroleum prices and the prospect of increased CO_2 emissions. There have been many studies attempting to incorporate bioactive biopolymers as an addition or replacement in the construction field. Furthermore, the use of gel-type natural polymers has just recently emerged for soil treatment [6]. This chapter provides a comprehensive overview of the characteristics of the biopolymers that can be used as construction materials, with an emphasis on the various kinds of biopolymers and their potential applications.

11.2 Characteristics of Biopolymers

The advantage of earthen building materials is found in their vast variety and low operational energy. To increase their tensile strength and mechanical performance, chemical stabilizers are frequently utilized. The primary downsides of these products are their significant environmental impact during manufacture, loss of recyclable nature, and potential deterioration in raw soil's hygrothermal capabilities. Traditionally, stabilization was accomplished by incorporating biopolymers, agricultural residues, and various fiber forms into the earthen matrix. Biopolymers are made from living organisms and can have a broad range of characteristics, including being of animal or plant origin, and having hydrophobic, hydrophilic, or amphiphilic properties. The polysaccharides, proteins, lipids, and other complex molecules can all be classified as biopolymers [7]. According to the studies evaluated, the increased amount for liquid enhancers ranges between 1 and 3 wt.% and can increase when the liquid replaces the water addition [8]. This is likely because the soil's low clay component had to be substituted by an auxiliary binder [9]. When stabilized with biopolymers, natural fibers like vegetable glues, excrement, juice, gums, and seagrass fibers can significantly improve the efficiency of the product. For instance, the stabilising procedure can significantly alter density by affecting both mechanical resistance and the thermal stability. Similar to this, porosity has an impact on a material's form as well as subsequently on its physical, hygrothermal, and durability characteristics. Additionally, the porosity has a direct impact on the system's capillary forces and barrier properties. As an illustration, the physical resistance is linked to the material's suction capabilities, which are indicated by the arrangement of its pores [10]. The biopolymers used might alter the porosity dispersion generally or just selectively, altering the material's physical properties. Filling the tiniest pores and enhancing the packing of soil particles with biopolymers can change the distribution of pore sizes. It is not always appropriate to utilize fibers to enhance a material's mechanical characteristics, and each situation calls for a case-by-case analysis of the process underlying the fiber's effectiveness. Major characteristics of biopolymers for civil engineering applications are presented in Figure 11.1. Laborel-Peron et al. reported the nature of the fiber and its structure, which can aid in the bonding with the particles, determine their impact on the mixture. Due to the space formed by inadequate bonding, the clay particles are less linked. The lower compressive strength occurs with increased porosity and less grain-to-grain friction. This impact could be intensified by the fibers expanding as a consequence of water retention during the manufacture of such mixture and the subsequent volume lowering while drying,

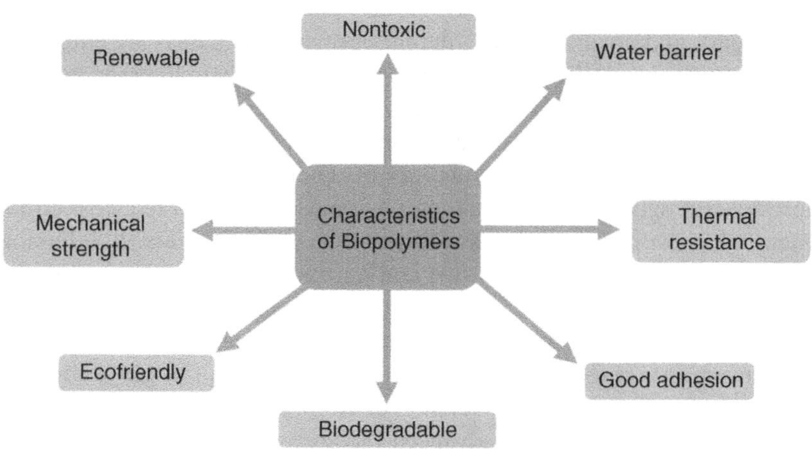

Figure 11.1 Main characteristics of biopolymers for civil engineering applications.

which leads in voids within the mixture [11]. Future research into biopolymer parameters may contribute to determining that different forms of reinforcements have varying characteristics [12].

11.3 Types of Biopolymers Used in Construction

Most of a biological cell is made up of biopolymers, which are typically composed of 90% dry cell content. Monosaccharides, amino acids, and nucleotides are the primary cell monomers that combine to make polymers like lignin, cellulose, starch, chitosan, gums, and proteins. These kinds of biopolymers are widely used in the construction fields.

11.3.1 Lignin

Lignin is a natural polymer that makes up between 20% and 30% of the cell walls in the plants. Water does not dissolve pure lignin. Chemically decomposing lignin yields lignosulfonates (LS), which are frequently used in building. They function as additives and can improve workability and processability by making concrete more flexible. By volume, it is the additive that is most frequently used in non-soil constructions [13]. Sulfate pulping is the primary technique used to produce LSs. In this procedure, dispersed lignin and hemicellulose are broken down using a hot sulfite solution. Sulfur dioxide is usually integrated with a base solution to produce raw liquor, where it develops sulfurous acid that aids in decomposition. It substitutes a sulfonate group for the hydroxyl groups. The used sulfite liquor contains LS, sugars, and organic acids. Different purifying techniques are employed for the different applications. LSs are employed largely to improve the processability and strength of the concrete, among other things. To enable LS to function as water-reducing concrete filler, sugars are eliminated. This aids in improving the mechanical qualities of concrete in addition to its compressive strength. The ethoxylation of hydroxyl groups can reduce retardation, which is necessary for implementation as plasticizers. Reinforcing additives were typically expensive and scarce. They may occasionally be harmful. Its toxicity can be reduced by at least half when used as a binder, and the finished material is more practical and water-resistant. They are employed as a reagent in digging petroleum and natural gas wells that retain the fundamental characteristics of drilling fluids in the oil extraction and oil refining sectors. The advanced LSs are available for reinforcement. As flotation additives, they are employed in the mining industry. They are also use in the chemical industries where they are employed in production of insecticides and other bioactive constituents as a dispersion and suspensions stabilizers.

The use of LS as a substitute for surface chemistry stabilizers is growing daily due to concerns about economic efficiency and sustainability [14]. As soil stabilizers during non-soil constructions, substances such as fly ash, gypsum, alum, lime, etc., are employed. Although these chemical stabilizers significantly increased compressive strength, they also made the soil more brittle and reduced groundwater quality, which polluted the ecosystem. Constructions may undergo seismic instability due to the soil's fragility. LS revealed similar characteristics to chemical stabilizers with minimal to no environmental toxicity. Moreover, LSs are less expensive than conventional chemical stabilizers because they are by-products. As an outcome of surface neutralization and the development of a stable particle cluster, LS-stabilized soil displays decreased double layer thickness, according to research into their behaviors. In this research, LSs are mostly employed to harden silty clay, low plasticized clay, sandy soil, etc. In research study of LS processed with highly plasticized clays and varied concentrations of the LS, found that synergistic interaction of LS and clay increased the soil's strength and lengthened its long-term life [14]. This might be as a result of the enhanced cohesiveness of soil particles. However, the lignin-modified silty clay had major limitations and could not be suitable for all types of uses. They discovered that for lignin to improve the silty clay's characteristics, the ideal moisture concentration should be 12%. In addition, it took four weeks to solidify and could change the physicochemical behavior of the elements in the sands [14]. A combination of LS and superplasticizers or pure plasticizers is indicated to achieve the most desirable performance.

11.3.2 Cellulose

The construction industry chemicals as a whole are entirely dominated by cellulosic materials. It is thought to consume more than 100,000 tonnes of derivatives annually, which, at the manufacturer's level, would be worth about $500 million. Among construction-related uses, carboxymethyl cellulose (CMC) and methyl are the two most common cellulose derivatives. The nonionic cellulose ethers dominate this industry, and have provided the nonionic cellulose derivatives, including their chemical, manufacturing, efficiency, and market research data. However, one such substance is micronized cellulose fibers, which are infrequently employed as a thickening in dry-mix mortar compositions, such as tile adhesives. Very small amounts of cellulose derivatives other than ethers are used. Usually, cellulose materials are used in construction materials to add viscosity and water retention. The usage of cellulose ethers is prevalent throughout the industry, and some possible applications are discussed to emphasize their unique benefits and performance qualities. The primary chemical additive used to promote water retention in dry-mix mortars and grouts is methyl cellulose (MC). Tile adhesives, gypsum, plaster with a cement, or lime base, and joint fillers for wall boards are a several significant applications [15].

Researchers have proven that cement composites with a high aspect ratio of nanocellulose (NC) can permit higher fracture strength. In preliminary experiments, the inclusion of nanofibers increased fracture energy by up to 50%, while the incorporation of a mixture of microcellulose increased fracture energy by 3%. According to recent studies, adding 0.2 vol % of cellulose nanofiber (CNF) to concrete mixture can increase flexural strength by 20% when compared to unreinforced cement paste. Nanoscale fibers are inserted into tiny pores inside concrete to increase its flexibility and tensile strength. The reduced amount of the concrete required, the lower the cost of labour and materials, and the gain from the decrease of greenhouse gas emissions by using NC as a high-performance cement filler are all benefits [16]. Cellulose fibers have a higher mechanical strength than typical fibers. Because they are frequently closely packed and have large aspect ratios, cellulose fibers are effective at stabilising fissures. Given that cellulose fibers have a wide variety of hydrophilic surfaces they aid in integrating cellulose effectively into the encased concrete building and produce in a lighter, consistent microstructure. The incorporation of cellulose fibers into to the cement matrix creates a significant link between the material's chemical and mechanical characteristics. The excellent aspect ratio and tightly woven fibers effectively stabilize crack sites, preventing cracks from forming. The ultra-high efficiency fracture qualities of concrete might be enhanced by the use of cellulose fibers. Additionally reported was the effect of various cellulose reinforcement methods on the processability and fracture behavior of composites. According to the study, adding 5% cellulose reinforcing to cement had an excellent crack-filling performance [17].

11.3.3 Starch

Starch from various sources is made up of one or more glucans. The polysaccharides amylopectin and amylose are the mainly two molecular parts of starch. All green plants contain starch, a type of carbohydrate that consists of the polysaccharide's amylose and amylopectin. Amylopectin is made up of longer, branch polymer chains of α-glucose units as opposed to amylose, which is made up of linear connected polymer chains of glycosidic linkages [18]. Because it is easily biodegradable, recyclable, easily accessible, and inexpensive, starch has been given a lot of consideration in the development of sustainable products. Starch derivatives allow fresh concrete to acquire stable conditions and flow characteristics. As a possible replacement for aggregates, corn starch displayed remarkable qualities. It creates thin concrete with similar shrinkage characteristics to those of conventional concrete.

Although pregelatinized and functionalized starches are frequently employed in constructions, native starch is rarely used. The pregelatinization is a modifying method for starch that significantly improves its cold-water solubility and viscosity yield – two characteristics that are undeniably crucial in the building industry. The method involves heating aqueous native starch slurries, which enlarges the granules and increases viscosity. The resulting viscous gel of pregelatinized starch is then dried to a powder using a drum dryer. High robustness, identical shrinkage, and creep qualities were shown by starch admixtures. Fresh concrete hardens and solidifies as a result of starch adhesion and absorption to its surface. The qualities of the starch-based concrete can be modified by a number of process variables, including temperature, pore volume, morphological, grain fineness, and adherence [19]. They are created by a process called pregelanitization that gives the starch the ability to gel with cold water. They can maintain nearly all of the functional qualities and base material viscosity, which is crucial in construction. Moreover, these biomaterials can be employed as a low-cost, low-temperature fluid loss intermediary for drilling fluids that consist of water. It typically operates best at low temperatures, particularly below 100 °C. Pregelatinized starch 1%–2% is used to achieve an API fluid loss of 2–5 ML at ambient temperature and 7 bar relative pressure. Additionally, it helps in the production of plasterboard. It assists in attaching the cardboard to the gypsum surfaces. Pregelatinized starch is transformed into carboxymethyl starch by raising the temperature. It has a carboxyl group that is negatively charged. The addition of carboxymethyl starch resulted in starch with reduced gelatinization temperature, greater solubility, and adequate storage stability, which altered the starch's ordered structure. Because it is alkaline and can substitute CMC, it is employed as an addition in the drilling of oil wells. Its thermal stability is higher than pregelatinized starch, reaching 120 °C. Different stress levels lead in strain failure in conventional concrete. This occurred from pressures building up in the cement matrix. The interfacial barriers between the substrate and aggregates also are proximal to crack formation. These microcracks assist in triggering creep failure, which spreads as more loads is applied. Continuously applying stress causes shrinkage and creep failure as the resulting strain can surpass the elastic strain barrier of the concrete. Admixtures made of starch performed admirably preventing creep strain. Akindahunsi et al. reported the creep behavior of admixtures containing varying amounts of cassava and maize starch [20]. They created concrete prisms with varied starch contents, ranging from 0% to 2% by weight, and discovered that the starch-based concrete significantly outperforms the regular concrete in terms of the strength and distortion. Additionally, they showed that starch-based concrete had no long-term negative impacts [21].

11.3.4 Chitosan

A naturally produced polymer called chitosan is a substantially or completely deacetylated derivative of chitin. It is a polysaccharide with the structural composition of β-(1→4) 2-acetamido-2-deoxy-D-glucose, some of which has been deacetylated. Chitin is a class of N-acetylglucosamine polymers with varying crystal structures and levels of deacetylation rather than a single polymer with a consistent stoichiometry. By more thorough deacetylation, usually between 50% and 100%, chitin is converted to chitosan. Although chitosan and cellulose are physically similar, chitosan is distinct because it is a polymer that includes primary amine groups.

Ustinova et al. reported that frost durability of the cement mixtures with the chitosan addition has been developed using a cement–sand mortar with a volume ratio of 1:3 and varied quantities of the addition ranging from 0.2% to 2.0%. It expands and opens up pores when there is water present, making cement stones denser. Study on the frost durability of cement mixtures with the addition of chitosan has been carried out [22]. Chitosan is added to cement compositions, which lowers the overall volume of pores and improves the way they are distributed. It is revealed that, when compared to synthetic admixtures, the incorporation of chitosan does not lessen the cement mixtures strength qualities on a polyethylhydrosiloxane basis. It should be incorporated to cement in amounts ranging from 0.6% to 1.0% of cement mass in order to strengthen the mixtures resistance to repeated freezing and thawing. The chitosan produced shear thinning rheological behavior that was comparable to that of microbial biopolymers like as welan and xanthan gums. So that its impacts are stronger and more evenly dispersed throughout the cement matrix, the application of polymers must be followed by the implementation of a larger energy mixture. By monitoring developed hydrates of Portland cement and their distribution, the analysis of the microstructure of the concrete is a useful tool for determining how mechanical characteristics and other qualities will perform. It seems desirable to carry on with studies into the blending of polymers with various properties. Synthetic latex and chitosan were combined with two polymeric additives to improve the physical and durability characteristics of the concrete. Chitosan can be designed to make effective organic–inorganic hybrid ecofriendly materials that have a variety of uses in civil engineering [23].

11.3.5 Gums

The primary substance contained within most seeds is starch or a similar carbohydrate, which the embryonic plants will utilize to fuel its growth. Since early on in our history humans have gathered gums for sustenance from flax, psyllium, locust, and quince seeds. Seed biopolymers like tara, carob, guar, and locust bean gum were found to be excellent thickeners for a variety of industrial uses including explosives, mining, ceramics, textile processing, and oil recoveries. The guar and locust bean gum are the only two seed biopolymers that are commonly used, and construction uses for seed gums other than starch have so far been quite sparse. The legume *Cyamopsis tetragonolobus*, a perennial plant that primarily come from India and Pakistan, produces guar gum in the form of seeds. Although a desirable alternative hydroxypropyl guar (HPG), which is used in dispersion pigments to deliver viscosity and stabilization, is a good viscosifier for aqueous systems, guar gum's utilization in the construction field was minimal until recently [24].

Native guar is also employed in top-hole drilled as a viscosifier to a certain extent in the construction of wells. Here, a water-based liquid phase gets its initial viscosity from the biopolymer. However, from a technological standpoint, the primary environmental benefit of quick recyclability occasionally poses concerns. The biodegradation of guar is expedited in particular since the soil and rock are teeming with microorganisms that contribute to the problem. As a guar stabilizer in this instance, biocides must be introduced to the drilling fluid. Guar gum is also employed as a thickening in trenchless drilling process for the same environmental purpose. Here, mobile equipment is used to drill big holes so that cables or pipelines can be placed beneath rivers or roadways. A typical recipe for guar-based innovative construction drilling fluid includes soda ash, fresh water, and % guar gum that has been pH-adjusted. One class of biopolymers comprises natural terpenes, exudate gums and resins include oil of colophonium, turpentine, and other balsams. They are created in plant cells and function as both an insecticide and a defence towards wounds and dehydration. Without a doubt, the development of lacquers and coating for furniture and wood used in constructions was spurred on by observations of glassy layers made from wood resins on plant trunks. Depending on the producing plant, the content of exudate gums and resins obviously varies. They all have 2-methyl-1,3-butadiene, or isoprene, as a distinctive structural unit in their molecules, despite the fact that they are often complex combinations of many chemicals. Their primary applications in constructions include concrete air entrainment, wood coatings and adhesives [25].

11.3.6 Protein

In proteins, the polymerization step creates lengthy chains of amino acids. Oilseeds that include collagen, keratin, casein, and enzymes also possess protein. They are all employed in industrial environments. The Latin word for casein, which translates to "cheese", refers to the % protein content of cow's milk. It has a pH of 6.6 and is negatively charged. The curd and whey can be made from milk when it turns sour; however, if more water is removed, casein will develop. It can also be obtained from milk by lowering its pH by mixing in certain mineral acids. The quantity of acid used during manufacturing and the fat content in the milk determine the casein's quality. High-quality casein can be produced with a low lipid and acid content. It should be rinsed in water as the last step to get rid of any remaining acids. Casein, the substance found in cheese, is insoluble in water, whereas casein salt is soluble and creates a paste like solutions. Fatehi et al. reported the casein and sodium caseinate's impact on the structure and geotechnical characteristics of the sand. They studied many process variables, such as curing time, which should quick after processing is accomplished; otherwise, it could degrade and emit an undesirable odor [26].

Although there are several synthetic nondegradable paints on the market, casein-based paints are biodegradable, nontoxic, and more affordable. Protein hydrolyzates can be extracted by processing of animal hair, hides or hoofs, blood, etc., notably from cattle, in the presence of sulfuric acid or enzymes. It significantly helps to reduce water's surface tension, making it compatible for the manufacture of foam. Researchers also looked at additive concentrations, additive amount of fat, and biopolymer concentrations, and discovered that higher biopolymer concentrations and curing times produced high compressive strength sand. The ideal temperature for biopolymer-modified sand was determined to be 60 °C. The temperature also has favorable impact on the increase strength. Protein-based biopolymer molecules broke down at temperatures over 120 °C, which decreased the durability of the composite sands. Caseinates, which are made from casein by neutralizing it with alkali hydroxides, are another form of casein that is employed in the paint industry. They are employed there because they have two favorable features for paints. One is the additive's viscous character, and another is adding self-levelling capabilities to produce a coating of paint that is levelled, yet they are both concrete. Foam with a very low density is combined with concrete for this function. These are mostly employed in the construction of roadways as a basement by filling ditches and prefabricated barriers. When compared to hexagonal bubbles made by synthetic foamers, proteins hydrolyzates-produced round bubbles had a 20%–30% better durability [27].

11.4 Applications of Biopolymers in Construction

The construction materials, concrete and dry-mix mortars, are employing the most bioadmixtures globally. Interior and exterior paints and coatings, gypsum wallboards, biobarriers, joint fillers and grouts, biofilters, and well construction are further segments of the construction material field. The different types of biopolymers used in construction and civil engineering are depicted in Table 11.1. Numerous applications have been developed, such as in natural adhesives, foams, sealants, and coatings. The role of main types of biopolymers in construction field is depicted in Figure 11.2.

11.4.1 Biocementation

On the basis of various biogeochemical processes carried out by microorganisms, various types of biocementation are feasible.

Table 11.1 Types of biopolymers utilized in construction sector.

Types of biopolymers	Advantages	Applications	References
Lignin	Improving processability, enhance mechanical strength	Bioconcrete, biocementation	[9, 44]
Cellulose	Recyclability, high density, thickeners	Biofilter, biocoating	[15, 16]
Starch	Nontoxic, low cost, thermal stability	Bioconcrete, bioclogging	[18, 19]
Chitosan	Ecofriendly, good mechanical strength	Biocoating, biofilter	[23, 39]
Gum	Good adhesive, high water retention capacity	Bioadhesive, biobarrier,	[24, 25]
Protein	Biodegradable, good chemical resistance	Biocoating, biofilter	[26, 43]

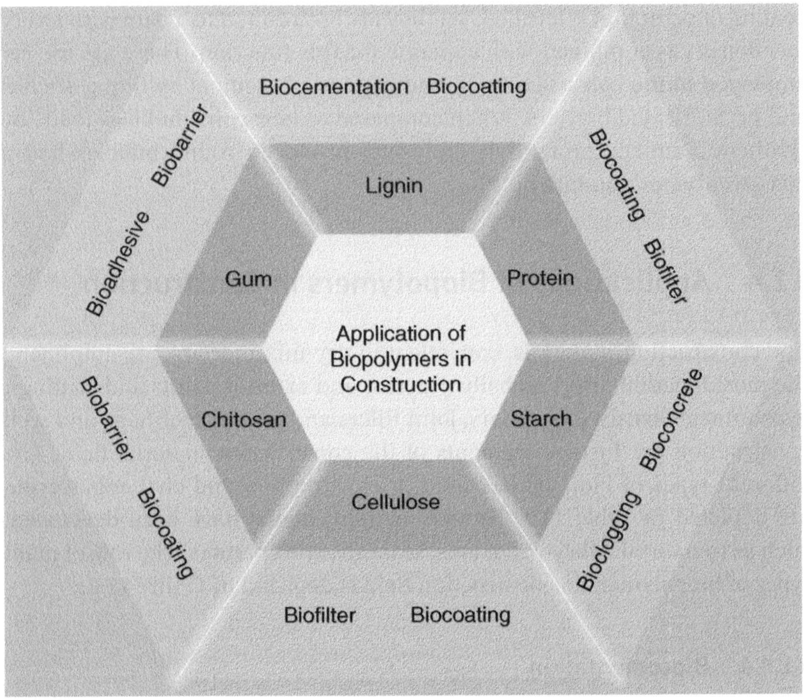

Figure 11.2 Role of various biopolymers as efficient construction materials.

11.4.1.1 Calcium- and Urea-Dependent Biocementation

The most widely used form of biocementation is based on a process known as microbially induced calcium carbonate precipitation (MICCP), which refers to the development of calcium carbonate minerals like as calcite and vaterite on the top of the soil particles and is a function of the attachment of cells of urease-producing bacteria (UPB) on the particles and the hydrolysis of urea by urease of UPB, which leads to the development of a microgradient of carbonate content and pH at the place of cell adhesion. For several geological and civil engineering fields, MICCP is being designed and field-tested [28]. Through the injection, surface spraying, and surface filtration, of calcium and urea mixtures with bacterial suspensions, biocementation can be carried out in bulk. To create biocemented monoliths, specified layers, or crusts on soil surfaces, various biotreatment methods are required. Such biocoating can also be employed to artificially create coral reefs that resemble the appearance of real coral reefs. Because such native coral reefs $CaCO_3$ framework takes a long time to build because of the relatively low calcium level in sea water, and even sometimes over decades, it may be crucial for the regeneration of coral reefs. Aquaculture and recreational marine zones can both benefit from using these artificial coral reefs. The lower viscosity of the mixture used in MICCP biogrouting/biocementation allows for comparatively long-distance insertion via the soil's tiny pores. MICCP must be enhanced or new types of biocementation should be developed. The MICCP has been developed for concrete self-repair [29].

11.4.1.2 Calcium Phosphate Biocementation

The calcium phosphate is precipitated from calcium phytate solution using the phytase activity of microorganisms produces a mixture of the crystal forms like as whitlockite $Ca_9Mg(PO_4)_6(HPO_4)$, hydroxyapatite ($Ca_5(PO_4)_3OH$), and monetite ($CaHPO_4$), all of which have a Ca-to-P weight ratio of 1.55 [30]. Because calcium phytate is poorly soluble, this method of biocementation requires pumping substantial amounts of solution into the soil.

11.4.1.3 Calcium Bicarbonate Biocementation

The precipitate of calcite employing the exclusion of CO_2 from calcium bicarbonate solution may be a significant biocementation approach. Calcium bicarbonate has a reasonably high solubility, around 1M, making biocementation and bioclogging practical. However, because of a rise in pH that prevents calcium carbonate from dissolving, the synthesis of calcium bicarbonate by dissolution of calcium carbonate during in the treatment with CO_2 only yields a low-concentration solution. Although biocementation of sand and rocks with deep fissures must be carried out at greater concentrations of calcium bicarbonate, low concentrations of calcium bicarbonate do not offer a barrier for the bioclogging or biosealing of rocks with thin cracks. The amount of CO_2 elimination from reaction determines the rate of precipitation of calcium carbonate from calcium bicarbonate in the environment. Because of the unstable nature of bicarbonate biocement, it is necessary to generate and store the solution at a high CO_2 pressure [31].

11.4.2 Bioconcrete

The most popular building material created by humankind is bioconcrete. Plank created a good description of current conventional concrete [32]. Concrete made with Portland cement and additional aggregates like steel or aluminium are the most often used building materials because of their affordability, ease of use in applications, and great compressive strength. Additionally, Portland cement and aggregates have certain drawbacks, including weak flexural strength, low chemical stability, barrier characteristics, and high greenhouse gas emissions [33]. Researchers worked to create ecofriendly and lighter construction materials to resolve these problems. The biopolymer concrete and aggregate components with strong structural qualities and rapid infrastructure maintainability are now being used in the construction field. The cement hydrate binders of traditional cement concrete are totally replaced with biopolymer binder or liquid resins in bioconcrete, a type of concrete–polymer composites [34].Similar to cement concrete, liquid resins are utilized as polymeric binders in the creation of bioconcrete, where they are polymerized in situ. The bioconcrete can be made using a variety of liquid resins, including phenol, epoxy, polyester styrene, polyurethane, acrylic, furan, vinyl ester, urea, phenol, and methyl methacrylate, among many others. Hardening is accomplished using hardeners or initiators in addition to liquid resin. The working and stiffening time can be changed and regulated by choosing the right kind of liquid resin and hardener in the proper amount. They were dried during selection because moisture affects with hardening. The strength of the particles affects the strength of bioconcrete. When making bioconcrete, aggregates and binders are employed instead of the typical cement and water. The lowest number of voids in the aggregate must be chosen to avoid conflicts. Silane coupling agents are applied during mixing to strengthen the connection between aggregates and polymeric binders. Different industrial wastes are receiving a great deal of attention now as a bioconcrete aggregate substitute. Recycled materials made of plastic and biomass are employed to develop bioconcrete. For instance, the preparation of the EPS solution for polyester concrete involves melting EPS in vinyl monomers, which serves as a liquid resin and aids in minimizing shrinkage.

Although they have widespread use, Portland cement concretes do have several problems that cannot simply be dismissed. Scientists were forced to develop bioconcrete as Portland cement concrete has internal and manufacturing restrictions. The bioconcrete is stronger, hydrophobic, chemically resistant, abrasive, and durable. However, bioconcretes have poor high thermal stability, are expensive, toxic, flammable, and have limited heat resistance. To avoid this, it is important to note the glass transition temperature of the polymeric phases as soon as they begin to exhibit undesired thermal characteristics. The bioconcrete is used presently all around the world. In the United States, bioconcrete is used for pipes, electrical insulators, underground tunnels, highways, dams, and other structures. It is utilized in the manufacture of goods in Japan, including piles, telephone cable ducts, septic tanks, tunnel liner segments, acid waste storage tanks, gutter covers, acid waste storage tanks, and panels.

11.4.3 Bioclogging

The method of bioclogging soil or porous matrix involves filling in the pores and channels in the material to substantially lower its hydraulic conductivity

[35]. In situ synthesis of microbial polysaccharides in porous soil to change its geological characteristics is another significant use of microbial biopolymers. Microorganisms frequently develop in soil as a biofilm that is adhered to solid surfaces and coated in slimy exopolysaccharides that the cells create. The soil permeability is reduced by this slime. Numerous studies have shown the impact of microbial biomass development on the decline in porosity, permeability, and hydraulic properties caused by bioclogging of the porous materials. The maximum hydraulic conductivity of the sands was reduced significantly by slime developers, based on a comparison study of four bacterial strains that were employed to inoculate quartz sand packed in columns with a regular flow of a matrix. It is common knowledge that practically all bacteria create exopolysaccharides when there are high volumes of carbohydrates or other water-soluble carbon sources compared to a nitrogen supply. Therefore, the commercial manufacture of bacterial water-insoluble polysaccharides uses wastes or subproducts from food manufacturing such as corn glucose syrup and cassava glucose syrup.

Various geological implementations, like as selective zonal environmental remediation, harbor and reservoir control, erosion prospective reductions, earthquake liquefaction prevention, building of responsive barriers, and long-term maintenance of polluted soil, can be accomplished by the growth of exopolysaccharide-producing bacteria in soil. The best types of microorganisms to use to bind soil particles and filled in soil pores are those that create impermeable extracellular polymeric substances. However, there are many more microorganisms that can be employed to bioclog soil in situ: the oligotrophic bacteria of the species *Caulobacter*, the aerobic Gram-negative bacteria of the genera *Acinetobacter*, *Agrobacterium*, *Alcaligenes*, *Arcobacter*, *Cytophaga*, *Flavobacterium*, *Pseudomonas*, and *Rhizobium*, the species of Gram-positive facultative anaerobic and aerobic bacteria, such as *Leuconostoc mesenteroides* producing water-insoluble exopolymer dextran. In anaerobic conditions and in a medium containing organic molecules and nitrate, the latter is capable of producing exopolysaccharides. Application of the concerns and supporting its active bacterium *Arthrobacter* sp. to lower the permeability of the sand column revealed that a bacterial mat formed at the inflow margin of the sand columns was responsible for the quick lowering of saturated hydraulic conductivity up to four orders of magnitude. *Microbacterium arborescens*, a facultative alkaliphile that grows best at a pH of 10.5, was shown to be able to grow in a solution of 12.5% sodium chloride and create significant amounts of a partly soluble exopolysaccharide that could agglomerate sand. The exopolysaccharides are crucial for sand dune stabilization and cell protection against the anaerobic bacteria that make up the majority of the bacterial strains employed to bioclog porous materials. Facultative anaerobic bacteria have become preferable route for media bioclogging due to the development of anaerobic environments in the permeable medium. The implementation of *Paracoccus denitrificans*, a facultative anaerobic bacterium that can create exopolysaccharides and can convert nitrate to nitrogen gas under anaerobic environments, might be the most appealing prospect for bioclogging of the permeable material. These microbes are not harmful to humans, plants, or other living creatures. The activation of native microorganisms can enhance biofilm development. Because they reduce the quality and volume of groundwater, biofilm growth and clogging are significant aspects in the process of producing drinking water hydration. The hydraulic conductivity decreased as an outcome of the formation of a clogging biofilm by exopolysaccharide-producing bacteria feeding the nutrient solution in compressed silty sand. Therefore, in-situ bioclogging of soil or cracks rocks can decrease their hydraulic conductivity and be utilized to minimize drain-channel erosion, create grout curtains to slow the progression of organic pollutants and heavy metals hold back the pipelines of dams and seawalls, as well as avoidable leakages in tunnels [31]. The durability of this form of bioclogging has also been challenged due to the fact that all microbials biopolymers are at least partly dissolved in water and can be removed from soils through groundwater flow. The microbial breakdown of soil exopolysaccharides by actinobacteria and fungi could also contribute to instability.

11.4.4 Biofilters and Biobarriers

The usage of specific biofiltration methods, which are using microbes to decompose putrid complexes to less rotten complexes, is a result of the social and health difficulties associated with the aromas from waste treatment plants. Such methods are commonly treated with desirable microbes to hasten the growth of a fast biofilm and shorten the time it takes for a biofiltration to start it up. These microorganisms can be trapped inside three-dimensional polymeric materials like as alginate beads to neutralize them [36]. The form of substrate surface used for immobilizing anoxic biomass might interfere with a bioreactor's efficiency, and the number of cells adhering to the matrices may change through one medium to another. A large number of organic materials have been used for immobilization. These matrices are available, simple to use, practical from an economic viewpoint, less damaging, and reliable. Several organic combinations have been used at wide scale and in the lab with a permeability barrier. A full-scale biobarrier technology exists that accomplished dispersed benzene toluene ethylenexylene remediation efficiency of greater than 99.9%. Solid organic materials like organic mulch and peat moss had usually been used in bioremediation and bioretention methods to increase the effectiveness of the permeability biobarriers, among other types of composites for use in permeability barriers. Among the several biopolymers found in celluose, lignin is known for its high efficacy and sorption capacity for nonionic organic complexes.

However, in addition to the sorption capacity of the biobarrier's sustaining materials, microbial activity and biofilm growth also have a substantial impact on the way that the permeability biobarrier performs. Nitrogen oxides (NOx) in low concentrations can be remedied using the biofilter treated with isolated nitrifying bacteria cells. According to the research, isolated nitrifying bacteria that used sodium alginate and poly (vinyl alcohol) as trapping matrixes were capable of eliminating up to 80.8% of the NOx. Under various operating circumstances, the cellulose showed the relevant advantages of the method for denitrification [36]. At optimum nitrate content and minimal hydraulic residence duration, 100% natural denitrification of the synthetic biofiltration was achieved. The immobilization of bacteria on the cellulose advanced regulation configuration performance, decreased cell leakage from the beads, increased immobilized cell performance, and enhanced adsorption efficiency. Chitin and chitosan are also used very effective at remove contaminants from petroleum refining industrial wastewater [37]. With the advancement of nanotechnology, biopolymers are becoming increasingly prominent as an odor control solution and as a recyclable substitute in biofiltration technologies. By modifying the microbe immobilization and adsorption in the biofiltration technology, biopolymers have shown that they have good removal capabilities. They could replace the significant usage of chemicals and polymer blends, though potential applications would probably develop in respective markets. They are good potential alternatives to synthetic polymers due to their constructed significant environmental advantages [38]. The construction of these innovative biofilters and biobarriers for large-scale wastewater treatments and application should also achieve a number of essential purposes, including the precise control of different by-products and the minimization of sludge production during the biofiltration process.

11.4.5 Bioadhesives

Bioadhesives are biopolymers that have the property of adhesiveness. The concept is usually used more widely to refer to a synthetic material intended to adhere to biological materials or glue made synthetically from natural monomers like carbohydrates. They can be made of a wide range of materials, but wood adhesive, protein, tannin, and lignin derivatives, as well as carbohydrates, have a significant impact. Although they are not yet in extensive commercial use, other adhesives discovered in nature are currently being developed. For instance, enhancing bioadhesives made by barnacles and mussels has been challenging despite the fact that they have great waterproof adhesive qualities. The adhesives made from biomass have been used extensively for a long time. Petro-based adhesives have been used more than bioadhesives, mostly because of the weak bonding capacities and inadequate resistance to water of most bioadhesives. The petro-based adhesives work well, but because they are made of nonrenewable raw resources and release hazardous byproducts, their uses are limited. As a viable substitute for petro-based adhesives, the development of ecologically friendly adhesives from sustainable biomass has received a great deal of attention in recent times. However, due primarily to their inadequate water resistance, biopolymers are not easily converted into adhesives. As a result, it could be essential to apply a tiny number of petroleum-based additives to achieve the desired qualities. A physically and chemically varied class of natural materials known as polysaccharides includes several items that have sticky qualities. In nature, polysaccharides are a commonly used kind of biopolymer. It is commonly known that these species, including mammals, plants, fungus, and bacteria, existed. These are involved in a variety of biochemical functions, including the storing of energy (starch), the construction of cell walls (cellulose), and cellular interaction (glycosaminoglycans). Such macromolecules have high molecular mass that can reach several million Dalton at times. The neutral sugars and anionic sugars that make up all of these polysaccharides can be combined to form homopolymers.

The polysaccharides that have been researched the most for the production of adhesives are gum arabic, starch, pullulan, chitosan, and certain microbial exopolysaccharides. Most starch- and dextrin-based adhesives are water-based and frequently used to connect paper and cardboard. The dextrin-starch is used extensively as an adhesive because it is more permeable in this form. A bioadhesive system based on mixtures of chitosan and modified starch has been created to reduce the starch adhesive's water solubility. Depending on different experimental configurations, mixtures of these polysaccharides can either produce viscoelastic liquids or hydrogels. For its adhesive qualities and water resilience, chitosan has also been employed on its own or in many polysaccharidic compositions [39]. In addition, these animal/fungal polysaccharides, which are the second most abundant following cellulose, are the one cationic polymer described as having positive charges at acidic pH values. Chitosan can only dissolve in this cationic form. Additionally, starch has been successfully transformed to enhance its adhesive capabilities. Hexa(methoxymethyl)melamine, for instance, has been used to cross-link starch and polyvinyl alcohol, producing an adhesive with improved characteristics. The combination of starch and tannins has also produced a wood adhesive with qualities comparable to those of conventional wood adhesives, which is promising for plywood uses [40]. Seed gums are polysaccharides that could be used in the production of adhesives. Many different forms of adhesives, including pressure-sensitive tapes, dentures and medicinal sealants, paper glues, pharmaceuticals tablet binders, and label pastes, employ gum dispersions like tamarind, locust bean, and guar gums. They exhibit high thermal stability and excellent adhesion. Additionally, locust bean gum shows excellent water-resistant properties. Wheat gluten (WG) has also attracted a lot of interest in this context. Traditionally, the confectionery sector and livestock feed are the main uses for WG, a by-product of the manufacturing of starch. For its distinct viscoelastic and adhesive qualities, WG is a form of elastomeric proteins that may be used in a variety of materials, including wood binders [41]. Gelatin, a naturally occurring water-soluble protein produced from collagen, is one of the most studied materials for tissue adhesives because of its ideal natural characteristics. Gelatin is thought to be immune-inducing, nontoxic, and biodegradable. The physically cross-linked gelatin adhesives have promising properties; however, they cannot act as an adhesive by themselves due to their low mechanical strength.

Ceramic tiles are adhered to walls or set on floors using grouts made of cement that are typically 5 mm thick. They must have adequate viscosity to prevent tile sagging. In countries wherever ceramic tiles are prominent, the market for tile adhesives is vast [42]. Tile adhesives need water-retention chemicals, just like superglue, to stop the water from escaping into the permeable wall or floor. Once more, the cellulose ethers described previously are employed for purpose. In order to offer what is referred to as greener strength and anti-sag characteristics, hydroxypropyl starch is also used. The plasters often call for lower concentrations of cellulose and starch ethers as tile adhesives.

11.4.6 Biocoatings

A macromolecule with a higher molecular weight is a polymer, which includes biopolymers. They are created by the bond formation of small, straightforward repeating units. The quantity of units is so great that changing just a few of these units will not modify the finished features of polymers. Polymers are classified according to the source, mechanism of creation, thermal responsiveness, and primary linkages by which they developed, physical characteristics, and type of monomer unit. The biopolymers are renewable because they originate from natural sources. When certain enzymes or catalysts are either present or absent, biological processes produce certain polymers. Biodegradation is the decomposition of substances into products that are harmless to the ecosystem, like carbon dioxide, water, and biomass, through the operation of natural sources microbes in a normal environment. Therefore, biopolymers employed in coatings and paints are often biodegradable polymers that are obtained from natural sources. This form of biopolymer is consequently more advantageous as a binder for paints and protective coatings.

However, cellulose is challenging to implement in paints and coatings owing to the inclusion of strong H-bonding, its crystalline form, large molecular mass, and hardness. As a result, cellulose is insoluble and inert. Thus, the hydroxyl groups in cellulose's structures must be altered into the suitable ester, ether, and acetal derivatives. The numerous cellulose derivatives, like as CMC, hydroxyl methyl propyl cellulose, hydroxyl ethyl methyl cellulose, hydroxyl ethyl cellulose, and others, are used as thickening agents for coatings and paints barrier protection colloids in emulsified paints, and film formers in hydrophobic materials. These derivatives come in a range of grades depending on their molecular mass, level of alteration, and water dispersion capability. In a variety of industries, such as water-based paints and construction additives, cellulose materials are becoming increasingly significant. A well-known hydrocolloid biopolymer is starch. It has an amorphous, branching amylopectin as well as linear, crystalline amylose. This is one of the most affordable biopolymers and a low-cost polysaccharide that is widely accessible. Hot water with starch mixed in it is used to thickening, stiffen, or glue coatings and paints. The materials manufactured from starch have dimensional stability characteristics, are brittle, and are sensitive to water. Acetylation of starch with acetic acid and pyridine helps to solve such difficulties. Gelatin is a transparent, odorless, fragile, and tasteless substance. Thus, it is a collagen product that has been permanently hydrolyzed. Dimethyl formamide is used to pulverize the dry gelatin to produce a hydrophilic, water-soluble product with a considerably large jelly consistency. Pulverized paints with gelatin particles and synthetic resin mixtures in dimethyl formamide are 30 mm or less in size. By eluting gelatin particles with hot water, it is possible to create paint layers with great water resistance. Chitosan-based polyurethane (PU) is used to create the self-healing coating, which is a highly desirable surface coating. This kind of coating could minimize the amount of maintenance and repairs on a range of products, costing consumers less and producing less waste. The PU utilized in exterior coatings to shield the surfaces of coated objects includes chitosan as one of its components. Chitosan reacts to the UV component of the employed sunlight, which damages the chemical composition of the polymer when a scratch forms on the outermost layer [43]. The other components in the coating mixture are then used to construct chemical chains, which probably cover the scratch. This method can produce a "scratch-free surfaces" in less than an hour and is incredibly effective.

11.5 Conclusion and Future Perspective

The demand for more ecofriendly polymers has emerged, and research to develop a wide range of biobased polymeric materials has been extensively explored. Depending on the specific polymerization processes used, biopolymers can possess a range of molecular structures. It is possible to create tailored biopolymers with the required strength or compressibility, inherent biodegradability, and minimal CO_2 generation during manufacturing. They have been considered a promising substitute for petrochemical products, which have rigorous environment protection regulations, and which currently represent roughly 60% of the worldwide biopolymer market. The innovation and manufacturing of biopolymers is being led by major international firms in the fields of chemical engineering and product manufacturing. Recently, a number of top businesses came to a deal to develop ecofriendly biopolymers, and they unveiled standards and a certification programme for biodegradable materials. Consumers can currently discover much about the product standards from them. This collaboration and the firms' joint interaction have increased interest for and enhanced the dependability of products made from biopolymers. Since 2010, the market for biopolymers has increased by more than 23% year on year. These substances are presently used in a variety of industries, including construction, medicine, food, and cosmetics. In the future, it is anticipated that this market expansion pattern will be maintained. The economic viability of biopolymers in contrast to cement has not yet occurred but given the trend of scientific advancements in the construction field, it is very possible that a demand for biomaterials for construction will emerge. With the increased biopolymer recovery ratio, the development and research of carbon free resources, the commercialization and mass manufacturing of biopolymers especially for construction industry, and other aspects, significant cost savings are projected. These developments will make it possible for nations with higher cement costs to receive locally produced construction adhesives at a lower price. Moreover, as the cost of concrete is higher in less developed nations, where the primary carbon resources for the manufacturing of biomaterials are cheaper, the local marketing of such biopolymers could aid in improving the strength and lifespan of geotechnical engineering in those regions.

References

1 Pacheco-Torgal, F. (2016). *Introduction to Biopolymers and Biotech Admixtures for Eco-efficient Construction Materials*, no. 2010. Amsterdam: Elsevier.

2 Clark, J.H. (2006). Green chemistry: today (and tomorrow). *Green Chemistry* 8 (1): 17–21. doi: 10.1039/b516637n.

3 Revuelta-Acosta, J.D., Garcia-Diaz, A., Soto-Zarazua, G.M., and Rico-Garcia, E. (2010). Adobe as a sustainable material: a thermal performance. *Journal of Applied Sciences* 10 (19): 2211–2216.

4 Maheri, M.R., Naeim, F., and Mehrain, M. (2005). Performance of adobe residential buildings in the 2003 Bam, Iran, earthquake. *Earthquake Spectra* 21 (SUPPL. 1): 337–344. doi: 10.1193/1.2098861.

5 Chang, I., Prasidhi, A.K., Im, J., and Cho, G.C. (2015). Soil strengthening using thermo-gelation biopolymers. *Construction and Building Materials* 77: 430–438. doi: 10.1016/j.conbuildmat.2014.12.116.

6 Chang, I., Jeon, M., and Cho, G.C. (2015). Application of microbial biopolymers as an alternative construction binder for earth buildings in underdeveloped countries. *International Journal of Polymer Science* 2015. doi: 10.1155/2015/326745.

7 Achenza, M. and Fenu, L. (2006). On earth stabilization with natural polymers for earth masonry construction. *Materials and Structures* 39 (1): 21–27. doi: 10.1617/s11527-005-9000-0.

8 Galán-Marín, C., Rivera-Gómez, C., and Petric, J. (2010). Clay-based composite stabilized with natural polymer and fiber. *Construction and Building Materials* 24 (8): 1462–1468. doi: 10.1016/j.conbuildmat.2010.01.008.

9 Zhang, T., Cai, G., and Liu, S. (2017). Application of lignin-based by-product stabilized silty soil in highway subgrade: a field investigation. *Journal of Cleaner Production* 142: 4243–4257. doi: 10.1016/j.jclepro.2016.12.002.

10 Chauhan, P., El Hajjar, A., Prime, N., and Plé, O. (2019). Unsaturated behavior of rammed earth: experimentation towards numerical modelling. *Construction and Building Materials* 227: 116646. doi: 10.1016/j.conbuildmat.2019.08.027.

11 Laborel-Préneron, A., Aubert, J.E., Magniont, C. et al. (2016). Plant aggregates and fibers in earth construction materials: a review. *Construction and Building Materials* 111: 719–734. doi: 10.1016/j.conbuildmat.2016.02.119.

12 Losini, A.E., Grillet, A.C., Bellotto, M. et al. (2021). Natural additives and biopolymers for raw earth construction stabilization – a review. *Construction and Building Materials* 304 (July): 124507. doi: 10.1016/j.conbuildmat.2021.124507.

13 Coban, O., Bora, M., Kutluk, T., and Özkoç, G. (2017). Mechanical and thermal properties of volcanic particle filled PLA/PBAT composites. *Polymer Composites* 16 (2): 101–113. doi: 10.1002/pc.24393.

14 Li, Y., Zhang, Y., Ceylan, H., and Kim, S. (2019). Laboratory evaluation of silty soils stabilized with lignosulfonate. *International Conference on Airfield and Highway Pavements*, Chicago, Illinois (21–24 July 2019), 531–540. American Society of Civil Engineers.

15 Reshmy, R., Philip, E., Paul, S.A., et al. (2020). Nanocellulose-based products for sustainable applications-recent trends and possibilities. *Reviews in Environmental Science and Biotechnology* 19 (4): 779–806. Springer Science and Business Media B.V. doi: 10.1007/s11157-020-09551-z.

16 Akhlaghi, M.A., Bagherpour, R., and Kalhori, H. (2020). Application of bacterial nanocellulose fibers as reinforcement in cement composites. *Construction and Building Materials* 241: 118061. doi: 10.1016/j.conbuildmat.2020.118061.

17 Peters, S.J., Rushing, T.S., Landis, E.N., and Cummins, T.K. (2010). Nanocellulose and microcellulose fibers for concrete. *Transportation Research Record* 2142: 25–28. doi: 10.3141/2142-04.

18 Datta, C. (2011). Starch as a biopolymer in construction and civil engineering. In: *Handbook of Bioplastics and Biocomposites Engineering Applications*, 317–344. doi: 10.1002/9781118203699.ch11.

19 Akindahunsi, A.A. and Uzoegbo, H.C. (2015). Strength and durability properties of concrete with starch admixture. *International Journal of Concrete Structures and Materials* 9 (3): 323–335. doi: 10.1007/s40069-015-0103-x.

20 Akindahunsi, A.A. (2019). Investigation into the use of extracted starch from cassava and maize as admixture on the creep of concrete. *Construction and Building Materials* 214: 659–667. doi: 10.1016/j.conbuildmat.2019.04.110.

21 Tusnim, J., Hoque, M.E., and Biswas, M.C. (2020). *Biopolymers in Building Materials*. Elsevier Inc.

22 Ustinova, Y.V. and Nikiforova, T.P. (2016). Cement compositions with the chitosan additive. *Procedia Engineering* 153: 810–815. doi: 10.1016/j.proeng.2016.08.247.

23 Bezerra, U.T., Ferreira, R.M., and Castro-Gomes, J.P. (2011). The effect of latex and chitosan biopolymer on concrete properties and performance. *Key Engineering Materials* 466 (May) 2014: 37–46. doi: 10.4028/.www.scientific.net/KEM.466.37

24 Chang, I., Im, J., Prasidhi, A.K., and Cho, G.C. (2015). Effects of Xanthan gum biopolymer on soil strengthening. *Construction and Building Materials* 74: 65–72. doi: 10.1016/j.conbuildmat.2014.10.026.

25 Pei, Y., Wang, L., Tang, K., and Kaplan, D.L. (2021). Biopolymer nanoscale assemblies as building blocks for new materials: a review. *Advanced Functional Materials* 31 (15): 1–30. doi: 10.1002/adfm.202008552.

26 Fatehi, H., Abtahi, S.M., Hashemolhosseini, H., and Hejazi, S.M. (2018). A novel study on using protein-based biopolymers in soil strengthening.

Construction and Building Materials 167: 813–821. doi: 10.1016/j.conbuildmat.2018.02.028.

27 Chevali, V. and Kandare, E. (2016). Rigid biofoam composites as eco-efficient construction materials. In: *Biopolymers and Biotech Admixtures for Eco-Efficient Construction* (ed. F. Pacheco-Torgal, V. Ivanov, N. Karak, and H. Jonkers), 275–304. Amsterdam: Elsevier.

28 Dejong, J.T., Soga, K., Kavazanjian, E., et al. (2013). Biogeochemical processes and geotechnical applications: progress, opportunities, and challenges. *Bio- and Chemo- Mechanical Processes in Geotechnical Engineering: Geotechnique Symposium in Print* 2013 (4): 143–157. doi: 10.1680/bcmpge.60531.014.

29 Pacheco-Torgal, F. and Labrincha, J.A. (2013). Biotech cementitious materials: some aspects of an innovative approach for concrete with enhanced durability. *Construction and Building Materials* 40: 1136–1141. doi: 10.1016/j.conbuildmat.2012.09.080.

30 Roeselers, G. and van Loosdrecht, M.C.M. (2010). Microbial phytase-induced calcium-phosphate precipitation – a potential soil stabilization method. *Folia Microbiologica (Praha)* 55 (6): 621–624. doi: 10.1007/s12223-010-0099-1.

31 Ivanov, V. and Stabnikov, V. (2016). Basic concepts on biopolymers and biotechnological admixtures for eco-efficient construction materials. In: *Biopolymers and Biotech Admixtures for Eco-Efficient Construction* (ed. F. Pacheco-Torgal, V. Ivanov, N. Karak, and H. Jonkers), 13–35. Amsterdam: Elsevier.

32 Plank, J. (2004). Applications of biopolymers and other biotechnological products in building materials. *Applied Microbiology and Biotechnology* 66 (1): 1–9. doi: 10.1007/s00253-004-1714-3.

33 Dosch, K. (2018). Resource efficiency in the building sector. In: *Factor X. Eco-Efficiency in Industry and Science*, vol. 32 (ed. H. Lehmann), 297–303. Springer, Cham. doi: 10.1007/978-3-319-50079-9_19.

34 Bulut, H.A. and Şahin, R. (2017). A study on mechanical properties of polymer concrete containing electronic plastic waste. *Composite Structures* 178 (June): 50–62. doi: 10.1016/j.compstruct.2017.06.058.

35 Chu, J. and Ivanov, V. (2014). Iron- and calcium-based biogrouts for soil improvement. *Geo-Congress 2014 Technical Papers.* 167: 1596–1601. doi: 10.1061/9780784413272.156.

36 Li, W., Lee, C., Wilkes, M. et al. (2014). Development of a rapid, effective method for seeding biofiltration systems using alginate bead-immobilized cells. *SCRIBD.* 5 (1): 1–4.

37 Ashraf, M.A., Maah, M.J., and Yusoff, I. (2012). Poly vinyl acetate (PVA) as fill material for land reclamation. *Chiang Mai Journal of Science* 39 (4): 693–711.

38 Ashraf, M.A., Batool, S., Ahmad, M. et al. (2016). Biopolymers as biofilters and biobarriers. In: *Biopolymers and Biotech Admixtures for Eco-Efficient Construction* (ed. F. Pacheco-Torgal, V. Ivanov, N. Karak, and H. Jonkers), 387–420. Amsterdam: Elsevier.

39 Mati-Baouche, N., Elchinger, P.H., De Baynast, H. et al. (2014). Chitosan as an adhesive. *European Polymer Journal* 60: 198–212. doi: 10.1016/j.eurpolymj.2014.09.008.

40 Moubarik, A., Charrier, B., Allal, A. et al. (2010). Development and optimization of a new formaldehyde-free cornstarch and tanninwood adhesive. *European Journal of Wood and Wood Products* 68 (2): 167–177. doi: 10.1007/s00107-009-0357-6.

41 Nordqvist, P., Lawther, M., Malmström, E., and Khabbaz, F. (2012). Adhesive properties of wheat gluten after enzymatic hydrolysis or heat treatment – a comparative study. *Industrial Crops and Products* 38 (1): 139–145. doi: 10.1016/j.indcrop.2012.01.021.

42 Mathias, J.D., Grédiac, M., and Michaud, P. (2016). Bio-based adhesives. In: *Biopolymers and Biotech Admixtures for Eco-Efficient Construction* (ed. F. Pacheco-Torgal, V. Ivanov, N. Karak, and H. Jonkers), 369–385. Amsterdam: Elsevier.

43 Karak, N. (2016). Biopolymers for paints and surface coatings. In: *Biopolymers and Biotech Admixtures for Eco-Efficient Construction* (ed. F. Pacheco-Torgal, V. Ivanov, N. Karak, and H. Jonkers), 333–368. Amsterdam: Elsevier.

44 Cai, G., Zhang, T., Liu, S. et al. (2016). Stabilization mechanism and effect evaluation of stabilized silt with lignin based on laboratory data. *Marine Georesources & Geotechnology* 34 (4): 331–340. doi: 10.1080/1064119X.2014.966217.

12

Biopolymers and Functional Biopolymers in Food Technology

Sanem Argin[1,*] *and Y. Martin Lo*[2]

[1] *Department of Food Engineering, Yeditepe University, Istanbul, Turkey*
[2] *Institute for Advanced Study, Shenzhen University, Shenzhen, China*
* *Corresponding author*

12.1 Food Hydrocolloids and Their Applications in Food Technology

Food hydrocolloids (food gums) are biopolymers obtained from plants, animals, and bacteria [1]. They swell by absorption of water or in some cases form a network structure, known as gels. Food hydrocolloids are added to food systems to modify texture, increase stability, and reduce the fat or calories of a product [1, 2]. They are used to thicken, gel, control syneresis, stabilize, emulsify or to aid the process (Table 12.1). Food hydrocolloids are also used in encapsulation of food ingredients and in gluten-free formulations to replace the function of gluten.

Xanthan gum, carrageenan, guar gum, locust bean gum, pectin, alginate, gum arabic, and gelatin are the most commonly used hydrocolloids in food systems. Their molecular weight, monomer type, the functional groups in their structure, type, number, and distribution of side chains, define their properties and their thickening and gelling ability. Polysaccharide-based food hydrocolloids either carry negative charge or neutral charge, which affects their function. On the other hand, protein-based hydrocolloids such as gelatin, can carry either negative or positive charge depending on the pH of the system.

Applications of Biopolymers in Science, Biotechnology, and Engineering, First Edition.
Edited by Arunima Reghunadhan, Akhina H., and Sabu Thomas.

Table 12.1 Food applications of hydrocolloids [1–6].

Biopolymer	Product	Function
Xanthan gum	Ice cream	Stabilizer, thickens and controls crystallization of mix
	Baked goods	Stabilizer, facilitates pumping
	Fruit gels	Adds body, prevents sticking
	Pizza sauces	Thickener, inhibits absorption by dough
	Microwave cakes	Stabilizer, helps moisture retention
	Beverages	Stabilizer, modifies mouthfeel
	Salad dressings	Stabilizer, suspending agent, emulsifier
	Dry mixes	Eases dispersion in hot or cold water, thickener
	Frozen foods	Improves freeze-thaw stability
	Syrups, toppings, relishes	Thickener, heat stability, uniform viscosity
	Cheese	Prevents syneresis
	Juice drinks	Suspending agent
	Gravies and sauces	Thickener, mouthfeel, flavor release, heat resistance
Guar gum	Ice cream, ice pops, sherbet	Improves body and texture, prevent unwanted ice crystals in freeze-thaw cycle
	Cold pack cheese	Prevent syneresis
	Pie fillings	Thickener, prevent syneresis
	Icings	Absorb free water
	Cake, donut mixes	Reduces mixing time, increases shelf life but producing moist crumb, increases water holding property
	Meat and gravy sauces	Allows reduction of the total solids
	Canned meat products	Prevent fat migration during storage, control syneresis
	Dietetic beverages and low carb products	Suspending ability, improving body

Table 12.1 (Continued)

Biopolymer	Product	Function
Alginate	Bakery creams	Instant gelling and thickening, heat stability, range of different textures, good mouthfeel, and flavor release
	Dressings	Thickening, stabilizing, emulsifying, good mouthfeel, acid stability
	Fruit juices	Stabilizer, emulsifier
	Fruit fillings and preparations	Gelling, thickening, stabilizing, prevents syneresis, excellent heat stability, range of different textures
	Ice cream and sorbet	Stabilizer, control of viscosity, prevents crystal formation and shrinkage, contributes to even and slow melt down
	Low fat spreads	Stabilizer, good mouthfeel, texture, and flavor release
	Restructured foods (e.g., onion rings, artificial fruits etc.)	Gelling, heat stability
	Yogurt	Stabilizer, good mouthfeel, texture, and flavor release
Carrageenan (κ-:kappa, ι-: iota, λ-:lambda)		
κ-&ι-, κ-&ι-&LBG	Dessert gels and low-calorie jellies	Gelling
κ-&LBG	Canned pet foods	Fat stabilization, thickening, suspension, gelling
κ-,λ-	Syrups	Suspension, improves body
κ-	Frozen concentrates	Pulping effect
κ-	Relishes and sauces	Improves body
κ-	Ice cream, ice milk	Prevents serum (whey) separation, controls meltdown
κ-, κ-&ι-	Pasteurized milk products	Suspension, improve body, emulsion stabilization
κ-, κ-&ι-	Custards	Gelling

(Continued)

Table 12.1 (Continued)

Biopolymer	Product	Function
λ-	Whipped cream	Stabilizes overrun
κ-&LBG	Yogurt	Improve body, suspension of fruits
Locust bean gum (LBG, carob gum)	Ice cream	Stabilizer, controls meltdown (κ-carrageenan needs to be added if LBG causes serum separation in the ice cream mix)
	Cultured dairy products	Prevents syneresis, improves spreadability, provides texture (with other gums).
	Soups, sauces, ready-to-eat meals	Provides texture and to improve freeze thaw stability (with other gums)
	Processed meat and fish products	Gelling (with carrageenan), water binder, provides texture
Pectin	Jams, jellies, desserts	Gelling
	Bakery fillings and toppings	Gelling
	Acidified milk and protein drinks	Thickening
Gelatin	Gelatin desserts	Gelling, improves texture
	Yoghurt	Prevents syneresis, thickening, gelling
	Aerated desserts	Stabilizer, improves texture, aeration
	Puddings and creams	Improves texture, thickening, gelling
	Soft and melted cheese	Stabilizer, improves texture
	Ice cream	Stabilizer, improves texture
	Icings	Thickening, gelling
	Wine and juice fining	Clarification
	Canned meat	Improves texture
	Pâtés	Stabilizer
	Frozen cooked meat	Binder
	Low fat spreads	Stabilizer, improves mouthfeel

Food hydrocolloids are used as thickening agents in canned foods, gravies, sauces, soups, dressings, low calorie foods, and beverages. When used in dry mix beverages, they provide instant viscosity once water is added to prepare the drink. They are used as suspending agents in salad dressings, fruit juice drinks, and chocolate milk. They stabilize emulsion type products such as salad dressings and cream sauces by preventing phase separation. They play a role in protein stabilization in products such as yogurt drinks, low pH milk beverages, and chocolate milk. Food hydrocolloids are used as gelling agents in jams, jellies, preserves, dessert gels, and canned pet food. They help controlling syneresis in sauces, low calorie preserves, pie filling, and desserts. They inhibit crystallization in confectionary and in ice cream up to several freeze/thaw cycles. In baked goods, icings and candy, food hydrocolloids are used as water binding agents. They also have film forming and fat holdout ability, which find applications in french fries, batter, and breaded foods.

In order to understand how food hydrocolloids provide all these functions, the structure and the nature of different hydrocolloids have to be well understood. Without knowing the structure and properties it is difficult to choose the correct hydrocolloid for a specific food application.

12.2 Biopolymers as Thickening Agents, Gelling Agents, and Processing Aids

Food hydrocolloids are excellent thickening and gelling agents that are used in many different food formulations to give the desired texture and mouthfeel (Table 12.1). Although there are many different types of hydrocolloids, the common property they posses is that they absorb water and swell. When they are dissolved in suspensions, the polymer chains occupy a certain amount of space in the liquid and increase the viscosity, i.e., resistance to flow, of the food solution. The increase in viscosity depends on the molecular weight (MW) of the hydrocolloid. As the MW increases the amount of gum required to obtain a certain viscosity decreases [7]. All hydrocolloids alter the rheological properties of a food system in a desired and/or prescribed fashion. Not only do they thicken the product by increasing the viscosity, but also change the flow behavior [8]. Some of the hydrocolloids, such as xanthan gum [9], guar gum [6], locust bean gum [10], alginate [11], and pectin [12] have shear thinning (pseudoplastic) rheological characteristics. This means that, once shear is applied, the viscosity decreases and when shear is removed the initial viscosity is recovered [13]. This property, for example, makes xanthan gum a very important ingredient in salad dressings [14]. In the formulations with xanthan gum, when the salad dressing is poured and then the bottle is turned back to its original position, unlike olive oil or such, the dressing does not leak from the wall of the bottle, which keeps the outside of the bottle clean. The underlying reason is that while the salad dressing is poured, the viscosity decreased due to the gravitational force and once the bottle turned back, it is recovered (shear thinning behavior), which allows the dressing to be collected into the bottle without leaking. Pseudoplastic behavior also eases the processes by facilitating the pumping of viscous products such as baked goods.

While all hydrocolloids thicken the food formulations, not all are capable of forming gels. Gels are network structures composed of many polymeric chains. When polymer chains come closer, linkages are formed between them through many physical bonds so that they trap the water inside, thus a liquid (solution) is converted into a solid (gel). Among aforementioned hydrocolloids, pectin, alginate, iota-carrageenan, kappa-carrageenan, and gelatin can form true gels.

Pectin is obtained from cell walls of plants, commonly from citrus and apple [15]. Pectin is used in gel formation of jams, jellies, desserts, bakery fillings, toppings, and fruit preparations for yogurts [16]. The backbone is composed of smooth regions (homogalacturonic acid) and hairy regions are composed of blocks of neutral sugar molecules such as rhamnose, galactose, arabinose etc. High methylester pectins (HM) gel with sugar-acid gelling mechanism, which requires a soluble solid (sugar) content higher than 60% and a pH range of 2.8–3.6 (optimum) [15]. For this reason, in high sugar jams, HM pectins are used. On the other hand, low methylester pectins (LM) gel with both sugar-acid gelling mechanism and egg-box gelling model [12]. Thus, LM pectins gel in a broader pH range. In the egg-box gelling model, Ca^{2+} ions added to the formulation or already found in the food system react with de-esterified galactronic units [17]. Thus, low degree of methyl esterification enhances gelling.

Alginate and carrageenan are extracted from cell wall constituents of brown seaweed and red seaweed, respectively [3]. It is also possible to produce alginate from *Azotobacter vinelandii* by fermentation. Alginate is a heteropolysaccharide composed of blocks of mannuronic acid (M) and gluronic acid (G) [18]. Alginate gels in the presence of divalent cations such as Ca^{2+} [19]. Calcium ions bind to G-blocks of alginate to form gels. Thus, the M/G ratio is important. High G content in the structure forms strong but at the same time brittle gels, which are more prone to syneresis on freeze/thaw cycles [20]. On the other hand, weaker and more elastic gels can be achieved by high M content in the structure, which has good freeze–thaw stability. Alginate exhibits shear thinning behavior when in solution.

Carrageenan has a linear backbone of repeating galactose units. It is naturally sulfated and negatively charged [21]. Only kappa- and iota-carrageenan can form gels where as lambda-carrageenan can only be used as a thickener with pseudoplastic behavior. The preferred ion for gelling is K^+ for kappa-, and Ca^{2+} for iota-carrageenan [21, 22]. The 3,6 anhydrobridges in the structure of kappa- and iota-carrageenan play a crucial role in the gel formation. Cohesive/elastic gels and firm/brittle gels can be formed by iota- and kappa-carrageenan, respectively. Carrageenan is the most preferred hydrocolloid in milk-based products (Table 12.1). The underlying reason is that carrageenan can incorporate casein micelles into its structure, thus the amount of carrageenan used to form a gel network is less compared to other types of hydrocolloids, which makes carrageenan very cost effective in the dairy industry [3].

Different from the hydrocolloids discussed so far (all are polysaccharides), gelatin is a protein produced by hydrolysis of collagen, which is the major constituent of skin, bones, and connective tissues. Sources of gelatin (bovine, poultry, pork, and fish) and extraction methods (acid (A type) or base (B type)) may vary, which affects the properties of the polymer [23]. Gelatin easily forms 100% thermoreversible gels when cooled below 30 °C and gel forming does not require addition of any other substances (cations, sugar etc.), as required in the case of other gelling hydrocolloids. It is widely used in confectionery (gummy bears, marshmallows, liquorice) and desserts as a gelling agent (yogurt, puddings, creams, icings) [24]. In the wine and fruit juice industries gelatin is used as a processing aid for clarification purposes. Positively charged gelatin removes the negatively charged fragmented pectin chains and tannins to produce clear juice. Type A gelatin is more preferred compared to type B gelatin, since type A gelatin has an isoelectric point (IEP) around 8.5–9.0, which makes it more positively charged at the pH of fruit juice, which is around 3.0–4.0 [23, 25, 26].

Locust bean gum (LBG) and guar gum (galactomannas) are the seed gums. Both are composed of mannose backbone and galactose side units. What differs between them is the mannose:galactose ratio and the distribution of the side chains [27]. Both do not form gels, but LBG can form strong synergistic interactions with other gums owing to its low galactose content and the uneven distribution of galactose units. Although both xanthan gum and LBG cannot form true gels on their own, when used together, they form thermoreversible elastic gels that are insensitive to cations [28–31]. One example to this synergistic interaction is the gelling system in desserts and mousses. The desired texture and shelf-life stability of water jellies are also achieved by mixing LBG with xanthan gum or carrageenan.

12.3 Biopolymers as Stabilizers and Emulsifiers

As can be seen in Table 12.1, biopolymers can stabilize emulsions/suspensions, increase freeze–thaw stability and prevent syneresis in many different types of products. For example, xanthan gum contributes to even distribution of spices and herbs in salad dressings; LBG in combination with other gums (carrageenan and/or guar gum and/or CMC) control the growth of sugar and ice crystals in ice cream [32]; and whey protein isolate and gum arabic stabilize nanoemulsions in beverages [33]. LBG in combination with xanthan gum, guar gum and carrageenan prevent syneresis and improve spreadability in cultured dairy products [31].

Another good example of biopolymers used as stabilizers is the antifreeze proteins (AFPs) added to low fat products. AFPs are glycoproteins found in animals, plants and bacteria that can survive in subzero environments such as arctic fish and wood frogs. Antifreeze proteins bind to small ice crystals to inhibit ice growth and recrystallization of ice [34]. The polar amino acid threonine in the structure of AFP, binds to ice crystals through H-bonding and converts the flat ice surfaces into highly curved forms. Crystallization on curved surfaces requires a greater driving force (i.e., lower temperature) than on flat surfaces, thus the nuclei stop growing [35, 36]. AFPs find applications in low fat ice creams, popsicles and frozen yogurt products to control the ice crystal growth and give a smooth texture to the product [37, 38].

12.4 Biopolymers Used in Functional/Dietetic Products and Gluten-Free Formulations

As the consumer's awareness of the relationship between food and health, as well as food and diseases such as high blood pressure, cholesterol, obesity, cancer, and cardiovascular problems, increase, the market share of functional foods worldwide also increased [39]. While functional foods and beverages are formulated, either some functional ingredients such as vitamins or dietary fibers are added or less healthy ingredients such as sugar and fat are reduced. Biopolymers play a role in functional foods in two ways: either as a healthy ingredient or as an ingredient that gives the desired texture to the product. Dietary fibers are a good example of the biopolymers with functional properties [1, 40, 41]. They are polysaccharides derived from plants that are not digested and absorbed in the upper gastrointestinal tract (i.e., stomach and small intestines) of human [42]. Dietary fibers are added to food products for weight management, due to their preventive effects against bowel cancer and for their prebiotic nature. One other example of functional biopolymers is the whey protein, a byproduct of cheese manufacturing, which is added to sports drinks to increase the protein content of the product.

On the other hand, some biopolymers are added to the functional products for practical reasons. The reduction of ingredients, especially carbohydrates and fat, affects the texture, viscosity, and mouthfeel of the healthy products. To overcome these problems, biopolymers are used in dietetic beverages and low carb products for their suspending and stabilization ability as well as their ability to improve the body. For example, in functional nondairy beverages, thickeners such as LBG or guar gum are added to give body to the product [27, 43]. These two hydrocolloids are also used in combination with pectin in acidified diary beverages such as yogurt drinks and fruit juice containing milks, and with carrageenan in neutral dairy beverages to increase the viscosity [21, 44]. As discussed in the previous subsection, antifreeze proteins stabilize low fat frozen products by inhibiting the ice growth resulting from the high water content. Biopolymers can also act as emulsifiers in vitamin fortified beverages. Whey protein isolate and gum arabic has been shown to stabilize nanoemulsion-based vitamin E delivery systems [33]. Gelatin is used to stabilize the emulsion in low fat spreads.

In baked goods, gluten is responsible for the viscoelastic properties of dough and creates the network structure that holds the gas inside. Low gas holding capacity of gluten free formulations causes low volume and weak structure [45–47]. In gluten-free products, hydrocolloids play a major role mimicking the network formed by gluten [46]. Xanthan gum, guar gum, locust bean gum, κ-carrageenan, pectin, hydroxylpropyl methyl cellulose (HPMC), and their mixtures are used in gluten free formulations for crumbly texture of baked goods and increase the emulsion stability of the batters/doughs [48–52].

12.5 Biopolymers Used in Encapsulation of Active Food Ingredients

Encapsulation is entrapment of active ingredients within a matrix or a capsule. Biopolymers are used as wall material to encapsulate food ingredients such as probiotic bacteria, vitamins, nutraceuticals, antioxidants, flavors, acids and bases, colorants, leavening agents etc. [53–55]. Encapsulation increases the stability and shelf life, masks unwanted odors/tastes, provides controlled or targeted release, prevents unwanted interactions, and eases the handling of food ingredients. Depending on the purpose, the properties of active material and desired release mechanism, different techniques and different wall materials are preferred for encapsulation [55, 56]. Table 12.2 lists the biopolymers used for encapsulation of food ingredients. Amylose and amylopectin content of starch vary depending on the source (maize, tapioca, potato, corn etc.) [57]. High amylose starches have good extrusion and film forming abilities and their solubility is low in water, which is important in controlled release. Chemically modified starches such as the hydrophobic octenylsuccinate (OSA) starch, acid modified tapioca starch or enzymatically modified starches such as cyclodextrin and maltodextrin are also used for encapsulation [58–61]. Modified cellulose derivatives such as methylcellulose (MC), hydroxypropyl methylcellulose (HPMC), hydroxypropyl cellulose (HPC) and ethylcellulose are commonly used cellulose derivatives for encapsulation purposes [62]. Proteins are generally used for encapsulation of lipophilic ingredients since they are surface active and excellent stabilizers of emulsions [63, 64]. They show good functional properties for encapsulation without chemical modification. However, one drawback is their instability at low pH. Partially hydrolyzed proteins are also used in encapsulation since they have improved physical properties and reduced allergenic potential. Among the hydrocolloids, gum arabic is usually preferred for encapsulation of lipophilic ingredients since it can act like an emulsifier [65].

12.6 Biopolymers Used for Food Packaging

Given that the food industry is the largest industry where plastic packaging materials are used [66], there is a strong need to develop packaging materials that are biodegradable, renewable, and environmentally friendly [67]. Composed of polymers that can be degraded by microorganisms,

Table 12.2 Microencapsulating biopolymers [53–65].

Origin	Carbohydrates	Proteins
Plant	Starch and derivatives, cellulose and derivatives, hydrocolloids (gum arabic, locust bean gum, guar gum, carrageenan, alginate, pectin)	Gluten (corn, wheat), Pea protein isolate, soy protein isolate
Animal	Chitosan	Caseins, whey protein, gelatin
Microbial	Dextran hydrocolloids (xanthan gum, gellan gum, alginate)	

biodegradable films made of biopolymers are ecologically safe materials [68]. The monomer structure of biopolymers affects their functional properties such as thermal stability, barrier properties against gases, moisture resistance and flexibility. Depending on their nature and structure, biopolymer-based coatings may be edible and consumed with the food product. They enhance organoleptic properties of food, and act as vehicle for active agents such as antimicrobials, antioxidants, nutraceuticals or colorings [66–71]. Starch and its derivatives, cellulose derivatives, pectin, chitosan, carrageenan, alginate, curdlan, gellan, pullulan, xanthan and some other polysaccharide based gums are used for biodegradable films/packages [66–75].

Biodegradable films made of proteins generally exhibit better mechanical and barrier properties compared to those made of polysaccharides [76, 77]. Plasticity, elasticity, the ability of network formation and high intermolecular binding potential are the advantageous properties of the proteins for packaging applications. Protein based films also have a unique structure that provide a wide range of functional properties. They can make bonds at different positions, and they have higher potential to form various linkages [78]. Soy protein, pea protein, corn zein, wheat gluten, rice bran protein and sorghum protein are the common plant-based proteins whereas gelatin, collagen, casein, whey protein, keratin, egg white and fish myofibrillar proteins are the most common animal based proteins used for biodegradable film formation [73, 76–82]. On the other hand, when compared to synthetic polymers, lower water vapor resistance and low mechanical strength limit the packaging applications of protein-based polymers [23, 78, 81].

References

1 Phillips, G.O. and Williams, P.A. (eds.) (2009). *Handbook of Hydrocolloids*. 1–948. Oxford: Elsevier.

2 Goff, H.D. and Guo, Q. (2020). The role of hydrocolloids in the development of food structure. In: *Handbook of Food Structure Development* (ed. F. Spyropoulos, A. Lazidis, and I.T. Norton), 3–28. Cambridge: RSC Publishing.

3 Alba, K. and Kontogiorgos, V. (2019). Seaweed polysaccharides (agar, alginate carrageenan). In: *Encyclopedia of Food Chemistry*, vol. 1 (ed. L. Melton, F. Shahidi, and P. Varelis), 240–250. Amsterdam: Elsevier.

4 Laaman, T.R. (2011). *Hydrocolloids in Food Processing*. Oxford: Blackwell Publishing, Ltd and Institute of Food Technologists.

5 Imeson, A.P. (1997). *Thickening and Gelling Agents for Food*. Boston, MA: Springer.

6 Mudgil, D., Barak, S., and Khatkar, B.S. (2014). Guar gum: processing, properties, and food applications—a review. *Journal of Food Science and Technology* 51: 409–418.

7 Razavi, S.M.A. and Irani, M. (2019). Rheology of food gum. In: *Bioactive Molecules in Food* (ed. J.M. Mérillon and K. Ramawat). Switzerland: Springer.

8 Foster, T. and Wolf, B. (2011). Hydrocolloid gums-their role and interactions in foods. In: *Practical Food Rheology: An Interpretive Approach* (ed. I.T. Norton, F. Spyropoulos, and P. Cox), 61–84. Oxford: Blackwell Publishing Ltd.

9 Meyer, E.L., Fuller, G.G., Clark, R.C., and Kulicke, W.M. (1993). Investigation of xanthan gum solution behavior under shear-flow using rheooptical techniques. *Macromolecules* 26: 504–511.

10 Yaşar, K., Kahyaoglu, T., and Şahan, N. (2009). Dynamic rheological characterization of salep glucomannan/galactomannan-based milk beverages. *Food Hydrocolloids* 23 (5): 1305–1311.

11 Onsøyen, E. (1997). Alginates. In: *Thickening and Gelling Agents for Food* (ed. A.P. Imeson), 1–24. Boston, MA: Springer.

12 Chan, S.Y., Choo, W.S., Young, D.J., and Loh, X.J. (2017). Pectin as a rheology modifier: origin, structure, commercial production, and rheology. *Carbohydrate Polymers* 161: 118–139.

13 Rao, M.A. (2007). *Rheology of Fluid and Semisolid Foods: Principles and Applications*, Food Engineering Series. New York: Springer.

14 King, A.H. (2010). Hydrocolloids in salad dressings. In: *Hydrocolloids in Food Processing* (ed. T.R. Laaman), 19–33. Oxford: Blackwell Publishing, Ltd., and Institute of Food Technologists.

15 Hoefler, A.C. (2004). *Hydrocolloids*. Eagan, MN: Eagan Press.

16 Nussinovitch, A. (1997). *Hydrocolloid Applications: Gum Technology in Food and Other Industries*. London: Chapman & Hall.

17 Burey, P., Bhandari, R.P., Howes, T., and Gidley, J.M. (2008). Hydrocolloid gel particles: formation, characterization, and application. *Critical Reviews in Food Science and Nutrition* 48 (5): 361–367.

18 Ali, A. and Ahmed, S. (2018). Recent advances in edible polymer-based hydrogels as a sustainable alternative to conventional polymers. *Journal of Agricultural and Food Chemistry* 66 (27): 6940–6967.

19 Draget, K.I. (2009). Alginates. In: *Handbook of Hydrocolloids* (ed. G.O. Phillips and P.A. Williams), 807–828. Oxford: Elsevier.

20 Szekalska, M., Puciłowska, A., Szymańska, E. et al. (2016). Alginate: current use and future perspectives in pharmaceutical and biomedical applications. *International Journal of Polymer Science* 2016: 1–17.

21 Imeson, A.P. (2009). Carrageenan and furcellaran. In: *Handbook of Hydrocolloids* (ed. G.O. Phillips and P.A. Williams), 164–185. Oxford: Elsevier.

22 Morris, E.R., Rees, D.A., and Robinson, G. (1980). Cation-specific aggregation of carrageenan helices: domain model of polymer gel structure. *Journal of Molecular Biology* 138 (2): 349–362.

23 Mariod, A.A. and Adam, H.F. (2013). Review: gelatin, source, extraction, and industrial applications. *Acta Scientiarum Polonorum, Technologia Alimentaria* 12: 135–147.

24 Schrieber, R. and Gareis, H. (2007). *Gelatine Handbook: Theory and Industrial Practice*. Weinheim, Germany: Wiley-VCH.

25 Djagny, K.B., Wang, Z., and Xu, S. (2001). Gelatin: a valuable protein for food and pharmaceutical industries: review. *Critical Reviews in Food Science and Nutrition* 41: 481–492.

26 Gómez-Guillén, M.C., Giménez, B., López-Caballero, M.A., and Montero, M.P. (2011). Functional and bioactive properties of collagen and gelatin from alternative sources: a review. *Food Hydrocolloids* 25: 1813–1827.

27 Wielinga, W.C. (2009). Galactomannas. In: *Handbook of Hydrocolloids* (ed. G.O. Phillips and P.A. Williams), 228–251. Oxgford: Elsevier.

28 Copetti, G., Grassi, M., Lapasin, R., and Pricl, S. (1997). Synergistic gelation of xanthan gum with locust bean gum: a rheological investigation. *Glycoconjugate Journal* 14: 951–961.

29 Schreiber, C., Ghebremedhin, M., Zielbauer, B. et al. (2020). Interaction of xanthan gums with galacto- and glucomannans. Part I: molecular interactions and synergism in cold gelled systems. *Journal of Physics: Materials* 3: 034013.

30 Garcia-Ochoa, F., Santos, V.E., Casas, J.A., and Gómez, E. (2000). Xanthan gum: production, recovery, and properties. *Biotechnology Advances* 18 (7): 549–579.

31 Philp, K. (2018). Polysaccharide Ingredients. *Reference Module in Food Science*. Amsterdam: Elsevier.

32 Pegg, A.M. (2012). The application of natural hydrocolloids to foods and beverages. In: *Natural Food Additives, Ingredients and Flavourings* (ed. D. Baines and R. Seal), 175–196. Oxford: Elsevier.

33 Ozturk, B., Argin, S., Ozilgen, M., and McClements, D.J. (2015). Formation and stabilization of nanoemulsion-based vitamin E delivery systems using natural biopolymers: whey protein isolate and gum arabic. *Food Chemistry* 188: 256–263.

34 Cao, H., Zhao, Y., Zhu, Y.B. et al. (2016). Antifreeze and cryoprotective activities of ice-binding collagen peptides from pig skin. *Food Chemistry* 194: 1245–1253.

35 Griffith, M. and Ewart, K.V. (1995). Antifreeze proteins and their potential use in frozen foods. *Biotechnology Advances* 13 (3): 375–402.

36 Gaukel, V., Leiter, A., and Spieß, W.E.L. (2014). Synergism of different fish antifreeze proteins and hydrocolloids on recrystallization inhibition of ice in sucrose solutions. *Journal of Food Engineering* 141: 44–50.

37 Clemmings, J.F., Hans, F., Zoerb, H.F. et al. (2000). Method of making frozen compositions. US Patent US6017574A.

38 Fenn, R.A., Needham, D., and Smallwood, S.E.K. (2004). Frozen food product. US Patent: US67939552B2.

39 Argin, S., Eskinazi, B., and Tavlı, D. (2019). A consumer perception study on functional dairy products among consumers in Istanbul, Turkey. *Turkish Journal of Agriculture Food Science and Technology* 7 (7): 963–970.

40 Fallourd, M.J. and Viscione, L. (2009). Ingredient selection for stabilisation and texture optimisation of functional beverages and the inclusion of dietary fibre. In: *Functional and Speciality Beverage Technology* (ed. P. Paquin), 3–38. Oxford: Elsevier.

41 Mehta, N., Ahlawat, S.S., Sharma, D.P., and Dabur, R.S. (2015). Novel trends in development of dietary fiber rich meat products—a critical review. *Journal of Food Science Technology* 52: 633–647.

42 Cui, J., Lian, Y., Zhao, C. et al. (2019). Dietary fibers from fruits and vegetables and their health benefits via modulation of gut microbiota. *Comprehensive Reviews in Food Science and Food Safety* 18 (5): 1514–1532.

43 Paquin, P. (2009). *Functional and Speciality Beverage Technology*. Oxford: Elsevier.

44 Wusigale, L. and Luo, Y. (2020). Casein and pectin: structures, interactions, and applications. *Trends in Food Science and Technology* 97: 391–403.

45 Capelli, A., Olivia, N., and Cini, E. (2020). A systematic review of gluten-free dough and bread: dough rheology, bread characteristics and improvement strategies. *Applied Sciences-Basel* 10 (18): 6559.

46 Matos, M.E. and Rosell, C.M. (2015). Understanding gluten-free dough for reaching breads with physical quality and nutritional balance. *Journal of the Science of Food and Agriculture* 95: 653–661.

47 Arendt, E.K. and Dal Bello, F. (2008). Functional cereal products for those with gluten intolerance. In: *Technology of Functional Cereal Products* (ed. B.R. Hamaker), 446–475. Oxford: Elsevier.

48 Arendt, E.K., Morrissey, A., Moore, M.M., and Dal Bello, F. (2008). Gluten-free breads. In: *Gluten-Free Cereal Products and Beverages* (ed. E.K. Arendt and F. Dal Bello), 289–320. Amsterdam: Elsevier.

49 Baldino, N., Laitano, F., Lupi, F.R. et al. (2018). Effect of HPMC and CMC on rheological behavior at different temperatures of gluten-free bread formulations based on rice and buckwheat flours. *European Food Research and Technology* 244: 1829–1842.

50 Sciarini, L.S., Ribotta, P.D., León, A.E., and Pérez, G.T. (2010). Effect of hydrocolloids on gluten-free batter properties and bread quality. *International Journal of Food Science and Technology* 45: 2306–2312.

51 Lazaridou, A., Duta, D., Papageorgiou, M. et al. (2007). Effects of hydrocolloids on dough rheology and bread quality parameters in gluten-free formulations. *Journal of Food Engineering* 79: 1033–1047.

52 Sabanis, D. and Tzia, C. (2011). Effect of hydrocolloids on selected properties of gluten-free dough and bread. *Food Science and Technology International* 17: 279–291.

53 Gaonkar, A.G., Vasisht, N., Khare, A.R., and Sobel, R. (2014). *Microencapsulation in the Food Industry*. Oxford: Elsevier.

54 Gani, A., Masoodi, F.A., Shah, U., and Shah, A. (2019). *Food Hydrocolloids as Encapsulating Agents in Delivery Systems*. Boca Raton: CRC.

55 Zuidam, N.J. and Nedovic, V. (2010). *Encapsulation Technologies for Active Food Ingredients and Food Processing*. Switzerland: Springer.

56 Kuang, S.S., Oliveira, J.C., and Crean, A.M. (2010). Microencapsulation as a tool for incorporating bioactive ingredients into food. *Critical Reviews in Food Science and Nutrition* 50 (10): 951–968.

57 Taggart, P. (2009). Starch. In: *Handbook of Hydrocolloids* (ed. G.O. Phillips and P.A. Williams), 108–141. Oxford: Elsevier.

58 Loksuwan, J. (2007). Characteristics of microencapsulated β-carotene formed by spray drying with modified tapioca starch, native tapioca starch and maltodextrin. *Food Hydrocolloids* 21 (5): 928–935.

59 Krishnan, S., Bhosale, R., and Singhal, R.S. (2005). Microencapsulation of cardamom oleoresin: evaluation of blends of gum arabic, maltodextrin and a modified starch as wall materials. *Carbohydrate Polymers* 61 (1): 95–102.

60 Hoyos-Leyva, J.D., Bello-Pérez, L.A., Alvarez-Ramirez, J., and Garcia, H.S. (2018). Microencapsulation using starch as wall material: a review. *Food Reviews International* 34 (2): 148–161.

61 Gupta, C., Chawla, P., Arora, S. et al. (2015). Iron microencapsulation with blend of gum arabic, maltodextrin and modified starch using modified solvent evaporation method—milk fortification. *Food Hydrocolloids* 43: 622–628.

62 Wallick, D. (2014). Cellulose polymers in microencapsulation of food additives. In: *Microencapsulation in the Food Industry* (ed. A.G. Gaonkar, N. Vasisht, A.R. Khare, and R. Sobel), 181–193. Oxford: Elsevier.

63 Kim, Y.D. and Morr, C.V. (1996). Microencapsulation properties of gum arabic and several food proteins: spray-dried orange oil emulsion particles. *Journal of Agricultural and Food Chemistry* 44 (5): 1314–1320.

64 Can Karaca, A., Low, N.H., and Nickerson, M.T. (2015). Potential use of plant proteins in the microencapsulation of lipophilic materials in foods. *Trends in Food Science & Technology* 42 (1): 5–12.

65 McNamee, B.F., O'Riorda, E.D., and O'Sullivan, M. (1998). Emulsification and microencapsulation properties of gum arabic. *Journal of Agricultural and Food Chemistry* 46 (11): 4551–4555.

66 Lagaron, J.M. and López-Rubio, A. (2009). Latest developments and future trends in food packaging and biopackaging. In: *Innovation in Food Engineering: New Techniques and Products* (ed. M.L. Passos and C.P. Ribeiro), 485–508. Boca Raton: CRC Press.

67 Tharanathan, R.N. (2003). Biodegradable films and composite coatings: past, present, and future. *Trends in Food Science and Technology* 14: 71–78.

68 Espitia, P.J.P., Du, W., Avena-Bustillos, R.J. et al. (2014). Edible films from pectin: physical-mechanical and antimicrobial properties – a review. *Food Hydrocolloids* 35: 287–296.

69 Cerqueira, M.A., Bourbon, A.I., Pinheiro, A.C. et al. (2011). Galactomannans use in the development of edible films/coatings for food applications. *Trends in Food Science and Technology* 22 (12): 662–671.

70 Bierhalz, A.C.K., da Silva, M.A., and Kieckbusch, T.G. (2012). Natamycin release from alginate/pectin films for food packaging applications. *Journal of Food Engineering* 110: 18–25.

71 Penhasi, A. and Meidan, V.M. (2014). Preparation and characterization of in situ ionic cross-linked pectin films: unique biodegradable polymers. *Carbohydrate Polymers* 102: 254–260.

72 Martau, G.A., Mihai, M., and Vodnar, D.C. (2019). The use of chitosan, alginate, and pectin in the biomedical and food sector—biocompatibility, bioadhesiveness, and biodegradability. *Polymers* 11 (1): 28.

73 Gabor, D. and Tita, O. (2012). Biopolymers used in food packaging: a review. *Acta Universitatis Cibiniensis Series E: Food Technology* 16 (2): 3–19.

74 Vartiainen, J., Vähä-Nissi, M., and Harlin, A. (2014). Biopolymer films and coatings in packaging applications-a review of recent developments. *Material Science and Applications* 5: 708–718.

75 Rhim, J.W. and Ng, P.K.W. (2007). Natural biopolymer-based nanocomposite films for packaging applications. *Critical Reviews in Food Science and Nutrition* 47: 411–433.

76 Cuq, B., Gontard, N., and Guilbert, S. (1998). Proteins as agricultural polymers for packaging production. *Cereal Chemistry* 75: 1–9.

77 Wang, L.Z., Liu, L., Holmes, J. et al. (2007). Assessment of film-forming potential and properties of protein and polysaccharide-based biopolymer films. *International Journal of Food Science and Technology* 42: 1128–1138.

78 Hernandez-Izquireda, V.M. and Krochta, J.M. Thermoplastic processing of proteins for film formation—a review. *Journal of Food Science* 73 (2): R30–R39.

79 Argin, S., Gulerim, M., and Sahin, F. (2019). Development of antimicrobial gelatin films with boron derivatives. *Turkish Journal of Biology* 43: 47–57.

80 Flores, R.N., Gimenez, B., Martin, F.F. et al. (2013). Physical and functional characterization of active fish gelatin films incorporated with lignin. *Food Hydrocolloids* 30: 163–172.

81 Nur Hanani, Z.A., Roos, Y.H., and Kerry, J.P. (2012). Use of beef, pork, and fish gelatin sources in the manufacture of films and assessment of their composition and mechanical properties. *Food Hydrocolloids* 29: 144–151.

82 Liu, L., Liu, C.K., Fishman, M.L., and Hicks, K.B. (2007). Composite films from pectin and fish skin gelatin or soybean flour protein. *Journal of Agricultural and Food Chemistry* 55: 2349–2355.

13

Biopolymers in Food Packaging

Sneha Sara Varghese[1]*, Vaisakh P.H.*[1]*, Eapen Philip*[1]*, Tressia Alias Princy Paulose*[1,*]*, and Akhina H.*[2]

[1] *Post Graduate and Research Department of Chemistry, Bishop Moore College, Mavelikara, Kerala, India*
[2] *International and Inter University Centre for Nanoscience and Nanotechnology, Mahathma Gandhi University, Kottayam, India*
[*] *Corresponding author*

13.1 Introduction

Compounds known as polymers are made up of monomeric units that are covalently linked together. The majority of the polymers that are traditionally produced from fossil fuels like crude oil, coal, and natural gases are among the countless polymers that are used in daily life. Despite their many benefits, traditional fossil-based polymers should not be used because they are not biodegradable [1]. Finding alternative renewable resources for the sustainable manufacture of polymers has become necessary due to the rising environmental consciousness.

There are a variety of sources with biological origins that promise to replace fossil fuels in the future. The organic materials found in natural sources are known as biopolymers. Biopolymers are large macromolecules made up of a great number of repeating units. A single molecule is defined as a macromolecule by International Union of Pure and Applied Chemistry (IUPAC). Terms like "biopolymers", "bio-based polymers", "bioplastics", and "biodegradable polymers" are not entirely clear. Although biopolymers are biodegradable, they do not contain biodegradable polymers that have

Applications of Biopolymers in Science, Biotechnology, and Engineering, First Edition.
Edited by Arunima Reghunadhan, Akhina H., and Sabu Thomas.

been created artificially. Both naturally derived and synthetically produced polymers are included in the category of biodegradable polymers, sometimes referred to as bioabsorbable polymers. Materials made from renewable sources are known as biobased polymers, and they can either be biodegradable or non-degradable. Although biodegradable, bioplastics are not always made from biological materials. In contrast to some biobased polymers, some petroleum-based polymers, such as polycaprolactone and polybutylene succinate, can also be biocompatible. The biomaterials are found to be biodegradable and biocompatible, making them helpful in a variety of applications, including edible films, emulsions, packaging materials, and medical implants, including organs, wound healing, tissue scaffolds, and dressing materials in the pharmaceutical sectors. One of most common forms of macromolecules are biopolymers, which are made up of large non-polymeric molecules like lipid and macrocycles as well as nucleic acids, proteins, carbohydrates, and lipids. Examples of synthetic macromolecules include plastics, synthetic fibers, and research materials like carbon nanotubes [2]. Their molecular networks may include a wide range of chemical side groups that help the molecules function in addition to repeating units of nucleic acids, saccharides, or amino acids. Typical forms of biopolymers discovered in microbes or genetically modified organisms using conventional chemical procedures are polylactic acid (PLA) and polyhydroxyalkanoates (PHAs). These comprise proteins from milk or collagen as well as carbohydrates from cellulose. The genetic engineering of microbes enables the biotechnological synthesis of biopolymers with tailored properties suitable for high-value medicinal applications, such as tissue engineering and drug delivery.

Maintaining the quality of perishable food goods requires the use of packaging. For a very long time, packaging made of petroleum-derived materials has been the material of choice. However, these materials provide significant environmental dangers because they are not biodegradable. Utilizing biodegradable polymers is a good substitute. This chapter discuss the application of biopolymers in food packaging.

13.2 Classification and Advantages of Biopolymers

Biopolymers are categorized in different ways based on the type of monomers, biodegradability, raw materials and backbones. First, there are two basic categories for biopolymers: biodegradable and nonbiodegradable biopolymers. These are then alternately divided into biobased and non-biobased biopolymers depending on the origin. The polymer backbone of the biopolymer can also be used to categorize it, and this includes polyesters, polysaccharides, polycarbonates, polyamides, and vinyl polymers. Based on the type of constituents, the three different types of biopolymers can be further divided into polysaccharides, polypeptides, and polynucleotide [3]. Classification of biopolymers is depicted in Figure 13.1.

Packaging is a method of preparing products for use after travel, distribution, storage, and at the retail stage. It indicates that an item should be transported in a defined manner, at an affordable price, to the end retail market. All conventional polymers that are used to make various types of food packaging can be loosely classified using actual use, packaging functionality, and ability to confine contents. One of the main reasons biopolymers should be used instead of traditional plastics is the existence of various pollutants in them and the possibility of their migration into food. The environmental problem and lack of biodegradability, which are side effects of using plastic made from petroleum compounds, are another important factor. In comparison to biopolymers, the production of 1 kg of traditional plastics uses 65% more energy. Conventional plastic takes an excessively lengthy time to degrade, which has a significant negative impact on the environment. Despite all of this, utilising conventional plastic was for a long time seen to be a viable idea, despite the advantages of employing recycled

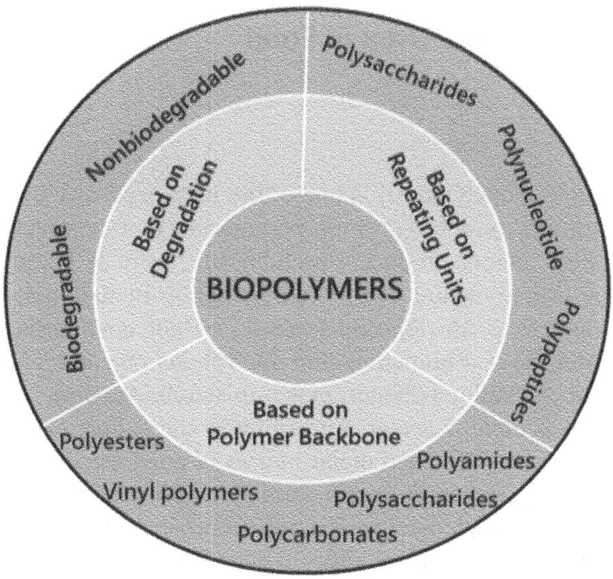

Figure 13.1 Classification of biopolymers.

material like biopolymers. Due to the necessity of using various substances, nearly all traditional plastics are poisonous and can only be used in food packaging in certain circumstances. Biopolymers have a longer shelf life than regular plastic and reduce greenhouse gas emissions by 30%–80% [4]. The majority of the products made using biopolymers are disposable and appropriate for compost manufacturing, which adds organic matter and nutrients to the soil. As a result, there is an improvement in nutrient and water retention, a decrease in chemical input, and a reduction in plant disease. It has been demonstrated that starch-based biopolymers degrade 10–20 times more quickly than conventional plastics. When conventional plastics are burned, poisonous smoke and gases are released into the air that are harmful to people, animals, and the environment. Although it was unnecessary to look into whether biopolymers could be burned and still degrade within a short amount of time, the fumes emitted if this happened would not be harmful to the environment. Because of their more varied structure, chemical makeup, and design, biopolymers provide researchers a wide range of chances to tailor the characteristics of the finished packaging material. The ability to incorporate nanoparticles with a variety of good traits, including heat and cold stability, impermeability to gases, strength, hardness, antibacterial qualities, impermeability to oxygen, etc., is a major benefit of using biopolymers [5, 6].

13.3 Major Biopolymers Used in Food Packaging

13.3.1 Starch

Starch is a hydrocolloid biopolymer. It is one of the least expensive biodegradable polymers and is a low-cost polysaccharide that is widely accessible. Agricultural plants produce starch in the form of hydrophilic granules. The primary sources of starch extraction include potatoes, corn, wheat, and rice. Amylopectin (poly–1, 4-Dglucopyranoside and –1, 6-Dglucopyranoside), a branching and amorphous polymer, and amylose, a linear and crystalline polymer, make up starch. Amylose has a helical shape and is soluble in water. According to the source of the starch, the relative concentrations and molar masses of amylopectin and amylose change, resulting in materials with various mechanical and biodegradable properties [7]. The elongation of starch increases with amylose concentration. Starch does not have a significant degree of stress stability. At 150 °C, the glucoside linkages begin to break, and at 250 °C or higher, the granules collapse. Hydrogen bond restructuring, or retrogradation, can be seen at low temperatures during cooling.Starch can be blended, retained whole, and utilized as filler or melted for blending components in different resins. Starch whiskers are used as fillers with polymer resins. Typically, starch is utilized as a thermoplastic. It is plasticized by destructuration in the presence of a controlled amount of heat, water, or plasticizers, and is subsequently extruded. Thermoplastic starch (TPS) is extremely sensitive to moisture. It has been demonstrated that the water content of TPS has a greater impact on its thermal characteristics than does the molecular weight of the starch. The resulting TPS is practically amorphous. Starch is biodegraded by enzymes by hydrolysis of the acetal bond. Non-toxic degradation products are produced.

Starch has several benefits as an edible film, including ease of manufacture, low cost, and a strong barrier to lipids and oxygen. It offers mediocre water resistance, though. Candy and pastry goods can be packaged using edible films made of starch [8]. The combined plasticizers' impact and blends has an effect on the properties of starch-based edible films revealing that films with high concentrations of surfactant and glycerol behaved similarly to those with considerably greater plasticizer contents. Plastic made of starch that is used in packaging is very fragile. Except when combined with other materials, chemically altered, plasticized, or a combination of these treatments, starch lacks the capacity to produce films with predicted mechanical qualities such high elongation, flexural, and tensile strength. Glycerol and other low molecular weight polyhydroxy compounds, polyethers, and urea are often used plasticizers.

13.3.1.1 Blends

13.3.1.1.1 Starch-Polyvinyl Alcohol

Excellent compatibility exists between TPS and polyvinyl alcohol (PVOH), and their mixtures are of particular importance. Both of them can be mixed in a variety of ratios to customize the resulting material's mechanical qualities. Blends have better tensile strength, elongation, and processability than pure TPS material [9]. The rate of breakdown of starch is significantly influenced by the PVOH level, and adding more PVOH will slow this rate.

13.3.1.1.2 Starch-PLA

Due to incompatibility, mixes of PLA and starch produced using standard methods have poor mechanical characteristics. Utilizing plasticizers or reactive agents in the extrusion process will promote elongation. Use of coupling agents like isocyanates is commonly seen. The isocyanate group and the hydroxyl groups of starch may react, forming urethane linkages and making these systems compatible. Investigated was the impact of starch gelatinization. It has been demonstrated that starch can act as a nucleating agent in PLA/gelatinized starch blends, improving the crystallinity of PLA blends and improving their mechanical properties. Utilizing a compatibilizer is another technique to enhance compatibilization. For this, maleic anhydride might be applied. Maleic acid compatibilized PLA/starch blends produce mechanical qualities that are better than those of unmodified PLA/starch blends. A biodegradable copolymer (PLA-grafted amylose) that can be employed as a compatibilizer in starch/PLA blends has been created.

13.3.1.1.3 Starch-PCL

TPS and PCL were combined to modify the rheological properties of the melt prior to the process in order to manufacture films utilizing the film blowing technique. A kind of starch blend with various synthetic components is produced by Novamont (Italy). There are four grades available, and PCL is included in one of them. The fastest rate of PCL breakdown is possible with the maximum amount of starch. Modified starch increases the PCL's Young's modulus while decreasing the tensile strength and elongation at break values. The mixture loses its ductility [10]. Depending on the applications, some synthetic materials with reduced biodegradability are employed to regulate the pace of biodegradation. Blends of PCL with high-amylose maize starch (25% wt.) had a 50% greater modulus than PCL and a 15% lower tensile strength. Blends with LDPE were created to improve the PCL/mechanical starch's characteristics. The inclusion of starch can greatly increase PCL's very poor rate of biodegradation.

13.3.2 Cellulose

The most prevalent naturally occurring polymer on the planet, cellulose is a nearly linear polymer of anhydroglucose. It is a cellobiose-based linear polymer with extremely long macromolecular chains. In all organic solvents, cellulose is crystalline, infusible, and insoluble. It tends to form tightly hydrogen bonded crystalline micro fibrils and fibers due to its ordered structure and arrangement of hydroxyl groups, and is most well-known in the area of paper or cardboard in the packaging context. Although some parts of primary food packaging use paper that has been waxed or coated with polyethylene, most paper is utilized for secondary packaging. Although cellulose is an inexpensive raw material, its hydrophilic properties, insolubility, and crystalline structure make it challenging to employ. The resulting cellophane has good mechanical qualities but is moisture sensitive due to its high hydrophilicity. However, because the theoretical melt temperature is higher than the degradation temperature and cannot be heat sealed, it is not a thermoplastic material. To enhance its barrier capabilities, cellophane is frequently treated with nitrocellulose wax or PVDC (poly vinylidene chloride), and, in this form, it is used to package baked foods, processed meat, cheese, and chocolates. Commercial production of several cellulose derivatives occurs, with carboxy-methyl cellulose, methyl cellulose, ethyl cellulose, hydroxyethyl cellulose, hydroxypropyl cellulose, and cellulose acetate (CA) being the most popular. Only CA is used often in food packaging. CA needs to be plasticized in order to produce films since it has relatively poor gas and moisture barrier characteristics.

Cellulose molecules are made up of a chain of β-(1,4)-linked glucose residues and have an amphiphilic nature. They also feature a high density of hydroxyl group [11]. To make cellulose more processable, it is transformed into its derivatives depending on infusibility and insolubility. To create food packaging materials, for instance, sugarcane pulp fiber free of chlorine bleach or ECF is utilized, and the fiber is pounded. This material can be moulded both wet and dry, with an emphasis on hygienic practises and security. The product can resist temperatures of −40 to 250 °C because the components are naturally derived from sugarcane fiber. The item can withstand freezing and can contain 100 °C boiling water or 150 °C boiling oil without leaking. More critically, the fiber from sugarcane pulp degrades in landfills within 45 days. The substance can be used to make packaging components including plates, glasses, and compartment boxes.

13.3.3 Chitosan

After cellulose, chitosan is considered to be the second most abundant polymer. Chitosan is derived from chitin and is a natural polymer. For many years, it has been known that chitosan possesses strong antibacterial capabilities against fungi, bacteria, and viruses. The usage of it is restricted by the material's gas and water vapour permeability, which is one of its few mechanical limitations. Due to its cationic nature, chitosan is soluble in acidic solutions like citric, formic, and acetic acids but insoluble in water. It has been authorized for use in food and wellness products owing to its antibacterial, biodegradable, biomedical, and biocompatible qualities [12]. Because it can be made from industrial waste from shellfish, chitosan is today a substance that is widely accessible. Chitosan can be obtained from the fungal cell walls of *Histoplasma farciminosum* and *Aspergillus fumigatus*, which contains a chitin component. It is also derived from the yeast *Saccharomyces cerevisiae* [13].

One of the causes of chitin's low solubility is the hydrogen bonding in the substance. Chitin derivatives are frequently created and used to get around this solubility problem. Chitin is used to create colloids, gels, and membranes. Chitin by nature is crystalline, and indeed the amount of deacetylation has a direct correlation with the crystallinity. To increase its utility, chitin is typically transformed into chitosan. Chitosan is non-allergenic, biocompatible, and nontoxic to animals [14].Chitosan can be divided into two types by molecular weight: high molecular weight (HMW) (> 300 kDa) and low molecular weight (LMW) (< 300 kDa). Depending on where it comes from, chitosan's molecular weight specifies its physicochemical properties, including viscosity, solubility, elasticity, tear strength, and biofunctionalities. Depending on the source and manufacturing process, the average molecular weight of chitosan ranges from 50 to 1000 kDa, 3.8 to 2000 kDa, or 50 to 2000 kDa. Chitosan with a higher molecular weight creates very viscous solutions that are challenging to manage in industrial settings. Using glycerol molecules of varying molecular weights, the viscosity of a chitosan film was standardized. High molecular weight chitosan films display non-Newtonian behaviour. Chitosan having a lower molecular weight inhibits phytopathogens more potently [15].

Chitosan is an emerging biodegradable food packing material. Furthermore, because of its antibacterial action, non-toxicity, and biodegradability, which open up a wide range of uses, chitosan has tremendous potential as a packaging antimicrobial material. When mixed with some other film-forming substances used in food packaging technology, chitosan film's functional characteristics can be improved, steadily producing a solution to the society's expanding challenges. Chitosan is thought of as an ecofriendly material for packaging because it is non-toxic, disposable, and biocompatible. Additionally, chitosan exhibits high qualities for gas and odour barrier in dry settings and is a good regulator against the proliferation of a wide variety of fungus, bacteria, and yeasts. Chitosan is a strong candidate for active antibacterial usage for food packaging because of these qualities as well as the simplicity of film creation.

13.3.3.1 PCL/Chitin-Chitosan Blends

Melt blending is used to create PCL blends with chitin to create biodegradable composites [16]. The melting or crystallization temperature is unaffected by the amount of chitin present. This was ascribed to a mixture that was not miscible. Solvent casting is another method of blending. In the process of blending with chitin, PCL loses some of its crystallinity. The same outcomes can be achieved with PCL/chitosan mixtures. It is anticipated that these mixtures will have good mechanical qualities.

13.3.4 Polylactic Acid

PLA is typically produced by polycondensing D- or L-lactic acid or by lactide, a cyclic dimer of lactic acid, through ring opening polymerization. There are two types of lactide: D-lactide and L-lactide. L-lactide is the synthetically produced isomer, and DL-lactide is the natural isomer. There have also been studies of other synthetic techniques. Because of the -CH3 side groups, PLA seems to be a hydrophobic polymer. Because of the methyl side group's steric shielding effect, it is more resilient to hydrolysis than PGA. For representative commercial PLA, the typical glass transition temperature is 63.8 °C, the elongation at break is 30.7%, and the tensile strength is 32.22 MPa [17]. Racemization of the D- and L-isomers or the use of a hydroxy acid co-monomer component can be used to control the physical characteristics and biodegradability of PLA. L-lactide can be converted into a semi-crystalline polymer (PLLA) with a crystallinity of about 37% while poly (DL-lactide) (PDLLA) is found to be an amorphous polymer. Their deterioration rates and mechanical characteristics differ. The tensile strength of PLLA ranges from 45 to 70 MPa, and it is a strong, clear polymer with an elongation at break of 85%–105%. It has a glass transition temperature of 53 °C and a melting point of 170–180 °C. PDLLA has a Tg of about 55 °C and no melting point. Brittleness and inadequate thermal stability are drawbacks of PLA. To increase the chain's mobility and encourage crystallization, PLA can be plasticized. Oligomeric acid, citrate ester, or low molecular weight polyethylene glycol are used to achieve plasticization [18]. Ring opening polymerization is used to create high molecular weight PLAs. By altering the ratios of the two enantiomers, this method also enables control over the ultimate properties of PLA. Other methods include chain extension reactions, solution polymerization, and melt/solid state polymerization.

The level of crystallinity affects how quickly PLA degrades. Since PLA degrades much more slowly than PGA, several copolymers of glycoside and lactide have now been studied as bioresorbable implant substances. Grafting is another method that can improve PLA's biodegradability. Ring opening polymerization with a tin catalyst was used to carry out the graft copolymerization of L-lactide with chitosan. With higher grafting percentages the melting transition temperatures and thermal stability of graft polymers rises. The deterioration of the graft polymer declines as the lactide content rises [19].

High molecular weight, resistance to water solubility, good processability (easily processed by thermoforming), and biodegradability of PLA make it a good material for food packaging. The tensile strength modulus, taste and odour barrier, temperature stability and processability, printability, and grease tolerance of polyethylene are all present in PLA. It also has the flexibility of PET or flexible PVC. Films, containers, and coating for paper and paperboard boards are all made from processed PLA. Although PLA appears to be a possible biodegradable polymer that might be used in the packaging of different foods, it has some drawbacks in its unaltered state, such being brittle and degrading quickly at high temperatures.

Various investigations revealed that PLA is used for the development of gas barrier films to be utilized for the food and medical-packaging products because it has fairly low susceptibility to oxygen and water vapour permeability compared to traditional non-degradable polymer resins. Additionally, the likelihood of PLA's commercialization is diminished by its high cost and fragility. Therefore, combining PLA with other appropriate biodegradable polymers will change various features and help to lower the cost of the total material. These polymers have comparably better flexural capabilities and high impact strength. There have been a number of documented advancements in North America and Europe involving the utilization of PLA as packaging material for products found in supermarkets. Foods such as Biota™ PLA-bottled water, Noble™ PLA-bottled juices, and Dannon™ yoghurt have all found usage for PLA-based containers. These containers are food-grade compliant according to EU and German standards. Due to its unique qualities, including biodegradability, GRAS status, and bioresource status, PLA is well suited for use in the food industry. A PLA-based antimicrobial packaging system would be superior to other antimicrobial systems in terms of cost-effectiveness, lack of regulatory problems, and antibacterial activity that is effective and environmental friendliness. Even if the PLA capability for antimicrobial packaging hasn't been thoroughly studied yet, more research and development are required to fully realize the potential of these materials [20].

13.3.5 Polyhydroxyalkanoates

Under imbalanced growth conditions, numerous bacterial species produce and accumulate PHAs, a group of bacterial polyesters. These polymers are created naturally by bacteria fermenting lipids and sugar. PHAs are simple macromolecules made up of monomers of 3-hydroxy fatty acids. A little over 55,115.57 tonnes of PHAs were manufactured for market in 2008. PHAs resemble synthetic polymers like polypropylene in terms of their thermomechanical characteristics [21].

It is of great importance that recent research has demonstrated that additional to PHB, a wide variety of PHAs can be generated by microbial fermentation because the qualities of PHAs rely on their monomer composition. The type of carbon source and microbes utilized determine the monomer makeup of PHAs. The medium chain length PHAs are elastomers with lower melting temperatures and a significantly lower degree of crystallinity, in contrast to the conventional highly crystalline thermoplastic PHB. Regarding uses for food packaging, PHAs' low water vapour permeability, which is comparable to LDPE, is a particularly intriguing characteristic. PHB has been generated biotechnologically and carefully researched as biodegradable polyester since 1925. Methyl is the alkyl substituent group. PHB has a crystallinity of over 50%, making it extremely crystalline. It reaches a melting point at 180 °C. Pure homopolymers are brittle substances. It becomes glass at about 55 °C. It shares some mechanical properties with synthetic, biodegradable polyesters, such as PLA. The amorphous phase undergoes secondary crystallization while being stored at ambient temperature. As a result, the polymer becomes harder and more brittle, while the stress and elongation modulus also rise ($E = 1.7$ GPa). The elongation at break then drops significantly (10%) [22]. It has a smaller processability window than ordinary plastics, which is a drawback. At temperatures near the melting point, PHB is vulnerable to thermal deterioration. PHB may be plasticized with citrate ester to simplify the procedure. Regarding mechanical properties and melting temperature (175–180 °C), PHB is similar to isotactic polypropylene (iPP). PHBs Tg is approximately 9 °C, and the ultimate PHB has an elongation to break that is significantly lower compared to iPP (400%). When 3HV or 4HB co-monomers are added, the mechanical properties are significantly altered. While stiffness and tensile strength are reduced, toughness rises with the addition of more of the corresponding co-monomer.

Numerous microorganisms such as bacteria, fungus, and algae can breakdown PHB in a variety of environments [23]. Though at a very modest pace, the hydrolytic breakdown results in the creation of 3-hydroxy butyric acid, a typical component of blood. To create biodegradable polymers for application in wastewater treatment, several monomers have now been grafted onto PHB. The grafted molecules were either hydrophobic (styrene or methyl acrylate) or hydrophilic (acrylic acid or sodium-p-styrene sulfonate). The degree of grafting varied with the monomers, getting stronger with styrene, sodium-p-styrene sulfonate, methyl acrylate, and acrylic acid in that order. PHB-containing multicomponent polymeric systems have been made in two different ways. The first process requires the radical polymerization of an acrylic polymer when PHB is present. The second involves melting PCL and PHB and combining them. In both procedures, peroxide is utilized to create intergrafted species that are responsible for compatibilization. Reactive blending has been used to these techniques. It should be mentioned that alternative chemical methods have been established for the manufacture of PHB in addition to the bacterial synthetic method. PHB is also produced via the ring opening polymerization of -butyrolactone [24].

The synthesis route determines the different structures that are produced. A polymer with partially stereo regular blocks is produced by chemical synthesis, whereas an isotactic polymer with arbitrary stereo sequences is produced by a bacterial process. Applications for PHB and related polymers, such Biopol, have been developed in a variety of fields, including packaging, hygiene products, agricultural and biomedical applications. Pressure-sensitive adhesives, biodegradable cheese coatings, high-solid alkyd-like paints, and biodegradable rubbers are just a few of the recent application discoveries based on medium chain length PHAs. Technically speaking, PHAs have a lot of potential. The use of biopolyesters will become financially advantageous once the cost of these products can be further lowered.

PHA polyesters can be made from renewable resources and are biocompatible and biodegradable. In addition to being totally biodegradable, they also have a number of other desired qualities like the displacement of petroleum and the reduction of greenhouse gas emissions. PHA is used in a variety of packaging materials that degrade naturally, including sheets, containers, bottles, films, laminates, and coatings. PHAs can be used to create more than 100 monomers and copolymers. PHAs are well suited for the food sector due to their high tensile strength, printability, flavor, and odor barrier qualities, thermal sealability, grease and oil endurance, temperature stability, and simplicity of dyeing. For instance, the US-based company Metabolix creates "Metabolix PHA", a blend of polyhydroxybutyrate (PHB) and poly (3-hydroxyoctanoate) that has received FDA approval for the production of food additives and the creation of packaging that maintains all the functional properties of nondegradable plastics [25].

13.4 Application of Biopolymers in Food Packaging

In food packaging, plastics products made from oil are utilized. Such polymers are inexpensive and readily available. They can be quickly fused, have strong mechanical qualities, and act as effective barriers against oxygen and volatile aromatic compounds [26]. In general, plastic products are not recyclable and are not accessible to water vapor, making them toxic to the biosphere [27]. Because of this, the usage of plastic products should be restricted and eventually substituted by ones that have minimal negative effects on the environment. Applying edible coatings and coverings comprising biopolymers like as starch, cellulose, chitosan, PLA, and PHAs can significantly improve the storage life for packaged foods. They function as efficient barriers again for movement of gases and vapor. The consumers prefer to buy products composed of biodegradable components. They also prefer to buy high-quality goods. It also addresses concerns about the ecological issues brought on by the recycling of non-renewable substances used to package food. They were ready to use more agricultural residues and are enthusiastic in using renewable resources for the food packaging. Many researchers from all over the world have researched the utilization of the edible films for packaging materials [28]. The usage of these films expands the life span of packaged foods while also enhancing their safety and purity. Biopolymer-based films can contain a variety of additives and act as barriers to prevent the diffusion of moisture, gases, and aromatic compounds [29]. Various antibacterial substances, including lysozyme, nisin, and organic acids, can be found in biodegradable films. They are employed in the manufacturing of numerous food items. This technique is the greatest method for keeping food fresh and preventing deterioration.

13.4.1 Paper Packaging

In addition to providing robust packaging, paper biopolymers can be employed to cover food items. Nonbiodegradable plastic items would be unnecessary in this approach. The paper and biopolymers combined deliver an intriguing utility while preserving the ecologically benign nature of the substance. The most common materials studied and used as biomaterials for the coating different types of papers include starch, cellulose, chitosan, isolated soy protein, caseinates, corn zein, whey protein isolate, carrageenan, wheat gluten, and alginate. When the paper biopolymer film is created based on whey protein, the use of paper for food packaging exhibits a considerable performance increase [30]. Despond et al. reported the packaging products in the paper with such a chitin/carnauba wax coating produces excellent outcomes because the paper has enhanced gas and water barrier characteristics [31].

Furthermore, every biopolymer product packaging must satisfy all specifications for traditional food packaging. This refers to the optical, mechanical, and permeability characteristics [32]. The first totally plant-based, sustainable packaging materials have been introduced by Tetra Pak. It was the first container to use caps constructed completely of bio-high-density polyethylene and films composed largely of bio-low-density polyethylene. All of these items are made from sugars and are laminated on paperboard that has received Forest Stewardship Council certification. Since 2015, this type of package has been offered for sale.

13.4.2 Edible Films and Coatings

A thin layer of substance that can be consumed is known as an edible film. It can be seen on the outside of various packaged foods, it can wrap the item, or it can be folded in between food ingredients. In other terms, it is a substance used to preserve food that increases a product's storage life without requiring removal before consumption. The product must be physically sound and able to stand on its own. The thickness of an edible film ranges from 0.050 to 0.250 mm. Food coverings that are edible and recyclable can increase the food's quality while extending its storage life and preventing environmental contamination. The stresses applied to edible films during food preparation and handling should be robust and versatile [33]. Usually, edible coatings can take the role of synthetic packing films because they have excellent mechanical characteristics. Foods can be covered with edible materials that contribute nutrients, enhance sensory qualities, and have antimicrobial capabilities. Films that are edible are produced using natural polymers. Substances that are employed to cover edible food should adhere to the regulations governing food items. Polysaccharides and proteins are the abundant renewable films. The most extensively employed polysaccharides include starch, cellulose, and its derivatives, as well as isolates taken from seaweeds. Edible polysaccharide-based coatings offer items with a variety of qualities due to the large number of polysaccharides that can be employed as raw materials for coatings. Protein films are made using gelatin, zein from corn, whey proteins, collagen from animal, soy proteins, and gluten from wheat. Protein and polysaccharide-based films are transparent to moisture but not to oxygen. The primary property of these coatings is strong mechanical durability when humidity levels are elevated.

In the modern food sector, edible packaging is extensively utilized. The food industry employs coatings to preserve food and increase shelf life. Edible films are applied using conventional coating techniques as spray fluidization, spraying, falling and pan coatings, dipping, and brushing. After these procedures, humid materials need to be dried, whereas lipid-based coatings need to be chilled. The removal of the solvent utilized in the above operations results in the formation of the most edible films. For instance, when thin coatings of a biopolymer solvents are sprayed to food and then evaporate, a thin film develops. The food sector uses ethanol and water as safe solvents. The films that are edible and recyclable should satisfy a variety of criteria, including a moisture barrier, a solute or gas barrier, a water or lipid solubility barrier, a color and appearance barrier, rheological and mechanical characteristics, and non-toxicity [34]. The kind of substance used to create coatings and films and the way those are made and applied, affects these aspects. To further enhance their qualities, plasticizers, crosslinking substances, antioxidants, antimicrobials, and texture enhancers can be applied. The use of biopolymers is important and beneficial. In general, edible films made of proteins and carbohydrates (cellulose, starch, chitosan) are more durable and flexible than edible films made of vax, and under some circumstances, they exhibit better gas-barrier abilities [35]. Han and Gennadios reported that edible packaging films are polymeric films made of edible elements that can offer a protective barrier to a commodity and build biochemical, physiological, and biological barriers between food and its external environments [36]. Additionally, edible films can enhance the look of food, can assist in maintaining its nutritional content, transport biologically active materials, and participate in shelf-life extensions [37]. Due to their ability to delay moisture absorption and minimize lipid oxidation and discoloration, edible films could enhance the quality for fresh, preserved, and processed meats, poultry, and fish. They make a product look better in retail packaging. By using them, food products that have been battered or breaded lose less mass and flavor while being fried and use less oil. Making edible films requires the application of antimicrobial and antioxidant substances for food packaging [38].

13.4.3 Antimicrobial and Active Packaging

Food poisoning, especially the kind that has occurred in recent years, has made us more aware of the necessity to develop innovative methods to prevent the growth of microbes to preserve food safety and quality. The application of antimicrobial films to a product's surface is one of the newest methods of food preservation [39]. Numerous studies demonstrate the effectiveness of antimicrobial films and coatings in lowering the levels of pathogen microbes like as *Salmonella typhi, Escherichia coli, Staphylococcus aureus, and Listeria monocytogenes* [40, 41]. The antimicrobial agents are incorporated using various matrices. Antimicrobial agents used in polymer coatings include polysaccharides, organic acids, peptides, enzymes, bacteriocins, and essential oils. The ability to act as inhibitors is a significant aspect of antimicrobial edible films. They have an impact on potential contaminants that are on a product's surface and restrict them from penetrating inside the product. Natural polymers, like cellulose, and chitosan, for instance, make sturdy films that can transport significant amounts of antibacterial chemicals. While chitosan films made from lactic or citric acids are soft and elastic, those made of hydrochloric, formic, or acetic acids are tough and brittle. This allows for the development of multilayered coatings and coverings.

Coma et al. reported when applied to the exterior of cheese, chitosan wrappers with 1% acid prevent *L. monocytogenes* and *L. innocua*. In comparison to chitosan-coated samples, the presence of an *L. innocua* population on the top of cheese was ten times greater in chitosan-free samples [42]. Gram-positive bacteria were suppressed by these films when they were made with nisin and lysozyme in both liquid and solid conditions, while gram-negative organisms were also prevented when EDTA was applied. Cerisuelo et al. reported on the active biodegradable packaging [43]. This is important because it prolongs the fruit's storage period without requiring more processing. Today, a wide range of biopolymers are employed to create antibacterial films, containing proteins like soy, whey, pea, wheat gluten, triticale, and fish protein, as well as polysaccharides like starch, cellulose, chitosan, carboxymethyl-cellulose, and mixtures of the two [44]. Antimicrobial films should contain at least one element capable of forming an adequate, consistent, cooperative, and adhesive composite. A film-forming component, plasticizers, a solvent, a pH-adjusting agent, and an antibacterial constituent are all components of mixes used to fabricate antimicrobial films [45]. When choosing antimicrobial drugs, it is crucial to have a detailed understanding of various types of microorganisms in order for them to be as efficient as possible. The classification of microorganisms into categories can be incredibly beneficial. The groups of microorganisms include those that require oxygen, those that build cell walls, those that require specific conditions during their growth phase, and those that are resistant to acids or osmosis [46]. Instead of using one antimicrobial agent, you might use two or more. Because of their complementing effects, the combination of two or more inexpensive, weaker-acting medicines can be quite effective in inhibiting microorganisms. Independent of earlier preserving food and manufacturing techniques, microorganism activity following later contamination must be avoided by the application of antibacterial coatings and films. No matter how few they may be, these films prevent the remaining bacteria from doing anything. This is how food preservation can be made less expensive without compromising safety. The acceptance and use of antimicrobial films in the food sector will depend on current regulations and the relationship between the cost of the films and their effectiveness when they are applied.

13.4.4 Biomaterials in Food Packaging

Numerous areas of food systems could be impacted by nanotechnology. Currently, advances made in nanotechnology have improved food packaging. While active packaging and antimicrobial packaging are being using, they speed up and simplify pathogen detection and barrier formation. It might be possible to replicate many components, such as colors and odors, using non-organic particles. They can also be utilized as reservoirs for the controlled release of medications or fungicides. In addition, nanoparticles can be employed to strengthen barriers against gases, enhance mechanical characteristics, and boost the resistance to oxidation of conventional biomaterials. Future food packaging will be significantly impacted by the use of nanocomposites [47]. Instead of using conventional packaging, advanced, smart packaging with multiple uses will be used. The fruits, vegetables, beverages, wine, and other kinds of foods will be better preserved in the future due to innovative materials that are being employed for food packaging. By including the right nanomaterials, it should be feasible to create packages with superior barrier, mechanical, and thermal characteristics. The ability of nano-structured materials to prevent the entry of pathogens like bacteria and viruses will be crucial for maintaining food safety. Consumers will be alerted if the food in the container has gone rotten due to nanosensors implanted in the packaging. The nanomaterials will cut down on waste from packaging. Weiss et al. reported that nanotechnology will benefit the food sector. The use of nanomaterials will boost the safety of food items. During manufacturing and distribution, food items are exposed to pathogen and contaminant detection using biosensors [48]. There are tools for tracking shipments and following a certain product. Systems that allow for the data gathering, localization, monitoring, and remote control of food items are referred to as smart systems or intelligent systems. The efficiency and security of food manufacturing and transportation can be enhanced by these solutions [49]. The most crucial information about food products can be expressed and displayed, and delivery methods that carry, safeguard, and deliver nutraceutical components to a particular location in which they become active can be enhanced.

13.5 Conclusion

In industrialized nations, biopolymer composites are now widely utilized for food packaging, essentially replacing conventional polymers, particularly for the packaging of organic, natural, and nutritional foods. Biopolymers have already been established in numerous applications, ranging from high-tech to food packaging. Despite the advantages that biopolymers have, a number of limitations prohibit their widespread commercial usage, notably in the area of packaged foods. This is typically due to affordability and performance when compared to traditional materials. Biopolymers can be divided into many categories and quality groups depending on how they are produced and used in the food sector. Cellulose-based biomaterials have the most widespread commercial application in the food packaging industry; relatively few other biobased materials are used commercially in the packaging of traditional food products. Industrial products made from biomaterials based on starch and PLA are being developed at an accelerating rate. Biopolymers are becoming more important in the application of food in this way. The main causes of this are the fact that they are made from renewable resources and that recycling allows for their reuse. Researchers that have concentrated on biopolymer applications have shown that these materials not only possess the necessary qualities for widespread usage in the food business, but also that they may one day be available for a reasonable commercial price. When compared to recyclable petroleum products, packaging materials generated from renewable bioresources may very soon be affordable and have the right characteristics for use in food packaging.

References

1 Sivakanthan, S., Rajendran, S., Gamage, A. et al. (2020). Antioxidant and antimicrobial applications of biopolymers: a review. *Food Research International* 136: 109327. doi: 10.1016/j.foodres.2020.109327.

2 Baranwal, J., Barse, B., Fais, A. et al. (2022). Biopolymer: a sustainable material for food and medical applications. *Polymers (Basel)* 14 (5): 1–22. doi: 10.3390/polym14050983.

3 Pang, Y.L., Tan, J.H., Lim, S., and Chong, W.C. (2021). A state-of-the-art review on biowaste derived chitosan biomaterials for biosorption of organic dyes: parameter studies, kinetics, isotherms, and thermodynamics. *Polymers (Basel)* 13 (17). doi: 10.3390/polym13173009.

4 Babu, R.P., O'Connor, K., and Seeram, R. (2013). Current progress on bio-based polymers and their future trends. *Progress in Biomaterials* 2 (1): 8. doi: 10.1186/2194-0517-2-8.

5 Nguyen Van Long, N., Joly, C., and Dantigny, P. (2016). Active packaging with antifungal activities. *International Journal of Food Microbiology* 220: 73–90. doi: 10.1016/j.ijfoodmicro.2016.01.001.

6 Cazón, P., Velazquez, G., Ramírez, J.A., and Vázquez, M. (2017). Polysaccharide-based films and coatings for food packaging: a review. *Food Hydrocolloids* 68: 136–148. doi: 10.1016/j.foodhyd.2016.09.009.

7 Ratnayake, W.S., Hoover, R., Shahidi, F. et al. (2001). Composition, molecular structure, and physicochemical properties of starches from four field pea (*Pisum sativum* L.) cultivars. *Food Chemistry* 74 (2): 189–202. doi: 10.1016/S0308-8146(01)00124-8.

8 Bravin, B., Peressini, D., and Sensidoni, A. (2006). Development and application of polysaccharide-lipid edible coating to extend shelf-life of dry bakery products. *Journal of Food Engineering* 76 (3): 280–290. doi: 10.1016/j.jfoodeng.2005.05.021.

9 Fishman, M.L., Coffin, D.R., Onwulata, C.I., and Willett, J.L. (2006). Two stage extrusion of plasticized pectin/poly(vinyl alcohol) blends. *Carbohydrate Polymers* 65 (4): 421–429. doi: 10.1016/j.carbpol.2006.01.032.

10 Shin, B.Y., Il Lee, S., Shin, Y.S. et al. (2004). Rheological, mechanical and biodegradation studies on blends of thermoplastic starch and polycaprolactone. *Polymer Engineering & Science* 44 (8): 1429–1438. doi: 10.1002/pen.20139.

11 Rein, D.M., Khalfin, R., and Cohen, Y. (2012). Cellulose as a novel amphiphilic coating for oil-in-water and water-in-oil dispersions. *Journal of Colloid and Interface Science* 386 (1): 456–463. doi: 10.1016/j.jcis.2012.07.053.

12 Jayakumar, R., Nwe, N., Tokura, S., and Tamura, H. (2007). Sulfated chitin and chitosan as novel biomaterials. *International Journal of Biological Macromolecules* 40 (3): 175–181. doi: 10.1016/j.ijbiomac.2006.06.021.

13 Merzendorfer, H. (2011). The cellular basis of chitin synthesis in fungi and insects: common principles and differences. *European Journal of Cell Biology* 90 (9): 759–769. doi: 10.1016/j.ejcb.2011.04.014.

14 Puvvada, Y.S., Vankayalapati, S., and Sukhavasi, S. (2012). Extraction of chitin from chitosan from exoskeleton of shrimp for application in the pharmaceutical industry. *International Current Pharmaceutical Journal* 1 (9): 258–263. doi: 10.3329/icpj.v1i9.11616.

15 Meng, X., Yang, L., Kennedy, J.F., and Tian, S. (2010). Effects of chitosan and oligochitosan on growth of two fungal pathogens and physiological properties in pear fruit. *Carbohydrate Polymers* 81 (1): 70–75. doi: 10.1016/j.carbpol.2010.01.057.

16 Yang, A., Wu, R., and Zhu, P. (2001). Thermal analysis and miscibility of chitin/polycaprolactone blends. *Journal of Applied Polymer Science* 81 (13): 3117–3123. doi: 10.1002/app.1763.

17 Briassoulis, D. (2004). An overview on the mechanical behaviour of biodegradable agricultural films. *Journal of Polymers and the Environment* 12 (2): 65–81. doi: 10.1023/B:JOOE.0000010052.86786.ef.

18 Jacobsen, S. (1999). Plastcising PLA. *Polymer Engineering & Science* 39 (7): 1303–1310.

19 Luckachan, G.E. and Pillai, C.K.S. (2006). Chitosan/oligo L-lactide graft copolymers: effect of hydrophobic side chains on the physico-chemical properties and biodegradability. *Carbohydrate Polymers* 64 (2): 254–266. doi: 10.1016/j.carbpol.2005.11.035.

20 Jin, T. and Zhang, H. (2008). Biodegradable polylactic acid polymer with nisin for use in antimicrobial food packaging. *Journal of Food Science* 73 (3). doi: 10.1111/j.1750-3841.2008.00681.x.

21 Galego, N., Rozsa, C., Sánchez, R. et al. (2000). Characterization and application of poly(β-hydroxyalkanoates) family as composite biomaterials. *Polymer Testing* 19 (5): 485–492. doi: 10.1016/S0142-9418(99)00011-2.

22 El-Hadi, A., Schnabel, R., Straube, E. et al. (2002). Correlation between degree of crystallinity, morphology, glass temperature, mechanical properties, and biodegradation of poly (3-hydroxyalkanoate) PHAs and their blends. *Polymer Testing* 21 (6): 665–674. doi: 10.1016/S0142-9418(01)00142-8.

23 Kim, M., Lee, A., Yoon, J., and Chin, I. (2000). Biodegradation of poly(3-hydroxybutyrate), Sky-Green® and Mater-Bi® by fungi isolated from soils. *European Polymer Journal* 36 (8): 1677–1685.

24 Juzwa, M. and Jedliñski, Z. (2006). Novel synthesis of poly(3-hydroxybutyrate). *Macromolecules* 39 (13): 4627–4630. doi: 10.1021/ma0602369.

25 Fakirov, S. (2015). Biodegradable polyesters. *Biodegradable Polyesters* 42: 1–347. doi: 10.1002/9783527656950.

26 Schreiber, S.B., Bozell, J.J., Hayes, D.G., and Zivanovic, S. (2013). Introduction of primary antioxidant activity to chitosan for application as a multifunctional food packaging material. *Food Hydrocolloids* 33 (2): 207–214. doi: 10.1016/j.foodhyd.2013.03.006.

27 González, A. and Alvarez Igarzabal, C.I. (2013). Soy protein – poly (lactic acid) bilayer films as biodegradable material for active food packaging. *Food Hydrocolloids* 33 (2): 289–296. doi: 10.1016/j.foodhyd.2013.03.010.

28 Khwaldia, K., Ferez, C., Banon, S. et al. (2004). Milk proteins for edible films and coatings. *Critical Reviews in Food Science and Nutrition* 44 (4): 239–251. doi: 10.1080/10408690490464906.

29 Khwaldia, K., Arab-Tehrany, E., and Desobry, S. (2010). Biopolymer coatings on paper packaging materials. *Comprehensive Reviews in Food Science and Food Safety* 9 (1): 82–91. doi: 10.1111/j.1541-4337.2009.00095.x.

30 Han, J.H. and Krochta, J.M. (2001). Physical properties and oil absorption of whey-protein-coated paper. *Journal of Food Science* 66 (2): 294–299. doi: 10.1111/j.1365-2621.2001.tb11335.x.

31 Despond, S., Espuche, E., Cartier, N., and Domard, A. (2005). Barrier properties of paper-chitosan and paper-chitosan-carnauba wax films. *Journal of Applied Polymer Science* 98 (2): 704–710. doi: 10.1002/app.21754.

32 Bordenave, N., Grelier, S., Pichavant, F., and Coma, V. (2007). Water and moisture susceptibility of chitosan and paper-based materials: structure-property relationships. *Journal of Agricultural and Food Chemistry* 55 (23): 9479–9488. doi: 10.1021/jf070595i.

33 Jooyandeh, H. (2011). Whey protein films and coatings: a review. *Journal of Nutrition* 10 (3): 296–301.

34 Sadek, C., Pauchard, L., Schuck, P. et al. (2015). Mechanical properties of milk protein skin layers after drying: understanding the mechanisms of particle formation from whey protein isolate and native phosphocaseinate. *Food Hydrocolloids* 48 (August): 8–16. doi: 10.1016/j.foodhyd.2015.01.014.

35 R, R., Philip, E., Madhavran, A., et al. (2021). Advanced biomaterials for sustainable applications in the food industry: updates and challenges. *Environmental Pollution* 283. doi: 10.1016/j.envpol.2021.117071.

36 Hun, J.H. and Cennadios, A. (2005). Edible films and coatings: a review. In: *Innovations in Food Packaging* (ed. J.H. Han), 239–262. Oxford: Elsevier.

37 Arrieta, M.P., Peltzer, M.A., López, J. et al. (2014). Functional properties of sodium and calcium caseinate antimicrobial active films containing carvacrol. *Journal of Food Engineering* 121 (1): 94–101. doi: 10.1016/j.jfoodeng.2013.08.015.

38 Rajasekharan, R., Bahuleyan, A.K., Madhavan, A., et al. (2022). Neem extract–blended nanocellulose derived from jackfruit peel for antibacterial packagings. *Environmental Science and Pollution Research* 30: 8977–8986. doi: 10.1007/s11356-022-20382-z.

39 R, R., Madhavan, A., Arun, K.P. et al. (2022). Chili post-harvest residue-derived nanocellulose composite as a matrix for in vitro cell culture and Hemigraphis colorata blended nanocellulose extends antimicrobial potential. *Sustainable Chemistry and Pharmacy* 25 (December) 2021: 100584. doi: 10.1016/j.scp.2021.100584.

40 Arancibia, M.Y., López-Caballero, M.E., Gómez-Guillén, M.C. et al. (2015). Antimicrobial and rheological properties of chitosan as affected by extracting conditions and humidity exposure. *LWT* 60 (2): 802–810. doi: 10.1016/j.lwt.2014.10.019.

41 Sánchez Aldana, D., Andrade-Ochoa, S., Aguilar, C.N. et al. (2015). Antibacterial activity of pectic-based edible films incorporated with Mexican lime essential oil. *Food Control* 50: 907–912. doi: 10.1016/j.foodcont.2014.10.044.

42 Coma, V., Martial-Gros, A., Garreau, S. et al. (2002). Edible antimicrobial films based on chitosan matrix. *Journal of Food Science* 67 (3): 1162–1169. doi: 10.1111/j.1365-2621.2002.tb09470.x.

43 Cerisuelo, J.P., Bermúdez, J.M., Aucejo, S. et al. (2013). Describing and modeling the release of an antimicrobial agent from an active PP/EVOH/PP package for salmon. *Journal of Food Engineering* 116 (2): 352–361.

44 Sánchez-González, L., Quintero Saavedra, J.I., and Chiralt, A. (2013). Physical properties and antilisterial activity of bioactive edible films containing *Lactobacillus plantarum*. *Food Hydrocolloids* 33 (1): 92–98. doi: 10.1016/j.foodhyd.2013.02.011.

45 Zinoviadou, K.G., Koutsoumanis, K.P., and Biliaderis, C.G. (2010). Physical and thermo-mechanical properties of whey protein isolate films containing antimicrobials, and their effect against spoilage flora of fresh beef. *Food Hydrocolloids* 24 (1): 49–59. doi: 10.1016/j.foodhyd.2009.08.003.

46 La Storia, A. (2011). Development and application of antimicrobial food packaging. PhD thesis. University of Naples Federico II.

47 Diéguez, M.C.V., Pelissari, F.M., do Amaral Sobral, P.J., and Menegalli, F.C. (2015). Effect of process conditions on the production of nanocomposite films based on amaranth flour and montmorillonite. *Lwt* 61 (1): 70–79. doi: 10.1016/j.lwt.2014.11.017.

48 Weiss, J., Takhistov, P., and McClements, D.J. (2006). Functional materials in food nanotechnology. *Journal of Food Science* 71 (9): 107–116. doi: 10.1111/j.1750-3841.2006.00195.x.

49 Sobhan, A., Muthukumarappan, K., and Wei, L. (2021). Biosensors and biopolymer-based nanocomposites for smart food packaging: challenges and opportunities. *Food Packaging and Shelf Life* 30. doi: 10.1016/j.fpsl.2021.100745.

14

Nanofiber Composites for Packaging Applications

Vandana A. Mooss[1], *Nimitha K.C.*[2], *RinuTresia P.X.*[3], *Jiji Abraham*[2], *and Arunima Reghunadhan*[4,*]

[1] *Department of Chemistry, Dr. Viswanath Karad MIT-World Peace University, Pune, Maharashtra, India*
[2] *Department of Chemistry, Vimala College(Autonomous), Thrissur, Kerala, India*
[3] *Department of Chemistry, Christ College, Irinjalakuda, Kerala, India*
[4] *Department of Chemistry, TKM College of Engineering, Karicode, Kollam-5, Kerala, India*
[*] *Corresponding author*

14.1 Introduction

The day-to-day of almost every human involves consuming one or another packaging material. This statement indicates the importance of packaging materials and industry. Throughout the whole supply chain management cycle, packaging is essential. In a cutthroat consumer market, it serves as a key point of differentiation. Packaging materials are in high demand as the population is increasing rapidly. The customer today recognizes that packaging is a crucial and fundamental component in the development of an effective brand identity. Proper packaging is required for consumer products to ensure safety, hygiene, ease of handling, etc. The packaging not only ensures safe delivery but as a communication medium between the producers and customers. Packaging is essential in the promotion of a product as it directly enters the mind of a consumer through the eyes. The packaging industry is growing day by day around the globe. The market scans and surveys suggest an increase of more than 26.5 million in the next five years. With the increase in the market of packaging materials, the printing industry and associated fields are also growing very quickly.

Applications of Biopolymers in Science, Biotechnology, and Engineering, First Edition.
Edited by Arunima Reghunadhan, Akhina H., and Sabu Thomas.

14.1.1 Materials Commonly Employed

The simple term packaging isn't that simple. It requires very careful selection of materials, design, and storage. Different end-use demands different specifications of materials. The most common use of packaging is in food materials where a variety of materials are employed such as paper, plastic, bioplastics, natural materials (wood and leaves), metals, and so on. The different materials used in the packaging and their applied fields are given in Table 14.1.

Material	Product	Application
Paper	Corrugated boxes Folding cartons Carry bags Common foldings	Food and beverages Pharmaceutical Cosmetics and personal care
Glass	Bottles Jars	Food and beverages Cosmetics and perfumery pharmaceuticals
Plastics (PET, HDPE, LDPE, PVC, PP), etc.	Bottles, stretch films, shrink films, flexible sheets, bags, etc.	Food and beverages Cosmetics Pharmaceuticals, electronics, etc.
Biopolymers	Sheets	Food
Wood	Boxes	Food and beverages, heavy machinery and equipment, electronic materials, etc.

Polymer-based or plastic-based packaging materials are the most common ones. The key advantages of plastic materials are their light weight, durability, sustainability, ease of molding, versatility, cost-effectiveness, energy saving, freedom of design, and so on. As all are aware, the chief disadvantage is their impact on the environment. Most plastic packaging materials are nonbiodegradable and cause serious pollution to the environment. The alternatives to plastics are demanding and here we find the role of paper and biodegradable plastics. Even though conventional plastics cannot be replaced by these materials completely, serious research is going on related to the degradation of plastics and the development of alternatives. Recycling and reuse are also considered as a solution to reduce pollution. This chapter discusses the alternatives to conventional plastic packaging materials, especially biopolymer-based composite materials.

Table 14.1 Influence of nanocellulose integration on characteristics of nanocomposite films.

Source of nano cellulose	Type of nano cellulose	Method of nano cellulose isolation	Composition of films	Properties of nano cellulose composite films	Sector	Ref
Jackfruit wastes	Cellulose nanoparticles (CNPs)	Acid hydrolysis	Oil of fennel seed, CNP and PVA	• Higher thermal stability • Better antibacterial and antioxidants activities	Food industry	[22]
Chemical precursors	CNF (cellulose nanofiber)	Solvent casting method	Chitosan/ cellulose nanofiber/glycerol	• Increased mechanical properties • Better barrier properties	Food industry	[23]
Potato peel	CNP	Acid hydrolysis	PVA, CNP and fennel seed oil	• Enhanced flexibility, elasticity, and thermal stability • Higher antibacterial and antioxidants activities	Food industry	[24]
Rice straw	CNC (cellulose nano crystal)	Acid hydrolysis	Chitosan, PVA, CNC and PEG (polyethylene glycol)	• Improved thermal stability, crystallinity, tensile strength • Enhanced biodegradation rates • Better antibacterial and fungal activities. • No influence on transparency of film.	Food industry	[25]
Carrot pomace	Modified CNF with AgNPs (silver nanoparticles)	Ultrasonication process	CNF / AgNPs and PLA	• Better transparency • Enhanced flexibility and hydrophobicity • Shows some antibacterial activity against *Bacillius cereus* and *E. coli* on surface contact. • Reduced degradation temperature. • Presence of CNF with 2Ms silver nanoparticles in composites have better mechanical properties.	Food industry	[26]

(*Continued*)

Table 14.1 (Continued)

Source of nano cellulose	Type of nano cellulose	Method of nano cellulose isolation	Composition of films	Properties of nano cellulose composite films	Sector	Ref
Bacterial	BNC		BNC and *Althaea rosea* flower gum	• Increased mechanical, barrier, thermal properties, and also better surface roughness. • Anti-ultraviolet properties. • Reinforced maximum properties at 5 wt% bacterial nano cellulose (BNC)	Food industry	[27]
Pulp of unbleached hard woods	LCNF (lignocellulose nanofibers)	Mechanical treatment	Essential oil of *Carum copticum*, chitosan, LCNF, glycerol	• Higher LCNF loading results in higher opacity and crystallinity. • For chitosan with essential oil (aromatic oil) and nanofibrils, there is strong antibacterial and antioxidant activity against *B.cereus* and *E. coli*. • Enhanced solubility and also water vapor barrier characteristics. • Good mechanical characteristics of lignocellulose nanofibers (LCNF) and essential oil-infused chitosan films.	Food industry	[28]
OPMF (oil palm mesocarp fiber CNF)	CNF	Nano fibrillation	Polylactic acid (PLA) and CNF	• Enhanced Young's modulus and tensile strength • Better poly latic acid (PLA) wettability • Enhance the crystallinity of the polylactic acid matrix by 43% at five percent carbon nanofiber (CNF)	Food industry	[29]

Pulp of palmito sheaths	CNF	Mechanical defibrillation	Chitosan-cellulose nanofibers.	• Reduced water absorption • Enhanced tensile strength • Higher crystallinity and Young's modulus • 20% reduction in ultraviolet light transmission at a 1.5% carbon nanofiber (CNF)	Food and cosmetic industry	[30]
Cocoa shells	CNF	Mechanical treatment	Glycerol, CNF, and Alginate.	• Alginate/CNF-coated Andean blueberries had improved firmness and look. • Reduced water vapor permeability and transparency of alginate sheets. • Blueberries treated with alginate/CNF had less water loss and respiration over the course of few days (specifically 21 days) in the refrigerator.	Agro-food industry	[31]

14.1.2 Bionanocomposites: An Overview

Composites are multiphase materials where the mixing of two or more components is made to improve specific properties. When the matrix components in the composites are biobased materials or biopolymers, they are termed as biocomposites. In addition to that, if we incorporate nanosized materials of any dimensions such as particles, fibers, tubes, or clusters, they are known as nanocomposites. Bionanocomposite based research started in the late 1970s. Biopolymer components can have both natural and synthetic origins. The biooriginated materials are chiefly polysaccharides such as cellulose and its derivatives, starch, soy protein isolate, chitin and chitosan, keratin, collagen, alginate, etc., and the synthetic biopolymers are poly(l-lactic acid)(PLA), polyhydroxy alkanoates (PHA), poly-(ε caprolactone)(PCL), polyglycolic acid(PGA), etc.

14.1.3 Nanofibers

Nanofibers are considered one dimensional nanomaterials where their diameter is normally at the nanometer scale. Nanofibers can be synthetic or natural and are often derived from biopolymers or metals, carbon, glass, etc. The high surface area makes them useful in many applications including tissue engineering and reinforcement. Fibrous morphology or structure can be obtained by various techniques. These techniques are mainly of two categories – conventional and modern methods. Under conventional methods come from template synthesis, electrospinning, self-assembly, sonochemical synthesis, etc. Methods such as centrifugal jet spinning, electrodynamic writing, solution blow spinning, etc., are considered modern techniques to obtain nanofibers [1]. Among these techniques, electrospinning is a well-established process. The nanofibers of biopolymers like PCL, and PLA are prepared by this method. The fiber structure can often be tuned via spinning and results in highly porous materials. The conductivity, polarity, viscosity, and electrical potential are the key factors affecting the nanofiber properties during electrospinning [2, 3]. In the case of self-assembly, the peptides are linked together to induce gelation, followed by phase separation resulting in the creation of fibers. This morphology can be obtained by template-assisted synthesis. Fibrils and hollow fibers often result. Ultrasonic irradiation into polymer solutions is the procedure followed during sonochemical synthesis. All the modern technologies enable us to fabricate fibers with tunable properties and fine structure [1].

14.1.4 Nanofiber Composites

Nanofiber composite structures include a polymer matrix and the fiber filler material. Depending on the type of interaction between polymer and filler, the morphology and properties change. The fibers were traditionally used as reinforcing agents in the polymer to enhance the properties. Nanofiber composites can be broadly divided depending on the type of matrix used. They can be ceramic matrix–polymer composites, metallic matrix composites and metal matrix composites. The nanofiber composite materials normally possess a large surface area that make them suitable for various applications. They also have thermal stability, enhanced electrical and mechanical properties, load bearing capacity, structural stability, porosity, biocompatibility, and so on depending on the choice of fiber material [4].

Various authors have reported the use of aramid fibers and glass fibers [5–8]. These fibers are well known for their strength. Cellulosic nanofibers are a good material for versatile applications [9, 10]. The variety of applications affects the importance of a material and in that sense, almost all the biobased fiber materials are considered important. The use of nanofiber composites of polymers has been explored mainly in the areas of tissue engineering [11], wound healing [12] and packaging. Here the discussion focuses on the different types of nanofiber composites used in the packaging industry.

14.1.5 Packaging Materials: Requirements

When selecting a material for packaging there are certain things that need to be taken into consideration. The general requirements of a packaging material are that it should be compatible with the material with which it is packed. The material should withstand any variations in temperature and pressure during transportation, should not release any harmful chemicals into the food and only allowed colors or pigments should be used. The most important requirement is they have to be environmentally benign, reusable and recyclable. The common packaging materials are plastics that may not always satisfy the aforementioned conditions so alternatives are always in demand. Nowadays most packaging material-related research is concentrated on biopolymer related composite materials, especially fiber composites. They have the properties of compatibility, processability, biodegradability, etc.

14.1.6 Fiber Composites in Packaging

High strength, chemical and water resistance, lightweight resistance to corrosion, high durability, resistance to fire, and electrical resistance, are the advantages of fiber composites. Natural fibers are fibers that are obtained from plants and animals. Here, cellulose is considered the main natural fiber. These fibers have excellent properties when compared with synthetic fibers because natural fiber composites are ecofriendly, organic, biodegradable, strong, and lightweight [13]. Moreover, they also have strong mechanical characteristics. These kinds of fibers have found many applications, especially in the field of packaging [14].

14.1.6.1 Cellulose Fiber Nanocomposite

Cellulose is a natural biodegradable polymer that is extracted from algae, industrial residue, agricultural wastes, plants, wood, and bacteria. Other sources are mango seed, wheat, rice, citrus peels, peanut husk, and cotton stalks [15]. Tasteless, odorless, highly crystalline, large molecular weight and being hydrophilic in nature are the main properties of cellulose [16]. These natural biopolymers are widely used in a variety of fields including energy storage, waste treatment and biomedical applications [17]. Cellulose is a sustainable and renewable resource and because of this, nowadays, great attention has been given to nanomaterials originating from the cellulose. Low toxicity, excellent tensile strength, and aspect ratio are the specific properties of nanocellulose.

Nanocelluloses are usually in the form of cellulose nanofibrils (CNFs) or cellulose nanocrystals (CNCs) that are extracted from several bioresources like bamboo, rice straw, sugar palms, etc. Many researchers have focused on the characterization and isolation of nanocellulose [18, 19]. Nanocellulose possesses good physical and mechanical properties. Because of the excellence of these properties, nanocellulose has been successfully applied in the fields of food packaging, optical materials, cosmetics, biomedical, and electronic devices.

Nanocellulose-based composites are another important derivative of nanocellulose (NC). Methods for the preparation of these composites include coating method, solvent casting layer by layer assembly, melt intercalation, in-situ, and ring-opening polymerization. The quality and safety of food have been improved by the use of NC-based composites that incorporate active agents, namely antimicrobials and antioxidants. In response to demands from the food industry, these composites have demonstrated good functional capabilities for active food packaging [20, 21].

Recently, there have been a lot of nanocellulose-based composites reported, particularly in biodegradable film forms. This type of nanocellulose-based composite is prepared from biodegradable sources such as nonwood fibers, fruit shells, fruits waste, vegetable peels, etc. Table 14.1 shows of the type of nanocellulose, the nanocellulose isolation method, film composition and the nanocomposite film's properties.

14.2 Modified Cellulose Fiber Nanocomposites

Better mechanical properties are obtained from cellulose via fibers via functionalization/modification methods. There are two types of modification: physical and chemical. Physical modification can be classified into heat treatment, plasma treatment, corona modification, UV modification, and gamma irradiation. Chemical modification is subdivided into two: non-bonding and bonding interactions. The barrier and mechanical properties of nanocomposites are improved via these methods.

14.2.1 Physical Modification

14.2.1.1 Heat Treatment

With this technique, impurities are removed from cellulosic fibers, the ratio of oxygen and carbon is raised, and the acidity of the surface of cellulose is increased by heating it in a convection oven. The fiber's mechanical characteristics are improved, and its crystallinity is increased through heat treatment [32]. A carbon nanocellulose film's (CNF) porosity and the permeability of oxygen are said to be reduced by heat treatment [33]. The mechanical and barrier properties of carbon CNFs were enhanced via heat treatment as reported by Sharma and co-workers [34].

14.2.1.2 Plasma Treatment

Heavy particles, electrons and free radicals are among the highly reactive species found in plasma that can be exploited to change the film's and fiber's surface. The various types of plasma such as low temperature, atmospheric glow discharge (AGD), and APPJ (atmospheric pressure plasma jet) are most frequently employed to modify the fiber's surface [35]. Cold plasma is used for coating purposes in packaging material. For example, Lu et al. successfully used BOPP/LDPE (biaxially oriented polypropylene/low-density polyethylene) coated with nano fibrillated cellulose layer to increase the oxygen-barrier effects of packaging material [36].

14.2.1.3 Corona Modification

One technique for activating the surface oxidation of cellulose fibers used in polyolefin composite is the corona treatment [36]. The ratio of oxygen/carbon on fiber surfaces is increased during corona treatment utilizing air, which also improves the wettability and polarity of the fibers. Additionally, the etching impact of this method improves mechanical stability and interfacial bonding between the matrix polymer and filler, which results in better barrier and mechanical characteristics [37].

14.2.1.4 Ultraviolet Irradiation

Ultraviolet light makes it easier for carbonyl groups to develop on wood and the surface of other natural fibers, which boosts the polarity of the fiber's surface. Additionally, ultraviolet radiation can cause the nearby cellulose molecules to link, improving the fiber's barrier and strength characteristics [38].

14.2.1.5 Gamma Radiation Treatment

In this method, the interaction of cellulose with gamma radiation occurred via the Compton scattering effect. Within the cellulose, there is a localization of energy induced which forms trapped radicals of macrocellulose. Gamma radiation treatment is used to increase barrier and mechanical characteristics of poly(caprolactone/nanocrystalline)/nanocellulose [39].

14.3 Chemical Method

14.3.1 Modification of Bonding

14.3.1.1 Silylation

Silane coupling reagents are very effective reagents that are used for enhancing the extent of interfacial cross-linking. In the silane reaction, more reactive sites can be produced. Because of this, alkaline-treated fibers have higher silane treatment efficiency than unreacted fiber. In order to prepare fibers for silane coupling, sodium hydroxide was applied for around 30 minutes. Following several rounds of distilled water washing, the fibers were then dried. The amount of cellulose hydroxyl groups at the interface between the fibers and matrix may be decreased by silane coupling reagents. The production of silanols is triggered by the presence of a hydrolysable alkoxy group. A stable covalent bond is then formed by the fiber's hydroxyl groups and silanols, which are chemisorbed onto the surface of fiber [40].

14.3.1.2 Mercerization Method

Good quality fibers are produced using this method [41]. The fibriliation, which follows mercerization, results in the breakdown of the composite fiber bundle into individual fibers. Mercerization makes more potential reactive sites available and improves fiber wetting. By reducing fiber diameter, mercerization raises the aspect ratio that creates a rough(coarse/ irregular) surface topography, which improves the interface of fiber–matrix adhesion and boosts mechanical characteristics [42].

14.3.1.3 Esterification Process

The esterification process that affects cellulose fibers is acetylation. In this process, acetic anhydride used as an acetylating agent. On the surface of the fibers –OH groups are acetylated, and it became highly hydrophobic in nature, so increasing the resistance to dampness. Acetylation reduces the composite's capacity to absorb water and improves the fiber–polymer interfacial adhesion. The moisture resistance and mechanical characteristics of the composites are enhanced by this change [43].

14.3.1.4 Polymer Grafting Method

In this method, two types of grafting approaches were adopted. They are "grafting from" and "grafting to", methods. These methods are used for enhancing the mechanical and barrier properties of cellulose. By pre-functionalizing the fiber's surface with an activator (initiator) when a monomer is present, the "grafting from" approach causes the polymerization process to occur, which results in the formation of the polymer on the fiber's surface [44, 45]. In the "grafting to" approach, a polymer is directly grafted to the surface of the fiber [46].

14.3.1.5 Modification of Fiber Using Isocyanates

Here, organo (aromatic and aliphatic) mono- or diisocyantes are used for the modification of the cellulose surface. To make the cellulose fibers hydrophobic, mono isocyanates are used thereby enhancing the WVP (water vapor permeability) and mechanical properties of resultant films, and to make them more compatible with non-polar thermoplastics like PLA [47]. Diisocyanates can be utilized to bind functional groups to the surface of nanocellulose to shield it when preparing thermoset matrices.

14.3.2 Non-Bonding Modification

Here cellulose underwent 2,2,6,6-tetramethylpiperidine-1-oxy (TEMPO) mediated oxidation and periodate oxidation to provide carbonyl functional groups such as aldehyde or ketone groups to the surface of the cellulose fibers. The resultant carbonyl groups can be used for the cross-linking formation between cellulose chains. This increased the mechanical and barrier characteristics of the fibers. The same thing happened in the case of periodate oxidation when the hydroxyl group of cellulose oxidized to the corresponding aldehyde.

14.4 Chitin Nanofiber Composites

Chitin is a polysaccharide of glucosamine. It is considered as the second most abundant natural biopolymer. Chitin is found to possess several properties that include inhibition of the growth of microorganisms, prevention of infection of wounds and increase in the activities of various antioxidant enzymes. Chitin is considered a biocompatible material due to its low toxicity, biodegradability, and low antigenicity [48].

Chitin is existed as nanofibrillar composites in several marine organisms like shellfish, sponges, and some fungi. Composites of chitin with calcium and silica are also known. Chitin nanofibers (CHNF) are usually prepared from marine arthropods like crabs and prawns by removing the proteins and minerals from them. Usually, this can be done with the help of sodium hydroxide or hydrochloric acid [48].

Chitin nanofibers are comparable with several non-renewable materials like glass, Kevlar etc., due to their high mechanical potential. Nanochitin is very small in size, lightweight, and highly stable. These properties make CHNF is a good material that can be used in food packaging and biomedical applications. Chitosan is a linear polysaccharide compound that contains beta linkages between D-glucosamine and N-acetyl-D-glucosamine. Chitosan can be prepared from chitin shells by treating it with an alkaline substance like sodium hydroxide. Chitin/chitosan composite films and coatings are now used as materials for food packaging. They are more effective materials than the pristine polymers. Sustainable processes for the extraction of chitin and biological production of chitin and chitosan can be enhanced by the use of suitable plasticizers [49].

Chitin nanofibers alone, along with biopolymers like thermoplastic starch, gelatin, and polyvinyl alcohol (PVA), can be made into films. These films can be prepared by the casting technique. Incorporation of starch, gelatin and PVA leads to an increase in physical and mechanical properties. Food packaging materials should prevent the transfer of moisture from the surroundings to the food particles. The resistance to moisture can be verified by water vapor permeability (WVP). For a good packaging material, the value of WVP should be as low as possible. WVP is based on three factors. Water sorption, water diffusion and water desorption. It is found that chitin nanofibers possess a higher crystallinity index and thus the permeability of water molecules is hindered. Incorporation of thermoplastic starch again creates higher hindrance towards water permeability. The structure of nano chitin fibers is verified by the field-emission scanning electron microscope (FE-SEM) images and represented in Figure 14.1. FE-SEM images of chitin nanofibers (ChNF), nanochitin-thermoplastic starch (NC-TPS), nanochitin-gelatin (NC-Gel) and nanochitin-poly vinyl alcohol (NC-PVA) are included in the figure.

A green nanocomposite can be prepared from chitin nanofibers and bamboo cellulose nanocomposites by simple blending. An investigation into the morphology, mechanical, thermal, and chemical properties, biodegradability, and water contact angles reveals that the tensile strength, Young's

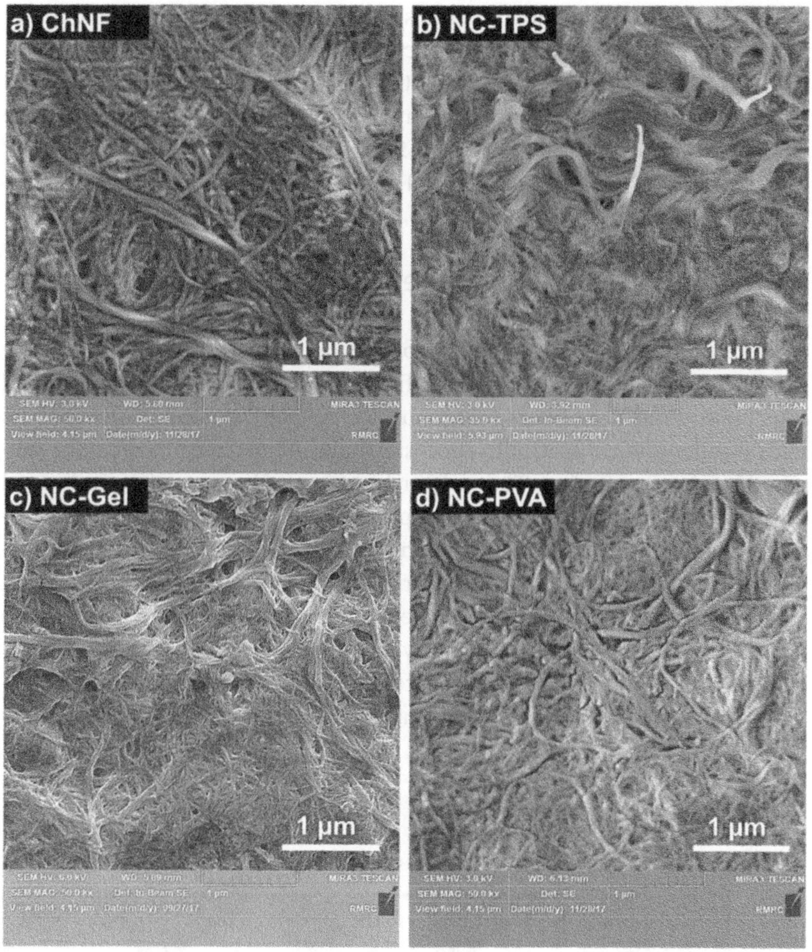

Figure 14.1 FE-SEM micrographs of (a) ChNF, (b) NC-TPS, (c) NC-Gel, and (d) NC-PVA [49]. *Source:* [49] / Springer Nature.

modulus, thermal stability, other mechanical and thermal properties, and biodegradability has tremendously improved when bamboo cellulose particles are blended with chitin nanofibers. These properties will make the material a potential candidate for food packaging applications. As the concentration of bamboo cellulose increases, the tensile strength increases up to three times and the Young's modulus increases up to 1.3 times. A food container should maintain its shape under food loading. High mechanical properties will make chitin–bamboo cellulose nanocomposites a promising material for food packaging. A comparison of the biodegradability of chitin nanofibers (CTNF), bamboo cellulose nanofiber (BACNF) and blends of CTNF and BACNF at different concentrations (chitin nanofibers at 94% (CT94:BA6), 75% (CT75:BA25) and 60% (CT60:BA40)) was carried out and improved biodegradable properties are exhibited by the blends when the concentration of BACNF increases [50]. Figure 14.2 shows the results of biodegradability test of CTNF-BACNF nanocomposites.

14.5 Electrospun Nanofiber Composites

Electrospinning is a top-down approach that can be used as an effective method for the preparation of nanofibers. Here, the polymer is stretched into micro- and nanofibers when an electric field is applied. Nanofibers with excellent biocompatibility, porosity and water permeability can be prepared by electrospinning. Thus, electrospun nanofibers are effectively used in the food packaging industry. The prepared fibers also exhibit good antioxidant and antimicrobial characteristics. The different types of electrospinning techniques are blend electrospinning, emulsion electrospinning and suspension electrospinning. Blend electrospinning can give rise to nanofibers with good hydrophobicity, spinnability, stability and mechanical and thermal properties. Gluten/zein nanofiber membrane can be prepared through this process and show a lower water vapor permeability than

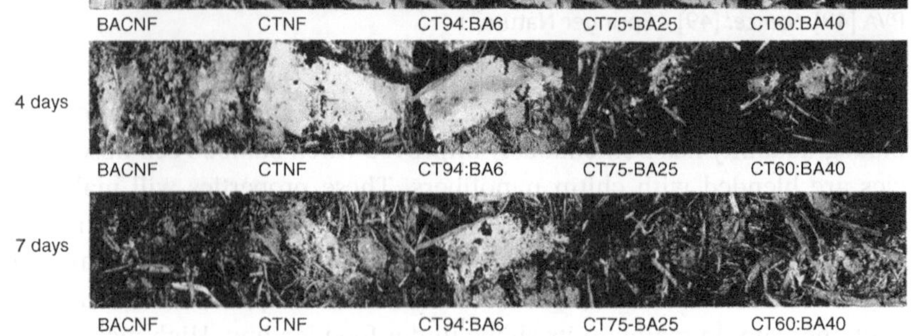

Figure 14.2 Biodegradability test of CTNF-BACNF nanocomposites [50]. *Source:* [50] / MDPI / CC BY 4.0.

PVA membranes and hence can substitute as a food packaging material. Emulsion electrospinning can make nanofibers and their composites with good encapsulation efficiency. Biopolymer-based antibacterial fiber membranes made of polylactic acid (PLA) and ultrasonically dispersed zinc oxide can be synthesized by the suspension electrospinning method. These composites show good protection against UV, visible light, and other environmental factors. Electrospun nanofibers with TiO_2 nanoparticles can be used to store bananas show delayed maturation of bananas and give good shelf life [50].

Zein, a protein present in corn, is renewable and possesses good biocompatibility. Electrospun nanofibers of zein can be used as a film coating for food preservation. Studies show that these fibers can act as edible packaging films. Electrospun poly vinyl alcohol nanofibers form packaging films that are water soluble. PVA nanowebs incorporated with vanillin/cyclodextrin can be synthesized by electrospinning and possess good thermal stability and durability. Hence, they can be used as packaging materials in the food industry. Studies have shown that electrospun PVA/cinnamon essential oil/b-cyclodextrin (PVA/cinnamon essential oil/b-CD) forms an antimicrobial film and exhibits a prolonged shelf-life of strawberries. Figure 14.3 represents changes that occurred with strawberries when kept in PVA/cinnamon essential oil/b-CD nanofilms.

More than 100 polymers can be electrospun and more can be produced by modified coaxial electrospinning. Electrospun nanofiber film of nano hydroxyapatite and gelatine provides a strong network structure, which improves the water resistance of the film. The properties of electrospun nanofibers can be improved by the addition of certain active materials. The antioxidant and antimicrobial characteristics of electrospun fibers are found to be improved by the addition of bioactive substances like essential oils and phenolic compounds. An electrospun nanofiber with compounded gelatine and two essential oils gives edible food packaging materials.

PLA is now becoming a good packaging material. It is completely biodegradable, recyclable, and environmentally friendly. Electrospun PLA nanofibers possess high transparency and good mechanical properties. Rigid porous PLA nanofibers have high tensile strength and small strain and are obtained by electrospinning.

Thus, one of the most important, promising, and future applications of electrospun nanofibers is in food packaging. It is because of the fact that the electrospun nanofibers can easily incorporate antioxidants, antimicrobials and other bioactive resources. This will enhance the shelf life and safety of the food materials. The applications of electrospun nanofibers in the food industry include encapsulation, food packaging, enzyme immobilization and filtration [51].

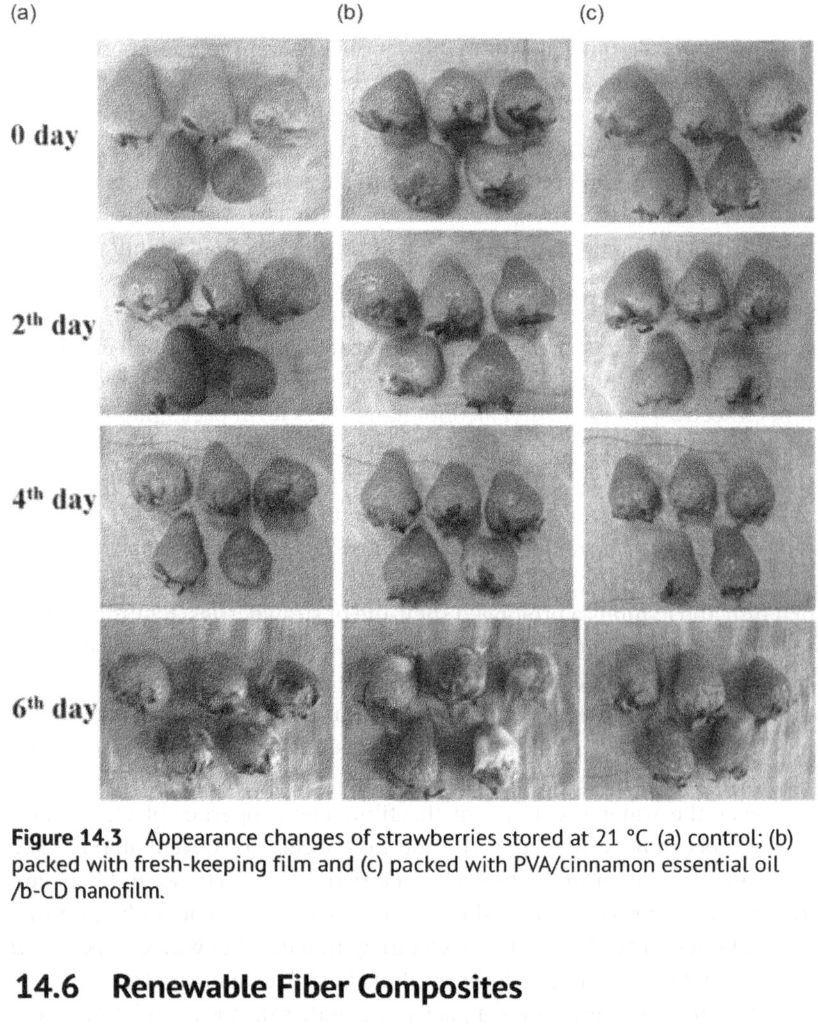

Figure 14.3 Appearance changes of strawberries stored at 21 °C. (a) control; (b) packed with fresh-keeping film and (c) packed with PVA/cinnamon essential oil /b-CD nanofilm.

14.6 Renewable Fiber Composites

Biopolymer composites perform best as biodegradable, renewable, and eco-friendly materials and are widely used in the food packaging industry. Cellulose nanofiber is one of the renewable fibers that are less expensive, less toxic and possess exceptionally good strength to weight ratio and are now widely used in the packaging industry. Cellulose nanofibers can be synthesized from lignocellulosic biomass via physical and chemical treatments. However, pristine cellulose nanofibers have poor barrier and mechanical properties and are not thermally stable. An alternative is produced by using cellulose as a filler to make composite materials. Cellulose fiber-based packaging materials do not cause any effect on the materials that are being packed. These materials are termed green materials and cause no hazards to human beings. Properly produced composites always give a cleaner environment and have found applications in sustainable packaging fields. Sustainable packaging is the manufacturing of packaging materials which are recyclable and biodegradable. They always provide a proper life cycle assessment and ensure a clean, safe, and non-hazardous environment for upcoming generations [52].

PLA is a biodegradable polymer that possesses a higher degree of crystallinity, melting temperature and glass transition temperature. Nanocomposites of PLA can be synthesized by methods like injection molding, extrusion and blending and act as a perfect substitute for thermoplastics used in food packaging. Antimicrobial packaging films provide a good shelf life. Starch, zein and PLA based films and composites, which are produced by the solvent casting technique, show antimicrobial properties. Renewable polymer packaging materials are stable at higher pressures and the food materials can then undergo high pressure processing (HPP) after packing. Figure 14.4 represents solid carrots packed with a multilayer zein-poly caprolactone (PCL) composite before and after high pressure (HP) treatment. It is observed that HP treatment up to a pressure of 700 MPa does not affect on the oxygen and water vapor barrier properties of composites [53].

Polymer composites of polyhydroxyalkanoates (PHAs), poly (b-hydroxy butyrate) (PHB) and poly(3-hydroxybutyrate-co23-hydroxyvalerate) (PHBV), bacterial cellulose and xanthan perform well as renewable packaging materials [54]. Bionanocomposites can provide good mechanical, thermal, barrier, and biodegradation properties along with new packaging

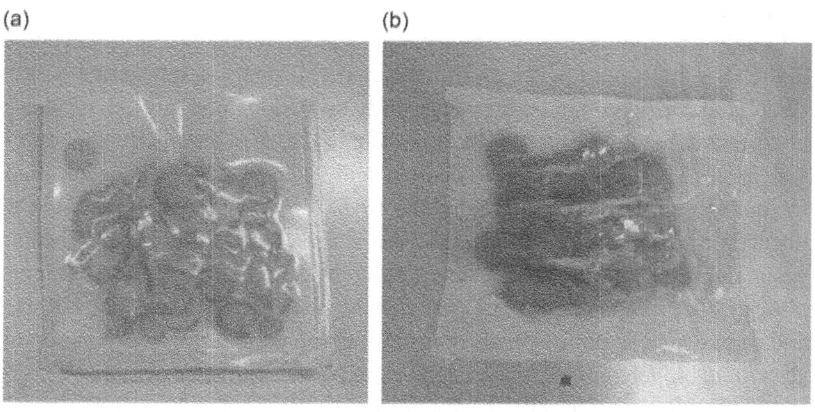

Figure 14.4 (a) Multilayer zein-PCL package containing solid carrots before HP treatment; (b) multilayer zein-PCL package containing solid carrots after HP pasteurization at 700 MPa [53]. *Source:* [53] / with permission of Elsevier.

functionalities. They can act as good oxygen scavengers and possess antimicrobial activities etc. [55]. More studies have to be carried out to fabricate such types of renewable nanocomposites to enhance the possibilities in the field of food packaging.

14.7 Characterization of Packaging Materials

14.7.1 Permeability

Permeability is one of the important properties of packaging films. It is a measure with which any transporting substance/gas passes through a film. Currently, developing biodegradable packaging materials with improved barrier properties is gaining considerable momentum. In food packaging films, a major emphasis has been given to developing superior barrier properties against oxygen, carbon dioxide, water vapors and any toxic materials from the environment to ensure safety and product-package shelf life. Oxidative rancidity of oil or fat in the food products and deterioration of texture and weight happen respectively due to oxygen and water permeation. The key factors such as crystallinity, orientation of fillers and polymorphism have been reported to influence on gas permeability [56]. There have been many experiments that paved the way to reduce moisture uptake or permeability and the inclusion of biofibers was found to be an effective strategy [57]. This, to a great extent, prevents moisture uptake by offering swelling constraints owing to the strong molecular interactions imparted after appropriate modification. Conventionally, the modification of natural fibers was attributed to alkali, silane, acetylation benzoylation reactions [58].

Cellulose nanofiber displayed better compactness compared to chitosan fibers in the PVA matrix and hence lowers oxygen permeability significantly [59, 60]. Chen et al. carried out a comparative study on various non-wooden cellulose fibers such as bamboo, cotton and sisal prepared by [2,2,6,6-Tetramethylpiperidin-1-yl] oxyl (TEMPO)-mediated oxidation and high-pressure homogenization methods to reinforce corn starch matrix [61]. The study revealed that among fibers, bamboo-reinforced films possessed the lowest diffusion rate of water vapor and oxygen. Although the fibers under investigation were of similar morphology, they exhibited different permeability rates owing to variations in size and degree of crystallinity. Wang et al. have observed that the incorporation of glycerin in biopolymers reduces the moisture permeability owing to their plasticizing effect [62]. Tarique et al. have studied the efficacy of thermoplastic arrowroot starch films as food packaging materials as it could enormously inhibit the water vapor permeability after loading of 10% arrowroot bagasse fibers [63].

Although numerous fiber reinforcements studies were reported in the context of biopolymer composites, arrowroot bagasse fiber has showed profound water holding capacity (>130%) due to its chemical composition. Furthermore, similar observations were noted by Muller et al. when cassava starch films were modified with fiber in the presence of glycerol [64].

The degree of permeability increases proportionally with respect to temperature. Additionally, the lack of a proper orientation of polymeric chains reduces the overall gas barrier properties. Furthermore, permeability to odor has also gained substantial interest as the packaging has to be protected against the permeation of any foreign odor. Carbon nano fibrils modified chitosan films showed improvement in opacity/light barrier effect ascribed to the compact arrangement of filler and matrix and hence light scattering [65].

14.7.2 Toxicity

Various types of natural fibers such as hemp, bamboo, and flax possess promising antibacterial activities [66]. The incorporation of natural fibers such as arrowroot in arrowroot starch films has demonstrated an ability to inhibit the colonization of a broad spectrum of pathogenic bacteria [63]. The arrowroot rhizomes consist of glycosides, terpenoids, tannins, flavones, and phenolic compounds, which play a key role in antibacterial or antifungal activities [67].

Chen et al. have investigated the efficacy of essential oil modified chitosan films for packaging films [65]. For the post-harvest packaging of fruits, biocomposities films made from pomegranate fibers and pectin exhibited better antifungal activity in the presence of thyme and clove essential oils. Thyme oil exhibited better antifungal activities than clove oil, which was ascribed to more phenolic constituents amounting to 52%–89% of its total composition. The studies unravelled the plausible mechanism that the inhibition of fungal/microbial growth was evidently due to the evolution of oxidized products from essential oils. Also, the inclusion of nanoparticles along with nanofibers imparted both antibacterial and anti-odor properties.

14.7.3 Mechanical Properties

The mechanical properties of fiber-modified films are governed by the composition, internal structure, the amount of cellulose, and the crystallite size of fibers [68]. The utility of natural fibers needs to overcome certain shortcomings such as water-insolubility, strong moisture absorption, and poor fiber-polymer compatibility. Plant-based fibers consisting hydrophilic groups show poor compatibility with hydrophobic matrices and often lead to processing defects. Therefore, the surface treatments are seen to be favorable to augment the compatibility issues. Alkali treatment is an effective process although an excess of alkali could lead to surface damage and thereby deteriorate mechanical strength. The mechanical performance of fiber-reinforced packaging materials lies in various factors such as (1) the nature of fiber, (2) the interface between matrix and fibers, and (3) the effective transfer of stress between matrix and fibers [69]. High modulus and strength for fiber-reinforced polymers are attributed to their superior percolation effect and interactions. When CNF was reinforced in polyvinyl alcohol films, they exhibited superior mechanical properties owing to the hydrogen bonding between fiber and PVA. Moreover, crystallinity of the cellulose fiber also was found to be contributing towards the enhanced tensile strength and elongation of PVA films [24]. Noteably, very high loading of cellulose nanofibers often does not contribute to high mechanical properties due to the tendency of agglomeration and consequently uneven stress distribution. The water absorption property also has an inverse effect on tensile strength and flexural strength [70]. Carboxyl cellulose nanofibrils incorporating chitosan films were evaluated for superior tensile strength and elongation at break. Chen et al. described the intermolecular hydrogen bonding and electrostatic force between matrix and cellulose nanofibrils [65, 71]. On the contrary, fibers possessing more polar groups lead to the inhomogeneous distribution of fibers in the matrix and fail to transfer load evenly. Dynamic mechanical analysis (DMA) of fiber-reinforced polymers provides essential information to evaluate the stress transfer and adhesive characteristics between fiber and matrices. Siaking et al. have identified that the optimum combination of coir fiber and pineapple leaf fiber in the polylactic acid matrix could enhance the storage modulus (stiffness) significantly, and the influence due to pineapple leaf fibers was intense [72]. Remarkably, individually both fibers did not impart any improvement in the tensile and flexural properties of polylactic acid films.

14.7.4 Morphology

The process of chemical and physical treatments involved, such as shearing of fibers, are seen to have direct correlation with the fiber length and morphology [73]. Ramesh and Radhakrishnan investigated whether the extraction method would considerably change the structure of cellulose nanofibers and hence the evaluation of the extent of agglomeration and degree of entanglement using SEM/TEM analysis [74]. The fiber-modified packaging film possessing a homogeneous morphology without any cracks as well as no sign of phase separation, which is desirable for packaging applications. A typical microstructure of fiber-reinforced packaging films exhibits minute cracks or pores in the presence of essential oil. The results reveal that even at low concentration, fillers could alter the surface properties of the polymer matrix. In another report, mechano-chemically treated hemp fibers consisting of crystalline cellulose fibrils were stated as promising biocomposities for packaging applications [75]. The resultant fibers after mechanical milling induced a reduction in mean diameter with a high degree of defibrillation (Figure 14.1). The nano zinc oxide in fibrous morphology imparted antibacterial properties to the polyhydroxybutyrate-co-3-hydroxyvalarate films due to the superior reinforcing effect in comparison to other morphologies. The ZnO fibers also improved the dispersion and distribution of themselves in the matrix [76]. The authors demonstrated the role of biofiber microstructure while governing the smoothness and uniformity of the polymer and consequently on mechanical properties. Uniform distribution and orientation of fibers justified the strong adhesion between matrix and fibers. Subramanian et al. have explained the role of the fiber architecture on moisture diffusivity and tortuosity of biocomposites to elucidate other desirable characteristics [77]. In a comparative study on wheat straw fibers and wheat straw/HDPE, Zhang et al. have reported that fiber reinforcement composites exhibited a substantial reduction in morphological defects and improved dimensional stability in presence of 0.4% nano silicon carbide [78]. Notably, the fibers before and after hydrolysis rendered similar morphology; however, the significant decrease in the thickness was manifested in an increase in hydrolysis time [79].

14.7.5 Thermal Properties

The thermal behavior of fiber-reinforced polymer materials provides additional information regarding the extent of matrix and filler interactions [80]. Enhanced thermal stability is one of the most desirable properties of polymer composites used in packaging applications. Inorganic fillers such as nano silicon carbide along with wheat straw fiber imparted better thermal stability than wheat straw-based biocomposites [78]. In this study, the modifications attributed to fibers were corroborated by unravelling the degradation profile during pyrolysis. Nguyen et al. reported the enzyme catalysis of hydrolyzed fibers of polylactic acid with noticeable differences in their melting enthalpy and surface softening temperature using differential scanning calorimetric analysis [79]. The selective hydrolysis of the amorphous region in the outer layers of the fiber was ascribed to the highest thermal characteristics. Further, nano cellulose fiber having difference in size, diameter, and aspect ratio from various sources displayed almost similar thermal degradation pattern reiterated the same chemical composition except the thermal stability [71]. Bamboo-based nano cellulose was superior in terms of thermal stability and had minimum weight loss; however, a proportionally higher concentration of nano cellulose could not augment the thermal stability of the composite. Similarly, a study on sugar palm fiber derived nanocellulose revealed that the percentage of lignin also played a major role in the thermal stability of biocomposities [81]. Optimization of the treatment process such as delignification (sodium hypochlorite and acetic acid) and mercerization (sodium hydroxide) were crucial to enhance thermal characteristics. Similarly, benzoylated fiber was successfully improved thermal stability owing to the better compatibility between fiber and matrix [82]. These results paved the way for finding numerous strategies of modifications of plant fibers for superior thermal behavior. A combination of bioadhesive with lotus fiber employed superior thermal stability along with thermally insulating properties enabling the scope of such materials for packaging applications. Flame retardancy could be achieved by specific fibers such a flax fibers or by employing inorganic metal oxide fillers [69].

14.7.6 Biodegradability

To witness green and sustainable development, biodegradability and composability of polymer packaging materials are highly crucial. Developing biosourced or biodegradable composite have gained considerable momentum and it is anticipated that due to the intervention of nanotechnology, composites should not be sacrificed to meet the requirement of biodegradability. Among biocomposities, the employing natural fibers was reported to be one of the effective strategies owing to high biodegradability, low cost, and ease of availability. The studies on lignocellulose fibers have been widely investigated due to its non-abrasive nature, low density, and low impact on health [83]. Arun et al. have examined the strengthening effect of CNF in polyvinyl alcohol matrix without compromising its characteristic biodegradability. The dissociation of CNF into disaccharide and cellobiose eventually increased the enzymatic degradation of polyvinyl alcohol films [84]. Similarly, the superior compatibility of starch and filler such as cellulose fibers were described by the virtue of glucose units present in both matrix and fillers. Additionally, they are fully biodegradable and also could be used to feed animals. Thermoplastic arrowroot films containing arrowroot fibers under a soil burial test showed that the weight loss of films was significantly enhanced in the presence of arrowroot fibers. A similar observation was reported by Hazrati et al. on biocomposites in the presence of *Dioscorea hispida* fibers, where the highest fiber loading enhanced the rate of biodegradation [61]. Bordon et al. have evaluated banana fiber (5 wt %) from banana waste creating protective fruit packaging bags with 90% biodegradability, making it a suitable candidate for a circular economy and promising sustainable development [83]. Polylactic acid packaging materials were not only biodegradable but also compostable. However, hydrolytic degradation is favored only at a higher temperature and desirable humidity. The biodegradation of cellulose fiber thermoplastic starch films was attempted using white rot fungi (*T. versicolor*). The unmodified starch films showed rapid weight loss within 30 days. However, in the presence of cellulose nano fiber or acyl cellulose nanofibers, the degradation rate has been retarded (Figure 14.2) High degree of crystallinity and the formation of a denser layer of water prevents the spread of degrading enzymes [60]. The role of nanofiber accelerating the degradation kinetics of polymer films also further sheds light on the scope of fiber reinforcement and its future prospects for various packaging industries [59].

14.8 Conclusions

In conclusion, nanofiber composites hold immense promise for revolutionizing packaging applications. Their unique combination of strength, flexibility, and barrier properties, along with their potential for sustainable and eco-friendly production, makes them a compelling choice for the packaging industry. As we continue to explore and develop new nanofiber composite materials and manufacturing techniques, we can anticipate even greater advancements in packaging solutions. These innovations have the potential to reduce waste, improve product shelf life, and enhance overall sustainability, ultimately benefiting both consumers and the environment. As research in this field progresses, we look forward to witnessing the continued evolution of nanofiber composites in the packaging industry, helping to address the challenges of modern packaging while minimizing its environmental footprint.

References

1 Alghoraibi, I. and Alomari, S. (2019). Different methods for nanofiber design and fabrication. In: *Handbook of Nanofibers* (ed. A. Barhoum, M. Bechelany, and A.S.M. Makhlouf), 79–124. London: Springer.

2 Wang, C., Wang, J., Zeng, L. et al. (2019). Fabrication of electrospun polymer nanofibers with diverse morphologies. *Molecules* 24.

3 Udomluck, N., Koh, W.-G., Lim, D.-J., and Park, H. (2020). Recent developments in nanofiber fabrication and modification for bone tissue engineering. *International Journal of Molecular Sciences* 21 (1): 99.
4 Ramalingam, M. and Ramakrishna, S. (2017). 1 – Introduction to nanofiber composites. In: *Nanofiber Composites for Biomedical Applications* (ed. M. Ramalingam and S. Ramakrishna), 3–29. Oxgford: Elsevier.
5 Yu, Z., Ait-Kadi, A., and Brisson, J. (1991). Nylon/Kevlar composites. I: mechanical properties. *Polymer Engineering & Science* 31 (16): 1222–1227.
6 Park, S.-J., Seo, M.-K., Ma, T.-J., and Lee, D.-R. (2002). Effect of chemical treatment of Kevlar fibers on mechanical interfacial properties of composites. *Journal of Colloid and Interface Science* 252 (1): 249–255.
7 Sathishkumar, T.P., Satheeshkumar, S., and Naveen, J. (2014). Glass fiber-reinforced polymer composites–a review. *Journal of Reinforced Plastics and Composites* 33 (13): 1258–1275.
8 Singh, J., Kumar, M., Kumar, S., and Mohapatra, S.K. (2017). Properties of glass-fiber hybrid composites: a review. *Polymer-Plastics Technology and Engineering* 56 (5): 455–469.
9 Coran, A.Y., Boustany, K., and Hamed, P. (1974). Short-fiber—rubber composites: the properties of oriented cellulose-fiber—elastomer composites. *Rubber Chemistry and Technology* 47 (2): 396–410.
10 Venkatarajan, S. and Athijayamani, A. (2021). An overview on natural cellulose fiber reinforced polymer composites. *Materials Today: Proceedings* 37: 3620–3624.
11 Mortimer, C.J. and Wright, C.J. (2017). The fabrication of iron oxide nanoparticle-nanofiber composites by electrospinning and their applications in tissue engineering. *Biotechnology Journal* 12 (7): 1600693.
12 Liu, M., Duan, X.-P., Li, Y.-M. et al. (2017). Electrospun nanofibers for wound healing. *Materials Science and Engineering C* 76: 1413–1423.
13 Keya, K.N., Kona, N.A., Koly, F.A. et al. (2019). Natural fiber reinforced polymer composites: history, types, advantages, and applications. *Materials Engineering Research* 1 (2): 69–85.
14 Nguong, C.W., Lee, S.N.B., and Sujan, D. (2013). A review on natural fiber reinforced polymer composites. *International Journal of Materials and Metallurgical Engineering* 7 (1): 52–59.
15 Sharmila, G., Muthukumaran, C., Kumar, N.M. et al. (2020). Food waste valorization for biopolymer production. In: *Current Developments in Biotechnology and Bioengineering* (ed. S. Varjani, A. Pandey, E. Gnansounou et al.), 233–249. Oxford: Elsevier.
16 Mishra, R.K., Sabu, A., and Tiwari, S.K. (2018). Materials chemistry and the futurist eco-friendly applications of nanocellulose: status and prospect. *Journal of Saudi Chemical Society* 22 (8): 949–978.

17 Li, Y.-Y., Wang, B., Ma, M.-G., and Wang, B. (2018). Review of recent development on preparation, properties, and applications of cellulose-based functional materials. *International Journal of Polymer Science* 2018: 8973643.

18 Khan, M.N., Rehman, N., Sharif, A. et al. (2020). Environmentally benign extraction of cellulose from dunchi fiber for nanocellulose fabrication. *International Journal of Biological Macromolecules* 153: 72–78.

19 Rashid, S. and Dutta, H. (2020). Characterization of nanocellulose extracted from short, medium, and long grain rice husks. *Industrial Crops and Products* 154: 112627.

20 Bideau, B., Bras, J., Adoui, N. et al. (2017). Polypyrrole/nanocellulose composite for food preservation: barrier and antioxidant characterization. *Food Packaging and Shelf Life* 12: 1–8.

21 Razavi, R., Molaei, R., Moradi, M. et al. (2020). Biosynthesis of metallic nanoparticles using mulberry fruit (*Morus alba* L.) extract for the preparation of antimicrobial nanocellulose film. *Applied Nanoscience* 10: 465–476.

22 Jancy, S., Shruthy, R., and Preetha, R. (2020). Fabrication of packaging film reinforced with cellulose nanoparticles synthesised from jack fruit non-edible part using response surface methodology. *International Journal of Biological Macromolecules* 142: 63–72.

23 Mohammadi Sadati, S.M., Shahgholian-Ghahfarrokhi, N., Shahrousvand, E. et al. (2022 August 24). Edible chitosan/cellulose nanofiber nanocomposite films for potential use as food packaging. *Materials Technology* 37 (10): 1276–1288. doi: 10.1080/10667857.2021.1934367.

24 Ramesh, S. and Radhakrishnan, P. (2019). Cellulose nanoparticles from agro-industrial waste for the development of active packaging. *Applied Surface Science* 484: 1274–1281.

25 Perumal, A.B., Sellamuthu, P.S., Nambiar, R.B., and Sadiku, E.R. (2018). Development of polyvinyl alcohol/chitosan bio-nanocomposite films reinforced with cellulose nanocrystals isolated from rice straw. *Applied Surface Science* 449: 591–602.

26 Szymańska-Chargot, M., Chylińska, M., Pieczywek, P.M. et al. (2020). Evaluation of nanocomposite made of polylactic acid and nanocellulose from carrot pomace modified with silver nanoparticles. *Polymers (Basel)* 12 (4): 812.

27 Yekta, R., Mirmoghtadaie, L., Hosseini, H. et al. (2020). Development and characterization of a novel edible film based on *Althaea rosea* flower gum: investigating the reinforcing effects of bacterial nanocrystalline cellulose. *International Journal of Biological Macromolecules* 158: 327–337.

28 Jahed, E., Khaledabad, M.A., Almasi, H., and Hasanzadeh, R. (2017). Physicochemical properties of *Carum copticum* essential oil loaded chitosan films containing organic nanoreinforcements. *Carbohydrate Polymers* 164: 325–338.

29 Ariffin, H., Norrrahim, M.N.F., Yasim-Anuar, T.A.T. et al. (2018). Oil palm biomass cellulose-fabricated polylactic acid composites for packaging applications. In: *Bionanocomposites for Packaging Applications* (ed. M. Jawaid and S.K. Swain), 95–105. London: Springer.

30 Franco, T.S., Amezcua, R.M.J., Rodrìguez, A.V. et al. (2020). Carboxymethyl and nanofibrillated cellulose as additives on the preparation of chitosan biocomposites: their influence over films characteristics. *Journal of Polymers and the Environment* 28: 676–688.

31 Medina-Jaramillo, C., Quintero-Pimiento, C., Gómez-Hoyos, C. et al. (2020). Alginate-edible coatings for application on wild Andean blueberries (*Vaccinium meridionale swartz*): effect of the addition of nanofibrils isolated from cocoa by-products. *Polymers* (Basel) 12 (4): 824.

32 Cao, Y., Sakamoto, S., and Goda, K. (2007). Effects of heat and alkali treatments on mechanical properties of kenaf fibers. *16th International Conference On Composite Materials*, 1–4.

33 Xia, J., Zhang, Z., Liu, W. et al. (2018). Highly transparent 100% cellulose nanofibril films with extremely high oxygen barriers in high relative humidity. *Cellulose* 25: 4057–4066.

34 Sharma, S., Zhang, X., Nair, S.S. et al. (2014). Thermally enhanced high performance cellulose nano fibril barrier membranes. *RSC Advances* 4 (85): 45136–45142.

35 Kalia, S., Thakur, K., Celli, A. et al. (2013). Surface modification of plant fibers using environment friendly methods for their application in polymer composites, textile industry and antimicrobial activities: a review. *Journal of Environmental Chemical Engineering* 1 (3): 97–112.

36 Lu, P., Guo, M., Xu, Z., and Wu, M. (2018). Application of nanofibrillated cellulose on BOPP/LDPE film as oxygen barrier and antimicrobial coating based on cold plasma treatment. *Coatings* 8 (6): 207.

37 Ragoubi, M., Bienaimé, D., Molina, S. et al. (2010). Impact of corona treated hemp fibers onto mechanical properties of polypropylene composites made thereof. *Industrial Crops and Products* 31 (2): 344–349.

38 Zaman, H.U., Khan, M.A., and Khan, R.A. (2009). Improvement of mechanical properties of jute fibers-polyethylene/polypropylene composites: effect of green dye and UV radiation. *Polymer-Plastics Technology and Engineering* 48 (11): 1130–1138.

39 Khan, R.A., Beck, S., Dussault, D. et al. (2013). Mechanical and barrier properties of nanocrystalline cellulose reinforced poly (caprolactone) composites: effect of gamma radiation. *Journal of Applied Polymer Science* 129 (5): 3038–3046.

40 Agrawal, R., Saxena, N.S., Sharma, K.B. et al. (2000). Activation energy and crystallization kinetics of untreated and treated oil palm fiber reinforced phenol formaldehyde composites. *Materials Science and Engineering A* 277 (1–2): 77–82.

41 Ray, D., Sarkar, B.K., Rana, A.K., and Bose, N.R. (2001). Effect of alkali treated jute fibers on composite properties. *Bulletin of Materials Science* 24: 129–135.

42 Faruk, O., Birat, K.C., Sobh, A. et al. (2017). Natural fiber reinforced thermoplastic composites. In: *Lightweight and Sustainable Materials for Automotive Applications* (ed. O. Faruk, J. Tjong, and M. Sain), 1–38. Boca Raton: CRC Press.

43 Shojaeiarani, J., Bajwa, D.S., and Hartman, K. (2019). Esterified cellulose nanocrystals as reinforcement in poly (lactic acid) nanocomposites. *Cellulose* 26 (4): 2349–2362.

44 Saedi, S., Madaeni, S.S., Seidi, F. et al. (2014). Fixed facilitated transport of CO2 through integrally skinned asymmetric polyethersulfone membrane using a novel synthesized poly (acrylonitrile-co-N, N-Dimethylaminopropyl acrylamide). *Chemical Engineering Journal* 236: 263–273.

45 Xu, K., Li, K., Tu, D. et al. (2014). Reinforcement on the mechanical-, thermal-, and water-resistance properties of the wood flour/chitosan/poly (vinyl chloride) composites by physical and chemical modification. *Journal of Applied Polymer Science* 131 (18).

46 Le Gars, M., Bras, J., Salmi-Mani, H. et al. (2020). Polymerization of glycidyl methacrylate from the surface of cellulose nanocrystals for the elaboration of PLA-based nanocomposites. *Carbohydrate Polymers* 234: 115899.

47 Espino-Pérez, E., Bras, J., Ducruet, V. et al. (2013). Influence of chemical surface modification of cellulose nanowhiskers on thermal, mechanical, and barrier properties of poly (lactide) based bionanocomposites. *European Polymer Journal* 49 (10): 3144–3154.

48 Serventi, L., He, Q., Huang, J. et al. (2021). Advances in the preparations and applications of nanochitins. *Food Hydrocolloids for Health* 1: 100036.

49 Mushi, N.E. (2014). *Chitin nanofibers, networks, and composites: preparation, structure, and mechanical properties*. Doctoral Thesis. KTH Royal Institute of Technology.

50 Souza, V.G.L., Pires, J.R.A., Rodrigues, C. et al. (2020). Chitosan composites in packaging industry—current trends and future challenges. *Polymers (Basel)* 12 (2): 417.

51 Senthil Muthu Kumar, T., Senthil Kumar, K., Rajini, N. et al. (2019). A comprehensive review of electrospun nanofibers: food and packaging perspective. *Composites Part B: Engineering* 175: 107074.

52 Qasim, U., Osman, A.I., Al-Muhtaseb, A.H. et al. (2021). Renewable cellulosic nanocomposites for food packaging to avoid fossil fuel plastic pollution: a review. *Environmental Chemistry Letters* 19 (1): 613–641. doi: 10.1007/s10311-020-01090-x.

53 Mensitieri, G., Di Maio, E., Buonocore, G.G. et al. (2011). Processing and shelf-life issues of selected food packaging materials and structures from renewable resources. *Trends in Food Science and Technology* 22 (2): 72–80.

54 Scarfato, P., Di Maio, L., and Incarnato, L. (2015). Recent advances and migration issues in biodegradable polymers from renewable sources for food packaging. *Journal of Applied Polymer Science* 132 (48). doi:10.1002/app.42597.

55 Ramos, Ó.L., Pereira, R.N., Cerqueira, M.A. et al. (2018). Chapter 8 – Bio-based nanocomposites for food packaging and their effect in food quality and safety. In: *Handbook of Food Bioengineering* (ed. A.M. Grumezescu and A.M. Holban), 271–306. Academic Press.

56 Cocca, M., Di Lorenzo, M.L., Malinconico, M., and Frezza, V. (2011). Influence of crystal polymorphism on mechanical and barrier properties of poly(l-lactic acid). *European Polymer Journal* 47 (5): 1073–1080.

57 Kanehashi, S., Kusakabe, A., Sato, S., and Nagai, K. (2010). Analysis of permeability; solubility and diffusivity of carbon dioxide; oxygen; and nitrogen in crystalline and liquid crystalline polymers. *Journal of Membrane Science* 365 (1): 40–51.

58 Syverud, K. and Stenius, P. (2009). Strength and barrier properties of MFC films. *Cellulose* 16 (1): 75–85. doi: 10.1007/s10570-008-9244-2.

59 Nazrin, A., Sapuan, S.M., Zuhri, M.Y.M. et al. (2020). Nanocellulose reinforced thermoplastic starch (TPS), polylactic acid (PLA), and polybutylene succinate (PBS) for food packaging applications. *Frontiers in Chemistry* 8: 213.

60 Babaee, M., Jonoobi, M., Hamzeh, Y., and Ashori, A. (2015). Biodegradability and mechanical properties of reinforced starch nanocomposites using cellulose nanofibers. *Carbohydrate Polymers* 132: 1–8.

61 Hazrati, K.Z., Sapuan, S.M., Zuhri, M.Y.M., and Jumaidin, R. (2021). Preparation and characterization of starch-based biocomposite films reinforced by Dioscorea hispida fibers. *Journal of Materials Research and Technology* 15: 1342–1355.

62 Wang, M., He, C., Chang, X., and Liu, D. (2016). Comparison on properties of composites prepared by four kinds of biological glues and wheat straw. *Acta Materiae Compositae Sinica* 2016 (11): 2625–2633.

63 Tarique, J., Sapuan, S.M., Khalina, A. et al. (2022). Thermal, flammability, and antimicrobial properties of arrowroot (*Maranta arundinacea*) fiber reinforced arrowroot starch biopolymer composites for food packaging applications. *International Journal of Biological Macromolecules* 213: 1–10.

64 Müller, C.M.O., Laurindo, J.B., and Yamashita, F. (2009). Effect of cellulose fibers on the crystallinity and mechanical properties of starch-based films at different relative humidity values. *Carbohydrate Polymers* 77 (2): 293–299.

65 Chen, S., Wu, M., Wang, C. et al. (2020). Developed chitosan/oregano essential oil biocomposite packaging film enhanced by cellulose nanofibril. *Polymers* 12 (8): 1780.

66 Sydow, Z. and Bieńczak, K. (2019). The overview on the use of natural fibers reinforced composites for food packaging. *Journal of Natural Fibers* 16 (8): 1189–1200. doi: 10.1080/15440478.2018.1455621.

67 Jayakumar, A. and Suganthi, A. (2017). Biochemical and phytochemical analysis of *Maranta arundinacea* (L.) Rhizome. *International Journal of Research in Pharmacy and Pharmaceutical Science* 2 (3): 26–30.

68 Le Digabel, F., Boquillon, N., Dole, P. et al. (2004 July 5). Properties of thermoplastic composites based on wheat-straw lignocellulosic fillers. *Journal of Applied Polymer Science* 93 (1): 428–436. doi: 10.1002/app.20426.

69 Mishra, T., Mandal, P., Rout, A.K., and Sahoo, D. (2022). A state-of-the-art review on potential applications of natural fiber-reinforced polymer composite filled with inorganic nanoparticle. *Compos Part C Open Access* 9: 100298.

70 Haameem, M.J.A., Abdul Majid, M.S., Afendi, M. et al. (2016). Effects of water absorption on Napier grass fiber/polyester composites. *Composite Structures* 144: 138–146.

71 Chen, Q., Liu, Y., and Chen, G. (2019). A comparative study on the starch-based biocomposite films reinforced by nanocellulose prepared from different non-wood fibers. *Cellulose* 26 (4): 2425–2435. doi: 10.1007/s10570-019-02254-x.

72 Siakeng, R., Jawaid, M., Ariffin, H., and Sapuan, S.M. (2019 May 1). Mechanical, dynamic, and thermomechanical properties of coir/pineapple leaf fiber reinforced polylactic acid hybrid biocomposites. *Polymer Composites* 40 (5): 2000–2011. doi: 10.1002/pc.24978.

73 Zini, E., Scandola, M., and Gatenholm, P. (2003 May 1). Heterogeneous acylation of flax fibers. Reaction kinetics and surface properties. *Biomacromolecules* 4 (3): 821–827. doi: 10.1021/bm034040h.

74 Ramesh, S. and Radhakrishnan, P. (2022). Areca nut fiber nano crystals, clay nano particles and PVA blended bionanocomposite material for active packaging of food. *Applied Nanoscience* 12 (3): 295–307. doi: 10.1007/s13204-020-01617-2.

75 Viscusi, G., Pantani, R., and Gorrasi, G. (2021). Transport properties of water vapor through hemp fibers modified with a sustainable process: effect of surface morphology on the thermodynamic and kinetic phenomena. *Applied Surface Science* 541: 148433.

76 Castro-Mayorga, J.L., Fabra, M.J., Pourrahimi, A.M. et al. (2017). The impact of zinc oxide particle morphology as an antimicrobial and when incorporated in poly(3-hydroxybutyrate-co-3-hydroxyvalerate) films for food packaging and food contact surfaces applications. *Food Bioprod Process* 101: 32–44.

77 Subramaniyan, S.P., Imam, M.A., and Prabhakar, P. (2021). Fiber packing and morphology driven moisture diffusion mechanics in reinforced composites. *Composites Part B: Engineering* 226: 109259.

78 Zhang, X., Duan, J., Zhuo, G. et al. (2022). Nano silicon carbide-treated wheat straw fiber reinforced high-density polyethylene composites. *Industrial Crops and Products* 182: 114834.

79 Nguyen, H.L., Bechtold, T., Fabbri, F. et al. (2022). Characterisation of enzyme catalysed hydrolysation stage of poly(lactic acid) fiber surface by nanoscale thermal analysis: new mechanistic insight. *Materials & Design* 219: 110810.

80 Peresin, M.S., Habibi, Y., Zoppe, J.O. et al. (2010 March 8). Nanofiber composites of polyvinyl alcohol and cellulose nanocrystals: manufacture and characterization. *Biomacromolecules* 11 (3): 674–681. doi: 10.1021/bm901254n.

81 Ilyas, R.A., Sapuan, S.M., Ishak, M.R., and Zainudin, E.S. (2017). Delignification of palm fiber. *BioResources* 12 (4): 8734–8754.

82 Shankar, S. and Rhim, J.W. (2020 January). Bionanocomposite films for food packaging applications. In: *Innovative Food Processing Technologies: A Comprehensive Review* (ed. K. Muthukumarappan and K. Knoerzer), 234–243. Oxford: Elsevier.

83 Bordón, P., Paz, R., Peñalva, C. et al. (2021). Biodegradable polymer compounds reinforced with banana fiber for the production of protective bags for banana fruits in the context of circular economy. *Agronomy* 11 (2): 242.

84 Arun, R., Shruthy, R., Preetha, R., and Sreejit, V. (2022). Biodegradable nano composite reinforced with cellulose nano fiber from coconut industry waste for replacing synthetic plastic food packaging. *Chemosphere* 291: 132786.

15

Polymers, Their Composites, Blends, and Nanocomposites for the Fabrication of Prosthetics

Md Enamul Hoque, Md Ushama Shafoyat, and Fatiha Tabassun*

Department of Biomedical Engineering, Military Institute of Science and Technology, Dhaka, Bangladesh
** Corresponding author*

15.1 Introduction

Macromolecules known as polymers are made up of monomers that repeat and are joined together by covalent bonds. Because of their distinctive qualities, including flexibility, toughness, and durability, they are perfect for use in the creation of prosthetics. Artificial devices known as prosthetics are intended to replace or improve missing bodily parts or functions as a result of illness, injury, or congenital deformity [1].

Polymer composites are substances made of two or more different phases that are mixed to produce a brand-new substance with enhanced qualities. The matrix phase is typically a polymer, while the reinforcing phase is typically a stiff, high-strength substance like carbon fibers, glass fibers, or nanoparticles [2]. Excellent mechanical qualities, such as high strength and stiffness, can be found in polymer composites, which are essential for prostheses [3].

Polymer blends are materials made up of two or more polymers that have been combined to produce a new substance with unique features. The material's mechanical qualities, which are crucial for prostheses and can be improved by blending polymers, can include tensile strength, elongation, and toughness [3].

Applications of Biopolymers in Science, Biotechnology, and Engineering, First Edition.
Edited by Arunima Reghunadhan, Akhina H., and Sabu Thomas.

Materials called nanocomposites are made of nanoparticles and a polymer matrix. The nanoparticles, which are typically smaller than 100 nanometers, can be formed from a variety of substances, including metals, ceramics, and polymers. The huge surface area and high aspect ratio of the nanoparticles in nanocomposites can provide the material with special mechanical, thermal, and electrical properties. Due to their improved characteristics, nanocomposites offer significant potential for use in the production of prosthetics [4].

This chapter contains a thorough analysis of the production of prosthetics using polymer composites, blends, and nanocomposites. For instance, researchers have looked into the production of dental prostheses using polymer blends, the production of prosthetic limbs using carbon fiber-reinforced polymers [5, 6], and the production of scaffolds for bone tissue engineering using nanocomposites [7].

Therefore, due to their distinctive features, polymers, and their composites, blends, and nanocomposites, have a significant amount of potential for use in the creation of prosthetics. The use of these substances may result in the creation of prosthetics with enhanced mechanical, thermal, and electrical characteristics, enhancing the quality of life for those who have lost limbs or have other problems.

15.2 Polymers for Prostheses Fabrication

15.2.1 Polyethylene

Since the middle of the 20th century, polyethylene (PE) has been extensively employed in knee arthroplasty. Over the past few decades, advancements in material production and processing have produced newer polyethylenes with various material characteristics. It is known that particle-induced osteolysis, which is the main cause of late failure and necessitates revision surgery, is caused by PE wear in knee arthroplasty. Despite the fact that there are many reasons for wear, the characteristics of PE have long been a subject of research as a contributing component [8]. However, the benefits have not been demonstrated to be to the same extent in knee arthroplasty. The introduction of newer, highly cross-linked PE has been found to improve wear rates in hip arthroplasty. The clinical and laboratory research to date is small, and only osteolysis and greater wear may result from the mechanism and liner fracture. The present highly cross-linked polyethylenes may be appropriate only for younger, more active people, according to the research. We are awaiting the findings of a recently finished randomized experiment comparing standard with high molecular weight PE in knee arthroplasty [8].

15.2.2 Polypropylene

The most popular sheet plastic used in orthotics and prosthetics is standard polypropylene (PP). The nonarticulated ankle–foot orthosis (AFO) is typically made of polypropylene and is known as a "MAFO" (molded AFO). It is important to keep in mind that not all MAFOs have rigid ankles, and not all non-articulated MAFOs have flexible ankle trims. We would highlight that the metal hinge be attached to the stiff polypropylene parts rather than, for example, the flexible polyethylene anterior shell, since metal joints are routinely employed when addressing particular disabilities. The most robust sheet plastic material now on the market, polypropylene, can be produced using a positive model. Common applications for orthotics include pelvic bands, MAFOs, knee–ankle–foot orthoses (KAFO), cervical–thoracolumbosacral orthoses (CTLSO), thoraco–lumbo–sacral orthoses (TLSO), and long-term use devices for the upper extremities where stiffness and durability are crucial, such as in joints, pelvic girdles, and other novel devices [9]. Orthopedic-grade polypropylene is regular-grade polypropylene that has had an additive added to it to increase its flexibility and tensile strength under stress. However, when maximal rigidity is required, this increased flexibility may not be desirable. To choose the best material and design for your orthotic therapy, a thorough evaluation is necessary at the beginning. Orthopedic-grade polypropylene is a great option when a custom-molded fracture orthosis is recommended for an active patient since it will reduce the likelihood of rupture in high-stress locations. If desired, a soft foam plastic interface can be placed on the positive model; during the vacuum-forming process, the hot polypropylene will cling to it [9].

15.2.3 Polycarbonate

Transparent sockets have long been acknowledged as having a potential value in both clinical and scientific settings. The Navy Prosthetics Research Laboratory used Plexiglas sockets in their studies of above-knee fitting during the 1950s, and the J.E. Hanger Co. of Atlanta found Plexiglas sockets to be very helpful in their groundbreaking work with the total-contact above-knee sockets in the early 1960s. However, making sockets out of this material takes a very long time because it has not been possible to form a socket from a single piece of the material. The Army Medical Biomechanical Research Laboratory suggested a technique for employing acrylic to cast a clear socket. The approach hasn't been widely used because it requires somewhat expensive materials and tremendous caution to achieve excellent results. A more straightforward method for casting transparent sockets using polyester resins was later developed at New York University, but it is so laborious and time-consuming that it has not been used on a regular basis in clinical settings. A technology for vacuum-forming polycarbonate sheet material that should make the use of transparent sockets feasible in everyday clinical practice has been discovered [10].

15.2.4 Polyurethane

Due to its excellent mechanical qualities, biocompatibility, and hemocompatibility, polyurethane has grown in popularity in the biomedical business. However, the material deteriorates throughout the course of polyurethane grafts' prolonged use. Novel polyurethanes having a siloxane segment, polycarbonate polyurethanes, and nanocomposite polyurethanes are available to improve biostability. Modern tissue engineering technologies are highly adapted for the production of polyurethane products with distinctive features, in addition to innovative polyurethanes. Modern techniques for creating polyurethane cardiovascular grafts using various polyurethanes are reviewed [11].

15.2.5 Polymethyl Methacrylate

Numerous studies have been conducted on bone, denture bases, and tooth substitutes as composite materials made of various particles and a polymer matrix because the composite materials have numerous significant features that make them appropriate for many dental uses [12]. According to estimates, acrylic resins make up 95% of the polymer materials used in dental applications as denture base materials. These materials typically include the thermoplastic polymer polymethyl methacrylate and have been the most widely used and accepted among all denture base materials [12].

Dental prostheses were the first application of polymethyl methacrylate (PMMA) in the 1930s [6]. PMMA was first made available as a heat-curable denture foundation material in 1936 [13]. This substance became the primary polymer utilized in the last few years as a denture base material when it was heated and pressed into the proper shape. The first time PMMA was used was as a cold denture base material to cure a dental sensation.

In 1938, denture base material was used [13]. In addition, several thermoplastic polymers have been utilized as denture foundation materials as alternatives to PMMA since 1960, but they have not been demonstrated to generate dentures that are more accurate or perform better than PMMA. As a result, PMMA resin continues to be the material of choice for denture base material [13]. The replacement of missing teeth, which may have been lost for a number of reasons, with fixed or removable dentures is known as prosthetic dentistry. A partial denture can be created to replace one or more missing teeth, and it has the advantage that retention and supports can come from neighboring, still-existing natural teeth and from design elements like occlusal rests and clasps. While a complete denture replaces all the teeth in a single dental arch, its goals would be to improve a patient's speech, attractiveness, chewing and biting, comfort, and capacity to participate in a variety of social and interpersonal activities [13].

15.2.6 Silicone

Polydimethylsiloxane (PDMS) is commonly referred to as silicone or silicone elastomer. These days, a wide range of industries, from personal care to automotive, use this well-liked polymer extensively [14]. The first silicones were released in 1946 [15], and maxillofacial prostheses started using them in the 1960s [16]. Although it cannot be altered or repaired, it is now the most popular material used in maxillofacial prosthetics due to its flexibility, heat resistance, transparency, and biocompatibility [17]. In prosthetic fabrication, different classes of silicone are used to cater to various needs and applications (Figure 15.1). Silicone elastomers are the primary class of silicone used in prosthetic fabrication. They offer excellent flexibility, durability, and biocompatibility, making them suitable for creating realistic and comfortable prosthetic components. Prosthetic silicone elastomers can be further classified based on their hardness and properties, such as high tear strength (HTV) or room temperature vulcanizing (RTV) silicones. It is important to note that these classes of silicone can have different formulations and variations within each category to meet specific prosthetic requirements. Prosthetists and manufacturers select the appropriate silicone materials based on factors such as desired flexibility, hardness, durability, and interaction with the wearer's body tissues.

15.3 Polymer Composites for Prostheses Fabrication

In both medicine and dentistry there are numerous biomaterials that fall under the categories of metals, polymers, ceramics, and composite materials. They were all created to treat bone and dental issues, and they each have benefits and drawbacks. Compared with bone and teeth, polymers are less mechanically strong. Metals are incredibly corrosive but also have exceptional mechanical qualities like wear resistance and biocompatibility. Ceramics have poor fracture toughness and are brittle. When combined,

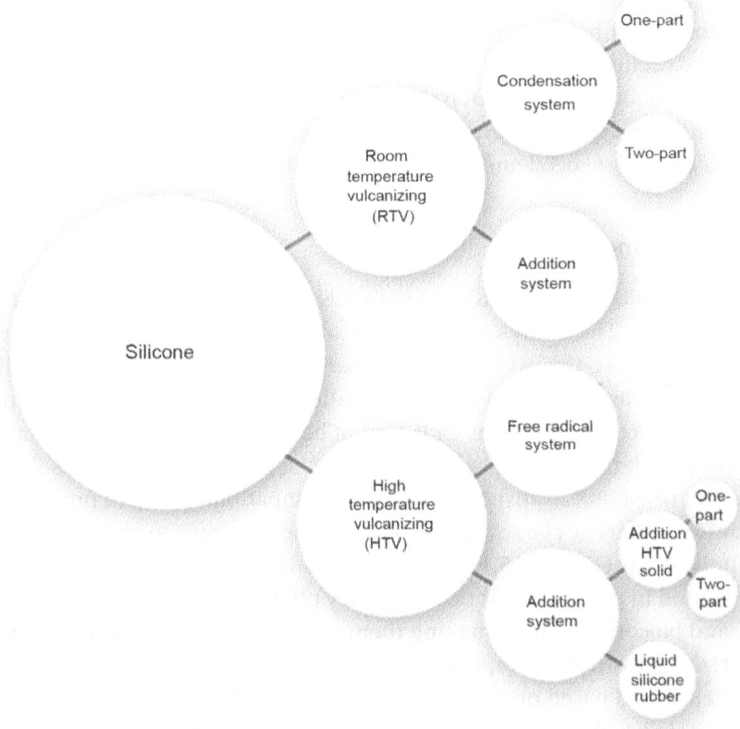

Figure 15.1 Diagram of the different classes of silicone.

these characteristics of metals, ceramics, and polymers are taken into account in the synthesis of composite materials. Polymer composite materials come in a wide range of compositions and possess unique properties that make them suitable for a number of orthopedic and dental applications, including dental implantation and denture base materials [18]. These materials also have superior properties to polymer materials, such as a good elastic modulus, high strength, low creep, and high heat deflection.

15.3.1 Carbon Fiber-Reinforced Polymer

One of the most expensive materials used in orthopedic applications is carbon. Carbon is extremely light, stiff, and capable of maintaining its shape under pressure. This is a result of its strong tensile properties under both tension and compression. In order to maximize fracture resistance and ensure consistent strength throughout the device, carbon has a bi-directional weaving pattern. This indicates that the brace (orthotics) or carbon socket (prosthetics) is exceptionally robust and can withstand significant pressure from patient use.

After the patient has approved the test socket, carbon fiber will be used in prosthetics to create the patient's final socket. The socket design depends on the type of prosthesis the patient is receiving, and the patient can personalize the socket to their preference. Depending on the type of brace and patient activity level, carbon fiber is utilized in orthotics. Carbon fiber enables the brace to be more rigid, which is necessary for braces that support an articulating ankle or for braces that require more stability [19].

15.3.2 Glass Fiber-Reinforced Polymer

A particular class of fiber-reinforced polymer composite is called glass fiber composites (GFC). Its favorable characteristics, including low density, high strength, and ease of processing, make it a popular material in the construction, automotive, and aerospace industries. Today, a combination of glass fiber [20] and plastic resin, which are distinguished by several exceptional qualities, can satisfy practically any requirement for structural material. Glass fiber-reinforced composites made their debut in 1942 as structural components for the aerospace industry. High-strength glass fibers, or S-Glass, were originally employed in cooperative projects between Owens Corning textile products and the U.S. Air Force in the early 1960s. S-2 glass started to be used in a range of commercial applications later in 1968. High-strength glass fiber offers a very favorable cost-to-weight ratio while combining high strength, high stability, transparency, and durability. By comparing their physical, mechanical, electrical, thermal, acoustical, optical, and radiation properties, high-strength glass fiber composites are examined for their utility. The strength/weight ratios of glass fiber composites are superior to those of the majority of other materials, and they have incredible impact resistance. Additionally, they are electrically efficient, resistant to moisture and outside weathering, and heat and chemically resistant. Along with these qualities, manufacturing is simple [21].

15.3.3 Aramid Fiber-Reinforced Polymer

The two types of fibers that are employed in polymer engineering composites are synthetic fibers and natural fibers. The most prevalent fibers are synthetic ones. Despite the fact that there are numerous varieties of synthetic fibers, glass, carbon, and aramid fibers are the most significant. DuPont first made Kevlar, an aromatic polyamide or aramid fiber, available in the early 1970s. Because of its high tensile strength and modulus, it was the first organic fiber suitable for use in advanced composites. It has a matching tensile modulus and roughly five times the tensile strength of steel. Originally created to replace steel in radial tires, aramid is currently utilized in a variety of settings. The aramid fiber is known by the brand name Kevlar. According to Vygotsky, the monomers of the aramid fibers are composed of terephthaloyl chloride and 1,4-phenylenediamine (para-phenylenediamine). The end product is a polymeric aromatic amide with modified amide groups and benzene rings. These polymer strands were made with a random alignment. Aramid fibers are technically long-chain synthetic polyamides. Due to their extraordinarily high tensile strength, aramid fibers are frequently employed in applications for armor and ballistic protection. Aramid fibers, which have a characteristic yellow tint, are frequently employed in high-strength and lightweight, innovative composite goods [22].

15.3.4 Hybrid Composites

In many engineering applications, hybrid composite materials are being used more and more because they have a number of improved features and advantages over classic composite materials. The n ($n > 2$) cooperatively working phases make up the mechanical properties of hybrid composites, which are particularly significant. Due to this, a linear coupling of numerical simulation models is used to predict the mechanical properties of hybrid composites [23]. However, in addition to the characteristics of the matrix and reinforcements, the mechanical behavior of hybrid composites also depends on the characteristics of the interface between these components and the matrix, which must be taken into account in the numerical modeling of the mechanical properties. Additionally, the impact of environmental aging should be considered in the numerical modeling of hybrid composite materials.

15.4 Polymer Blends for Prostheses Fabrication

15.4.1 Polyaniline/PUs Composites

In order to project soft films for flexible strain sensors, polyaniline (PANI), an extensively researched intrinsically conducting polymer incorporating dopants, is used. By in-situ polymerizing aniline and creating a double network, Wang et al. created a unique, flexible, sensitive, self-healing, and recyclable PANI/PU hydrogel sensor [24]. The produced hydrogel displayed mechanical properties similar to skin, including 1.1 MPa of strength, 0.3 MPa of Young's modulus, and 500% elongation at break, as illustrated in Figure 15.2. Additionally, the hydrogel sensor exhibits an exceptional conductivity of 7.87 S/m1 and a sensitivity of 2.89 [24].

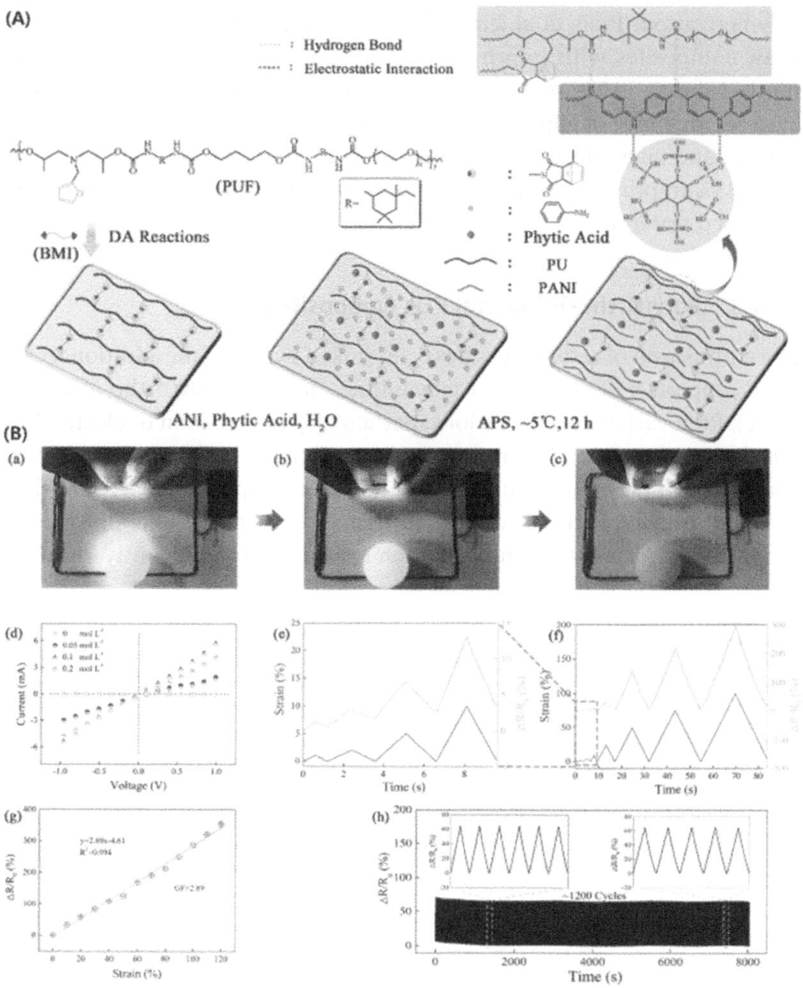

Figure 15.2 (A) The creation of a dual interpenetrating and synergistic network is performed. (B) By illuminating an LED bulb, the original sample is shown in (a), stretched to a strain of around 350% (b), and fractured at a strain of more than 500% (c). (d) PANI/PU hydrogel C-V curves were created using various aniline feed concentrations. relative resistance change (e, f) during the repeated stretching–releasing from 1% to 100% of the maximum strain. (g) fluctuation in the PANI/PU hydrogel's relative resistance as a result of the applied strain. (h) relative resistance fluctuation throughout about 1200 cycles of stretching-releasing deformation at 25% maximum strain [24]. *Source:* (B)(b)[24] / MDPI / CC BY 4.0.

15.4.2 Polypyrrole PUs Composites

From an electrical composite standpoint, polypyrole (PPy) has concentrated on promising materials and has demonstrated exceptional electrical capabilities among diverse conducting polymer fillers. By using vapor phase polymerization, Tyagi et al. created percolative polyurethane–polypyrrole–straw composites with improved mechanical strength and dielectric constant as shown in Figure 15.3 [24]. The fabricated composites were said to have a wide range of applications, including sensors, actuation, energy generation, and microelectronics, thanks to their 26% improved tensile properties, 133% improved dielectric constant, low dissipation factor, and 51.3% decreased water uptake.

15.4.3 Poly(Ortho-Toluidine) PUs Composites

The enthusiasm of researchers has been stoked by one of the additional electrical properties of substituted polyanilines, such as poly(ortho-toluidine) (POT), for a variety of applications, particularly the creation of electrochromic devices. In 2014 [24], a report on the performance of POT nanoparticles distributed in PU matrices using a solution mixing approach was published. The performance of these coatings in terms of thermal stability, physicomechanical characteristics, and corrosion resistance is significantly improved

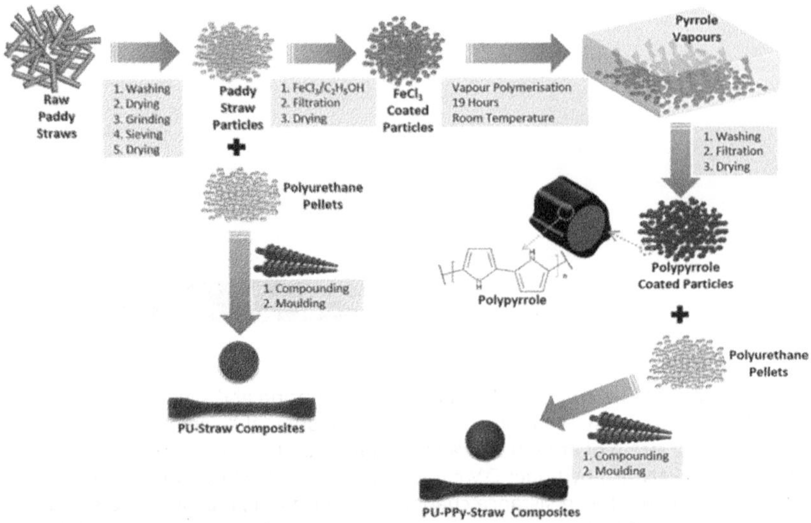

Figure 15.3 A schematic of how uncoated and coated PPy/PUs composite paddy straw particles are prepared [24]. *Source:* [24] / MDPI / CC BY 4.0.

by the minimum dispersion (0.25–1.0 wt%) of POT in castrol oil polyurethane (COPU). Electrochemical impedance spectroscopy (EIS) and potentiodynamic studies showed that the POT/COPU-coated MS effectively offers protection to the mild steel through a barrier mechanism in anticipation of acid and salt medium. The salt spray test also revealed that coatings behaved similarly to how they would in an acidic environment. In acidic and salty settings, the POT/COPU coatings have demonstrated superior corrosion protection over the COPU coatings.

15.4.4 Poly(ethylenedioxythiophene) /PUs Composites

Figure 15.4 illustrates the investigation of the mechanical and electrical properties of uniform and transparent ILPU gels made of PU elastomer and ionic liquid(IL).ThedevelopmentofIL/PU/PEDOT(poly(ethylenedioxythiophene): PSS composites, which involved sandwiching the IL/PU gel between two PEDOT: PSS films as flexible electrodes, was shown to exhibit rapid and intense bending to anode under the electric field. The bending displacement

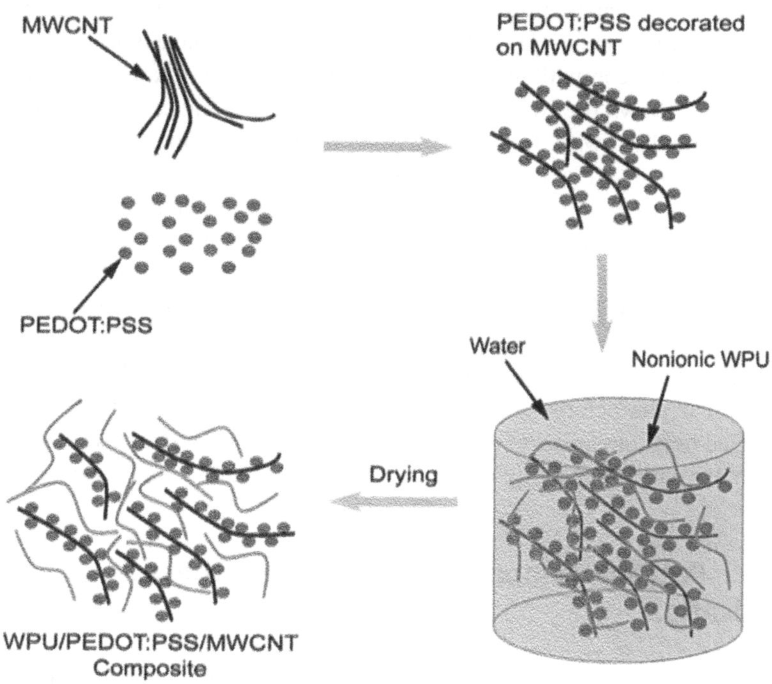

Figure 15.4 Shows a schematic diagram of the nonionic composite of WPU, PEDOT: PSS, and MWCNT film production [24]. *Source:* [24] / MDPI / CC BY 4.0.

of 3.8 mm was obtained at IL 40 wt.%, 2 V, which is appropriate for the strain of 0.32%. The electrical composite actuator that was created as a result has potential at frequencies higher than 10 Hz [25]. Another study by T. Chen et al. produced graphene–polyurethane dielectric elastomer composites with extremely high relative permittivity (350 at 1 kHz), low dielectric loss, noncovalent PEDOT: PSS functionalization (0.2 at 1 kHz), low loss tangent (0.4), and low loss modulus (200 Mpa) that are advantageous for the development of microcapacitors in the matrix and reducing current leakage. This outcome was described as a promising material for electromechanical applications involving micro-actuators.

15.4.5 Other Conducting Polymer/PU Composites

Due to its superior electrical conductivity of 6000 Scm1 and thermal conductivity of 5000 WmlK1, graphene, or reduced graphene oxide (rGO), a one-atom-thick two-dimensional carbon lattice, has recently received a lot of attention as a potential new material for multifunctional EMI shielding [26, 27]. However, numerous researchers have been working to develop conductive polymer-reinforced polymer matrix composites that are highly effective. In order to complete this objective, rGO fillers have typically been used to improve compatibility within or on a polymer matrix. For instance, Figure 15.5 shows the two-stage bath product that Wang et al. developed [28]. GO modified with 3-mercapto propyl trimethoxysilane (MPTES) was produced and mixed with a mechanical stirrer (300 rpm) for 1, 2, and 3 wt.% of the synthesized WPU, respectively, for 1 hour at ambient temperature. The cotton fabric was prepared, dipped into the solution, and dried. It was then exposed to UV light for 30 minutes. Finally, using a violent click reaction, an M-rGO-WPU/cotton fabric with a conductive linked network and low M-rGO loadings was created. The development of the conductive interconnected network allowed for the production of fabric with improved mechanical characteristics, increased EMI shielding performance, and increased heat transmission capabilities. This study demonstrated the material's potential for use in high-performance EMI shielding applications.

15.5 Polymer Nanocomposites for Prostheses Fabrication

Significant progress has been made in the development of materials for additive manufacturing (AM) applications in recent years. However, there are still a few composite or nanocomposite materials being used to improve

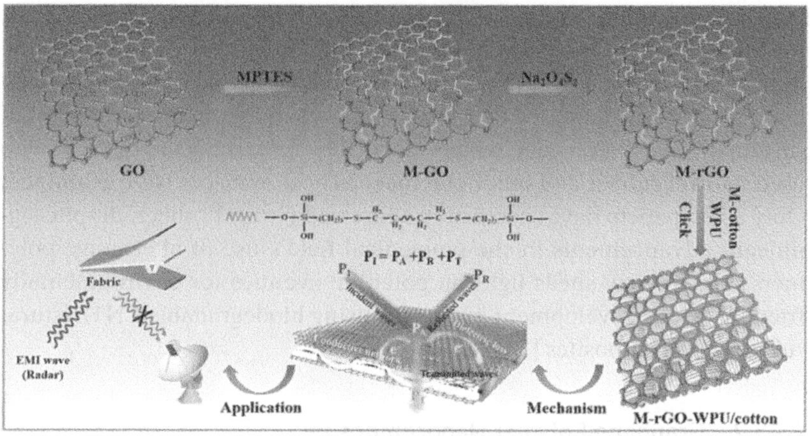

Figure 15.5 Shows a schematic depiction of the M-rGO-WPU/cotton fabric's construction and EMI shielding used [28]. *Source:* [28] / MDPI / CC BY 4.0.

performance and multifunctionality. This review makes an effort to offer a thorough analysis of both commercially available materials and current developments in high-performance polymer nanocomposites used in different AM processes. The discussion covers four AM processes: fused filament fabrication (FFF), selective laser sintering (SLS), multijet fusion (MJF), and stereolithography (SLA). The portfolio of AM materials is being rapidly expanded by the development of printable polymer composites, particularly polymer nanocomposites, which enable the manufacture of multifunctional parts with intricate architectures [29].

15.5.1 Carbon Nanotube-Reinforced Polymer Nanocomposites

The trend to adopt a sustainability strategy in the creation of polymer nanocomposites for prosthetic purposes is as significant as the hunt for workable substitutes for traditional materials in biomedical applications. Due to their unmatched strength-to-weight properties, carbon nanotube-(CNT) reinforced polymer nanocomposites have taken the lead in the current prosthetic sector. Researchers have disagreed about the types of polymers employed for this purpose and their long-term effects on the environment. Among the common worries are the suitability, affordability, and sustainability of materials for the creation of prosthetics. As a result, this research tackles issues with sustainable development goal (SDG) compliance in biomedical manufacturing, which emphasizes material selection taking into account environmental implications. Additionally, the contributions of earlier studies were evaluated using the number of papers on CNT-reinforced polymer nanocomposites, which has dramatically increased over the past ten years. From the standpoint of sustainability, several conclusions reached by researchers in the area who used natural rubber and other polymers as host matrices were examined. Only a few papers have specifically addressed natural rubber, despite significant advancements in the biomedical field's use of alternative polymers. This review sheds light on potential avenues for environmentally friendly device development and usage using biodegradable CNT/natural rubber nanocomposites [30].

15.5.2 Graphene Polymer Nanocomposites

Commercialization of a PMMA composite containing graphene is used in dental prosthetics. The various fiber ratios offer a wide spectrum of colors that enhance dental esthetics in prosthetics. There is no research that has addressed how graphene affects its mechanical properties. We looked at PMMA and PMMA-containing graphene fibers (PMMA-G) that were provided for machining in the shape of discs. Shore hardness and Vickers microhardness measurements were made after Raman spectroscopy was used to investigate the existence of graphene fibers. Pin-on-disc wear testing and mechanical compression tests were used to determine the maximum strength and Young's modulus (E) values. The precise wear rate and the coefficients of friction were calculated in accordance with recognized international standards. The samples were then examined using field emission scanning electron microscopy (FESEM) to determine the shape of the graphene embedded in the PMMA. The findings indicated that graphene was present in PMMA, with an estimated concentration of 0.1027% by weight in G-PMMA. There were no statistically significant differences between the Shore hardness and Vickers microhardness values. With a p value of 0.005, differences in the compression maximum strength (129.43 MPa for PMMA and 140.23 MPa for PMMA-G), E values (2.01 for PMMA and 2.89 GPa for PMMA-G), and reduced wear rate for the G-PMMA samples (1.93 107 for PMMA and 1.33 107 mm^3/Nm) were noted. In comparison to PMMA, the coefficients of friction for PMMA-G dropped from 0.4032 to 0.4001. According to the findings, adding a small amount of graphene to the prosthesis material significantly improved its mechanical characteristics. Therefore, we can say that this material's aesthetic appeal is what makes it so popular for dental prosthetics [31].

15.6 Fabrication Techniques for Polymer-Based Prostheses

While orthoses are assistive devices that restore the stability and mobility of a body part for people with neuromuscular and/or musculoskeletal disorders, prostheses are devices that restore the functionality of a missing body part. The standard prosthetic and orthotic devices are shown in Figure 15.6 [32]. Users' satisfaction with orthotic and prosthetic (O&P) equipment when treating patients is found to be strongly influenced by its fit and alignment. Additionally, giving the patient a pleasant O&P device within four weeks following an impairment or amputation considerably reduces the rate of device abandonment. Therefore, two essential elements that might considerably enhance the lives of patients are a quick fabrication method and a comfortable fit of the O&P devices. In order to assure a custom fit, the currently used conventional manufacturing method for O&P devices takes a lot of time and includes a lot of manual labor and design iterations, which results in a lot of waste. However, they have been empirically examined and shown to offer patients long-lasting, trustworthy, and safe O&P equipment. In order to create high-performance O&P devices, numerous academics and O&P practitioners have recently begun exploring innovative manufacturing technologies. AM technologies have been used to build a variety of

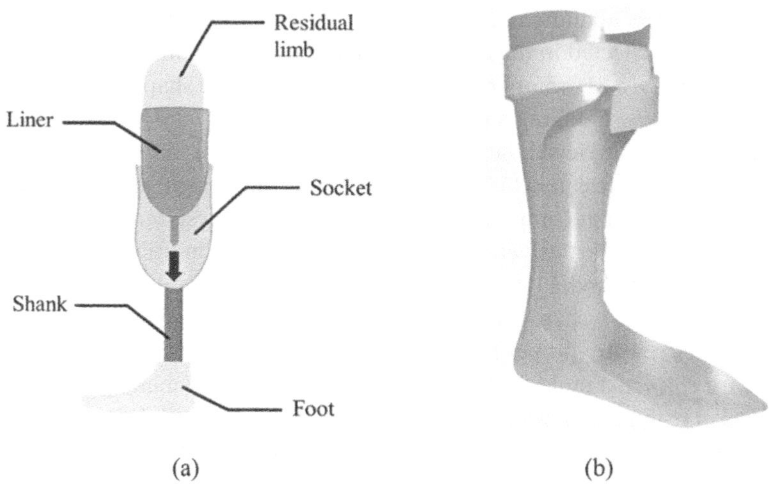

Figure 15.6 (a) Lower-limb prostheses; (b) ankle–foot orthoses [32]. *Source:* [32] / MDPI / CC BY 4.0.

customized foot orthoses. Using gait analysis and subjective measurements of fit and comfort, these orthoses have been compared to products that have been traditionally manufactured. These results show that it is possible to customize foot and ankle–foot orthoses using the AM method, and they also highlight the tremendous clinical potential of this methodology. Similar to the definitive prosthetic devices, prosthetic sockets created utilizing AM methods showed improved comfort, greater step symmetry, and equivalent lower extremity joint performance. Before AM can be extensively used in a service system for the manufacturing of personalized orthoses and prostheses, there are still some clinical, financial, and technological obstacles to be cleared.

15.6.1 Conventional Manufacturing Processes

Laura Trevelin, RTPc (June 2022), gave a detailed account of how prostheses are typically made, which begins with casting and bandaging the residual limb in plaster. The negative mold, which is removed after the cast has dried and is used as the basis for the positive mold, is then filled with plaster. The pressures on the limb are then manually adjusted on the positive mold to change their intensity and position. The patient's needs are taken into account when choosing the best suspension technique and equipment. The diagnostic "check" socket is then constructed using transparent thermoplastics that can be heated and molded. The "check" socket is tried on a patient to assess alignment and fit after it has been put together using the componentry selected in the proper standard neutral alignment.

Alternatively, a 3D scanner can be used to capture the shape of the residual limbs instead of taking a plaster cast of them. Afterward, pressure-sensitive parts are relieved, and necessary sections are reinforced using computer-aided design (CAD) software. Once the design is complete, a positive model is cut from foam, such as expanded polystyrene (EPS), and the check socket is then created utilizing the above-described processes from that carved model.

For four to six weeks, the patient wears the check device while being monitored to ensure proper fit, alignment, and comfort. If the fit and alignment are correct, a mold for the final socket is created by making a copy of the socket using an impression substance (such as Jeltrate alginate) and filling it with plaster. Alternatively, the check socket can be plastered over and severed. If the shape needs to be changed more drastically because the fit is not right, the check socket would be filled again with plaster. The suspension procedure might even be completely altered in specific circumstances. The technician would next construct a second check socket, necessitating the use of additional fittings. The initial fitting's back and forth is what takes the most time.

The technician would laminate the final socket shape once it had been checked, approved, and copied and the plaster mold had dried fully. Stretchable polyvinyl alcohol (PVA) is used to cover the plaster cast in a bag that serves as a separator. Then, layers of glass (a nylon-and-glass fiber composite), fiberglass, and carbon fiber are added depending on the prosthetist's practical expertise and suitability for the patients, taking into account their weight, degree of activity, and socket suspension style. After the layers are applied, a final PVA bag funnel is placed over it, and the bags are taped together to close them off. A vacuum hose is then connected to the sealed bags to remove any remaining air between them. The socket is thoroughly covered with acrylic resin after being added, being cautious not to generate any material wrinkles or air pockets. The funnel is shut off and given a full hour to cure. After breaking the plaster out, trimming, and grinding the edges, the trim lines are removed with a cast saw. After that, the device is put back together in its original configuration while being careful to maintain alignment. Particularly for new amputees, whose shapes change considerably in the first two years, the patient may occasionally need to have constructed a completely new socket within six to twelve months of the initial fit. The patient may also occasionally require minor alterations in alignment and fit. The typical production process for O&P devices is shown in Figure 15.7 [32].

The traditional manufacturing method for orthoses is relatively comparable to that for prostheses, with the difference that no "check" device is built during the creation of orthoses, unlike with prostheses. The process starts with the bodily component's measurements. The molding procedure follows,

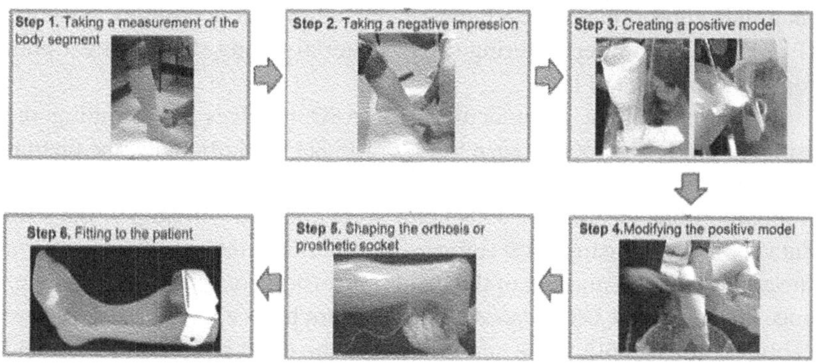

Figure 15.7 The conventional manufacturing process of O&P devices [32].
Source: [32] / MDPI / CC BY 4.0.

during which a casting technique is used to capture an impression of the orthosis. Then, to make a positive mold, liquid plaster is put into the negative mold. It could take two to 48 hours to completely dry the mold. The positive mold then needs to be polished and adjusted. The negative mold is eliminated when the positive mold has dried and become smooth. Following that, a thermoplastic (usually polypropylene (PP), polyethylene (PE), or other copolymers) is applied to the model and stapled to the plastazote or platform foam pulled over it [33].

The thermoplastic is then heated in the oven. The thermoplastic material becomes softer due to the increased temperature, allowing the technician to shape it over the model. Uneven edges are smoothed off when the mold is separated from the model. The first fitting of the orthotic device might begin at this stage. Adjustments are frequently necessary, which can take time. A completely new orthosis can be needed if large levels of modification are anticipated. In some situations, the process might need to be redone from scratch [33].

15.6.2 Additive Manufacturing

The fourth industrial revolution, known as Industry 4.0, contains several key elements, one of which is additive manufacturing (AM), sometimes referred to as 3D printing. A part is made via selective material addition in a 3D volume, which produces very little waste. Rapid design iteration, low-cost, lightweight parts, and wide personalization are also made possible by this digital but toolless manufacturing method, all of which are essential for O&P devices. Although the clinicians' understanding of patient care is invaluable, they must integrate advanced manufacturing to stay competitive. The four AM process types that can be utilized to 3D print polymer materials appropriate for O&P devices include fused deposition modeling (FDM), selective laser sintering (SLS), material jetting (MJ), and vat photopolymerization (VPP).

Step 1 of additive manufacturing of AFO involves 3D scanning the patient's ankle and foot, as seen in Figure 15.8(1). Additionally, the plantar surface of the patient's foot is scanned, either directly or utilizing the foam impression box (see Figure 15.8(1) for AFO creation). In Step 2, the scan data is processed using CAD software to stitch together two scans – one of the ankle/foot and another of the foam impression box – by lining up reference dots that have been placed along the foam box's edge. These dots were captured during both scans and utilized by the CAD program to align and sew the two scans together. CAD software is used to manually generate and smooth the timelines for AFO. The path and support structure for AM and AFO is designed in Step 4. Step 5 is where the AFO is constructed [34].

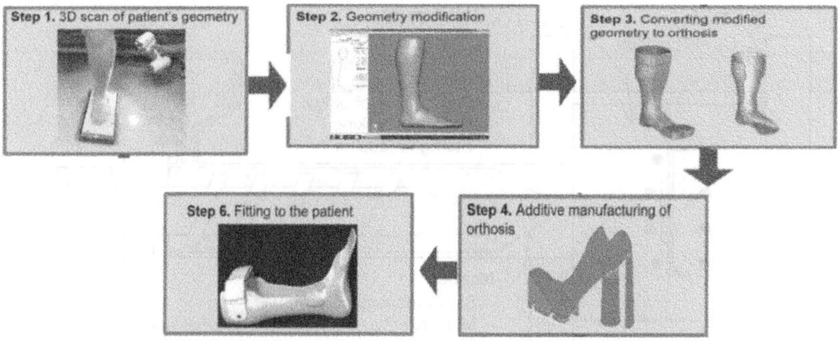

Figure 15.8 Manufacturing processes for additive manufacturing [34]. *Source:* [34] / with permission of Elsevier.

15.6.3 Injection Molding

Through an aperture (sprue), molten polymer is pumped under intense pressure into a mold cavity during the injection molding process. Pellets of polymer material are fed through a hopper into an injection molding machine. Through a hole (sprue), molten polymer is pushed through a high-pressure process into a mold cavity. Pellets of polymer material are fed through a hopper into an injection molding machine. The material is then driven into a split mold using a feeding screw and pressed into it, filling the cavity using runners and a sprue gate. Extruders and injection molding machines are similar.

The screw action of the two machines is where the biggest distinction lies. An output of continually lengthy products (pipe, rod, and sheet) is produced by the extruder's continuous rotation of the screw. Since the screw of the injection molding machine not only rotates but also moves forward and backward in accordance with the steps of the molding cycle, it is known as a reciprocating screw. When the molten polymer is pumped into the mold in the filling step, it functions as a ram, and it then retracts backward in the molding step. Heating elements are used to soften and melt the polymer over the barrel. A cooling system that provides regulated cooling and material solidification is built within the mold. When the polymer has solidified, the mold holds it there until it opens and the ejector pins release the part. Injection thermoplastics are mostly molded; however, elastomers and thermosets can also be extruded. In this instance, the heating and melting of the substance in the heated barrel cause cross-linking to take place. The main design of an injection molding machine is displayed in Figure 15.9 [35]. Injection molding is a very productive technology that offers excellent accuracy and control over the shape of the created pieces. The process is effective for mass-producing numerous similar pieces. Thermoplastics such as

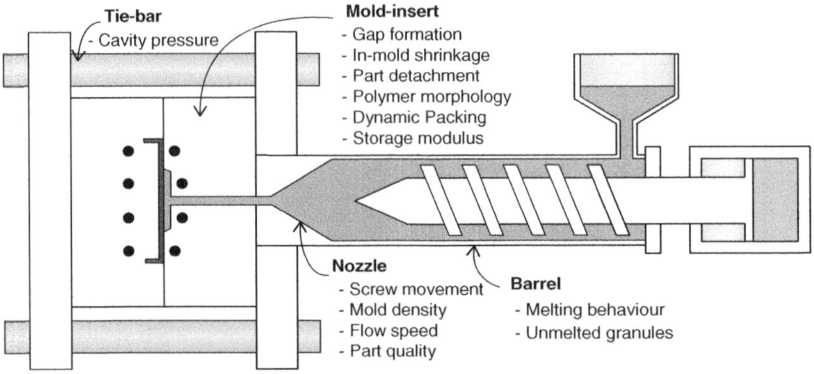

Figure 15.9 A simplified model of an injection molding machine [35].

polypropylene (PP), polycarbonate (PC), acrylonitrile-butadiene-styrene (ABS), and nylon 6 (N6) are frequently used in injection molding.

15.6.4 Compression Molding

For a long time, one of the most established processes for producing plastic parts has been compression molding. A typical compression molding procedure is shown in Figure 15.10 [36]. In a typical single or twin extruder, the resin and chopped fibers are combined to create the bulk charge [36]. The mold cavity tool receives the extruded bulk charge next. The heated bulk charge is compressed at high pressure in the mold cavity to create the finished item. The thermoset or thermoplastic resin used in compression molding is a choice. A thermoset substance often has a low viscosity, and as a result, it flows readily and fills the cavity with ease, resulting in nice, even coverage. Cross-linking between the polymer chains, which might take a while, is how thermoset materials solidify. The cycle time to produce the thermoset part may therefore be greater. Thermoplastic resins, on the other

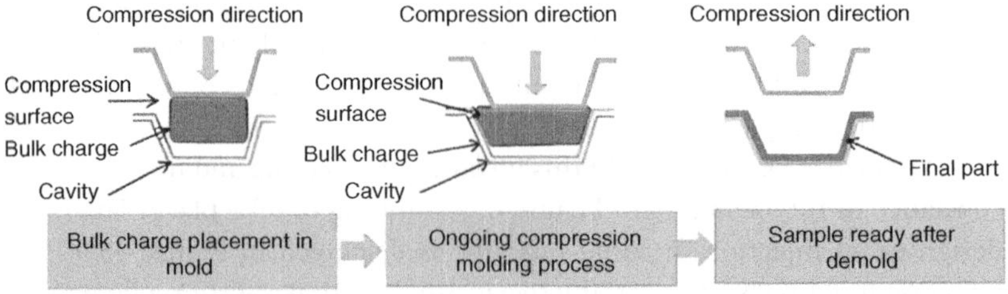

Figure 15.10 A typical compression molding process [36]. *Source:* Adapted from [36] / Elsevier.

hand, are heated until they melt and are then forced to flow into the cavity before the resin cools and hardens. The cycle time can be shortened because thermoplastic resins don't need to be cured. Filling the mold cavity can be difficult since thermoplastic polymers have a higher viscosity than other materials, and they have an even higher viscosity when combined with carbon fiber. The melt can be forced to fill the void by applying more pressure, but doing so also causes the fibers to be sheared at a faster rate. This implies that the final component's fiber length may be considerably shorter. Thermoplastic materials are favored because they have a bigger volume and a faster cycle time, which are crucial for automotive applications. Furthermore, innovative composite choices, sometimes known as hybrid techniques, can be applied to improve the performance of such parts. In the hybrid method, the fiber-reinforced resin bulk charge is compression molded with discrete continuous fibers, like woven unidirectional tapes, positioned at strategic intervals [36]. All of this necessitates careful mold cavity and process parameter design to allow for an even and smooth material flow. In this research, we offer a technique for including the starting fiber orientations and the viscosity effect of the thermoplastic resin in the compression filling simulation.

The alignment of the discontinuous fibers in compression- or injection-molded thermoplastic products depends on the melt flow during the production process. As a result, different parts of the completed item have different fiber orientations. The material characteristics of the completed pieces may be highly anisotropic because of the variation in the orientation distribution of the discontinuous fibers. Warping occurs often in most compression- or injection-molded items. This is mostly caused by anisotropy from fiber orientation and differential shrinkage brought on by changes in cooling rate. Additionally, because there are fibers present, the fiber-reinforced portions are naturally stiffer. As a result, controlling warpage in the fiber-reinforced molded item becomes more difficult.

Only recently have Moldex3D and other companies begun to address compression molding process simulation, notably for long fiber-reinforced thermoplastic (LFT) composite materials [36]. In order to replicate the compression molding of bulk material in a mold cavity, the original offering extended the injection molding formulation with changed boundary conditions depending on the compression surface and mold shape. The bulk charge is typically a complicated mixture of semi-solid and semi-liquid when it is ready for compression molding after being extruded from a twin or single screw extruder, adding to the challenge. The process-dependent constitutive characteristics of these bulk materials must be taken into account.

In other words, when simulating the deformation and stress relaxation in the compression molding process, the viscoelastic behavior of the plastic pieces cannot be overlooked. Stokes and Bushko [36] developed a model to analyze residual stresses and take injection molding's viscoelastic effect into account. Matsuoka [36] used viscoelastic simulation to foretell the warpage and residual stress of the injection-molded portion of an integrated circuit (IC). Bickerton [36] explored liquid compression molding, where the viscoelastic behavior is taken into account and the resin is infused into a pre-impregnated fiber. For compression molding of discontinuous fiber-reinforced composites where the fiber-reinforced material is considered a viscoelastic material, there is no prior simulation that is adequate. Implementing this method for the compression molding process is crucial and required.

15.7 Challenges and the Future Perspectives

The need for prosthetic devices is still not being fully addressed as the number of amputees increases in low-resource environments. Because of this, these amputees constantly deal with difficult physical, emotional, and financial problems like traveling hostile terrain, going without money, and facing social rejection. Although there have been attempts to accommodate this need, doing so presents challenging and complex difficulties. Governments, NGOs, and private organizations are among the groups attempting to provide amputees with affordable, long-lasting prostheses. To accomplish this, all of these institutions must overcome several infrastructure, technological, and commercial obstacles. There hasn't been much work done to offer a thorough analysis of the main distribution paths and their unique issues because these organizations use a variety of distribution tactics in low-resource situations.

According to the World Health Organization, there are 40 million amputees in the developing world, yet only 5% of them have access to any kind of prosthetic care [37]. The significance of building mechanisms for providing appropriate care grows as the number of patients needing amputations rises [37]. By 2030, it is anticipated that over 590 million people will have diabetes, a condition that increases a person's risk of amputation by 25 times [37], with low- and lower-middle-income nations experiencing the largest increases in prevalence [38]. Additionally, during humanitarian situations like the Haitian earthquake or the Sierra Leonean civil war, many people may need to have their limbs amputated in inhospitable locations with poor access. Several organizations, including hospitals, for-profit businesses, faith-based organizations (FBOs), and nonprofits, are working to make prosthetic devices more accessible. The Polus Center, Prosthetic Suppliers World Rehabilitation Fund, and Jumping Kids are a few nonprofit organizations that operate in areas with limited resources. These organizations typically make purchases and distribute goods to local hospitals, nonprofits, and workshops. Financial donations as well as new or used prosthetic materials and devices are collected by FBOs like Standing With Hope and Christian Blind Mission for distribution in low-resource areas. By producing and marketing their own devices, two instances of for-profit and hybrid for-profit/nonprofit organizations, Nonspec and 3D Life Prints, work to increase access to cheap prostheses. 95% of the amputee population in underdeveloped nations still lacks access to appropriate prosthetic care and cost-effective devices, despite the efforts of all of these different groups. This chapter offers a thorough analysis of the existing procedures and difficulties associated with providing amputees with prosthetics. Amputees can start to overcome common problems like finding employment, being forced to rely on others for aid, and the social stigma that comes from living in their own communities if they have access to the right prostheses.

It is an exciting time to be working on prosthetic technology since science and our understanding of how the human body and brain function have made enormous leaps. Prosthetics prototyping, customization, and production costs have decreased due to additive manufacturing, or 3D printing, while developments in neuroscience have provided intriguing hints about how the brain interprets touch, pressure, temperature, and pain. Scientists have been looking for ways to make prosthetics feel and behave like the real thing, from exoskeletons that can react to changing terrain in real time to prosthetic hands that can replicate 100 different feelings. People's lives also get better when these helpful technologies advance.

i) <u>3D-printed prosthetics improve lives in the world's poorest regions.</u> Although limb loss has afflicted millions of people, new World Health Organization surveys show that just 10% of those individuals have access to prosthetic care. Many patients struggle against exorbitant fees and social stigma even with access. The nonprofit LimbForge steps in to help with that and in order to make high-quality prosthetic limbs more affordable for people in impoverished countries, the organization plans to adopt 3D printing. According to Sam Hobish, LimbForge's creative director and main UX designer, "this is not just about building a device. Learning how to utilize it, accepting it as a part of your body, and feeling at ease displaying it to others are all important."

ii) New thought-controlled prosthetics restore the sensation of touch. More than 100 sensations can now be restored to amputees by the University of Utah researchers using a prosthetic arm they've dubbed Luke. The researchers have developed an information loop that is then converted into signals that the brain interprets as touch sensations by implanting a device into a person's residual nerves and placing electrodes in muscles.

iii) Neuroprosthetics and the future of artificial touch. According to Dr. Sliman Bensmaia, assistant professor of organismal biology and anatomy at the University of Chicago, "in the end, we will have a prosthesis that roughly approximates the dexterity of the human hand." To get there, researchers will need to learn more about how the human brain interprets signals, such as those that rush to your head when you unintentionally bang your fingers on the vehicle door. Well, it's hard enough to manage a robotic hand; why would you want to make it feel, too? one would question. It turns out that controlling something is pointless if you can't make it feel that way, according to Bensmaia.

iv) Australia's first 3D-printed paw helps Millie the greyhound run again. When Australian architect Ed Dieppes and his wife, Nora, adopted Millie, a greyhound puppy missing one of her front paws, they vowed to help her live a full and comfortable life but after spending nearly $10,000 on prosthetic paws that kept needing to be replaced, the Dieppeses knew they had to find a different solution. By using Autodesk Fusion 360 and teaming up with a 3D-printing shop in Queensland, the Dieppeses were able to experiment with designs without shouldering expensive manufacturing costs. Most importantly, the end result has been a hit with Millie, who can now run to her heart's content.

v) Brain activity mapped to improve prosthetic design. University of Houston (UH) scientists have been learning more about the function the brain plays in walking, while researchers at universities in Utah and Chicago examine hand prostheses and the role of touch sensation. Researchers think that incorporating brain activity into the mix will give patients more range of motion (and perhaps the ability to play sports), as current state-of-the-art prosthetics cannot adapt to changing terrain in real-time. Jose Luis Contreras-Vidal, a Cullen University professor of electrical and computer engineering at UH, adds, "It looks so simple to walk." But the truth is otherwise. According to Contreras-Vidal, the discoveries may lead to powered exoskeletons and wheelchairs that would increase the freedom and quality of life for patients who have had spinal cord injuries.

vi) Ambionics Tests 3D-printed prosthetics for children. Anyone with kids will attest that because children outgrow their clothes so rapidly, using hand-me-downs is frequently the only practical method to reduce costs. What if, though, your child needs a prosthetic limb? In order to solve this problem for his son Sal, one father, Ben Ryan, used 3D printing, digital design tools, and his skills as an inventor to produce prosthetic limbs that can be produced quickly and inexpensively.

15.8 Conclusions

The main focus of this chapter has been on various fabrication methods and procedures for prosthetics. In the fabrication of materials and polymers, engineering plays a crucial role in the creation of tools and machinery for fundamental studies of human motion, but its preeminence lies in the second stage of any prosthetics program, the design of the prosthetic device. Safety, function, control, efficiency, beauty, comfort, simplicity, and durability are just a few of the numerous aspects one must take into account when designing a prosthetic device. Since they are all interrelated, it is difficult to rank the importance of these elements, hence, the design will almost always be a compromise.

Knowing the mechanical, pneumatic, hydraulic, or electrical means by which the desired performance can be achieved, as well as the forces available, the forces applied, and the proper distribution of masses in the device, is necessary for safety, function, control, efficiency, and appearance. Understanding the range and distribution of pressure that the body's tissues and blood vessels can withstand without suffering harm or causing the amputee pain is necessary for comfort. Knowledge of the durability and simplicity, the deterioration that may result from perspiration and body acids, continuous use, temperature changes, abrasion, and chemicals from external sources is necessary for both cost and maintenance of the device, as well as knowledge of what materials and material combinations may be used to minimize such deterioration.

It has been demonstrated that, depending on the type of prosthetic device a patient receives, expectations and daily goals might vary. According to one experiment, a user's opinion of the gadget affected how important other features, such as comfort, were to them. A gadget with improved hand articulation is presumably going to alter the user's perception of the required comfort, lowering its relative weight to the user. The promise of more control creates a level of expectation and significance for strong performance.

Additionally, it has been observed that the type of amputation affects the variables that lead to rejections.

Technique, age, and gender are important; women with acquired limb loss are more likely than men to reject devices, while adult patients with congenital limb loss are more likely to reject and quit using a device. All populations are impacted by the major issue of abandoning prosthetic equipment, which has several root causes. Sensory feedback, aesthetics, function, control, comfort, and durability have all been cited as significant issues requiring more investigation in terms of prosthesis design and adoption in the adult population.

References

1 Barrios-Muriel, J., Romero-Sánchez, F., Alonso-Sánchez, F.J., and Salgado, D.R. (2020). Advances in orthotic and prosthetic manufacturing: a technology review. *Materials (Basel)* 13: 295. doi: 10.3390/ma13020295.

2 Sapuan, S.M., Haniffah, W.H., Al-Oqla, F.M. et al. (2016). Effects of reinforcing elements on the performance of laser transmission welding process in polymer composites: a systematic review. *International Journal of Performability Engineering* 12: 535. doi: 10.23940/ijpe.16.6.p535.mag.

3 Alameri, I. and Oltulu, M. (2020). Mechanical properties of polymer composites reinforced by silica-based materials of various sizes. *Applied Nanoscience* 10: 4087–4102. doi: 10.1007/s13204-020-01516-6.

4 Wahab, M.A., Islam, N., Hoque, M.E., and Young, D.J. (2018). Recent advances in silver nanoparticle containing biopolymer nanocomposites for infectious disease control – a mini review. *Current Analytical Chemistry* 14: 198–202. doi: 10.2174/1573411013666171009163829.

5 Türk, D., Einarsson, H., Lecomte, C., and Meboldt, M. (2018). Design and manufacturing of high-performance prostheses with additive manufacturing and fiber-reinforced polymers. *Production Engineering* 12. doi: 10.1007/s11740-018-0799-y.

6 Zafar, M.S. (2020). Prosthodontic applications of polymethyl methacrylate (PMMA): an update. *Polymers* 12: 2299. doi: 10.3390/polym12102299.

7 Wang, M. and Duan, B. (2011). Nanocomposite scaffolds for bone tissue engineering: design, fabrication, surface modification and sustained release of growth factor. *MRS Online Proceedings Library* 1301: 99–110. doi: 10.1557/opl.2011.470.

8 Chakrabarty, G., Vashishtha, M., and Leeder, D. (2015). Polyethylene in knee arthroplasty: a review. *Journal of Clinical Orthopaedics and Trauma* 6: 108–112. doi: 10.1016/j.jcot.2015.01.096.

9 Showers, D.C., and Strunck, M.L. (1984). Sheet plastics and their applications in orthotics and prosthetics. *O&P Virtual Library*. 38 (4): 41–48. https://www.oandplibrary.org/op/1984_04_041.asp (accessed 6 August 2023).

10 Mooney, V. and Snelson, R. (1972). Fabrication and application of transparent polycarbonate sockets. *O&P Virtual Library*. 26 (1): 1–13 https://www.oandplibrary.org/op/1972_01_001.asp (accessed 6 August 2023).

11 Cruz, R.L.J., Ross, M.T., Powell, S.K., and Woodruff, M.A. (2020). Advancements in soft-tissue prosthetics part B: the chemistry of imitating life. *Frontiers in Bioengineering and Biotechnology* 8: 147. doi: 10.3389/fbioe.2020.00147.

12 Hamad, Q. (2018). Studying the mechanical properties of denture base materials fabricated from polymer composite materials. *Al-Khawarizmi Engineering Journal* 14: 100–111. doi: 10.22153/kej.2018.01.006.

13 Sheng, T.J., Shafee, M.F., Ariffin, Z., and Jaafar, M. (2018). Review on poly-methyl methacrylate as denture base materials. *Malaysian Journal of Microscopy* 14: 1–16.

14 Hatamleh, M.M. and Watts, D.C. (2010). Mechanical properties and bonding of maxillofacial silicone elastomers. *Dental Materials* 26: 185–191. doi: 10.1016/j.dental.2009.10.001.

15 Chalian, V.A. and Phillips, R.W. (1974). Materials in maxillofacial prosthetics. *Journal of Biomedical Materials Research* 8: 349–363. doi: 10.1002/jbm.820080415.

16 Barnhart, G.W. (1960). A new material and technic in the art of somato-prosthesis. *Journal of Dental Research* 39: 836–844. doi: 10.1177/00220345600390041001.

17 Montgomery, P.C. and Kiat-Amnuay, S. (2010). Survey of currently used materials for fabrication of extraoral maxillofacial prostheses in North America, Europe, Asia, and Australia. *Journal of Prosthodontics* 19: 482–490. doi: 10.1111/j.1532-849X.2009.00538.x.

18 AL-Oqla, F.M., Sapuan, S.M., Anwer, T. et al. (2015). Natural fiber reinforced conductive polymer composites as functional materials: a review. *Synthetic Metals* 206: 42–54. doi: 10.1016/j.synthmet.2015.04.014.

19 Campbell, A.I., Sexton, S., Schaschke, C.J. et al. (2012). Prosthetic limb sockets from plant-based composite materials. *Prosthetics and Orthotics International* 36: 181–189. doi: 10.1177/0309364611434568.

20 Stickel, J. and Nagarajan, M. (2012). Glass fiber-reinforced composites: from formulation to application. *International Journal of Applied Glass Science* 3. doi: 10.1111/j.2041-1294.2012.00090.x.

21 Sathishkumar, T.P., Satheeshkumar, S., and Jesuarockiam, N. (2014). Glass fiber-reinforced polymer composites - A review. *Journal of Reinforced Plastics and Composites* 33: 1258–1275. doi: 10.1177/0731684414530790.

22 Denchev, Z. and Dencheva, N. (2012). *Manufacturing and properties of aramid reinforced polymer composites*. 465–530. doi: 10.3139/9781569905258.008.

23 Alam, R., Alimuzzaman, S., Shahid, A. et al. (2023). Collagen/*Nigella sativa*/chitosan inscribed electrospun hybrid bio-nanocomposites for skin tissue engineering. *Journal of Biomaterials Science, Polymer Edition* 34 (11): 1–22. doi: 10.1080/09205063.2023.2170139.

24 Choi, S.-M., Shin, E.-J., Zo, S.-M. et al. (2022). Revised manuscript with corrections: polyurethane-based conductive composites: from synthesis to applications. *International Journal of Molecular Sciences* 23: 1938. doi: 10.3390/ijms23041938.

25 Okuzaki, H., Takagi, S., Hishiki, F., and Tanigawa, R. (2014). Ionic liquid/polyurethane/PEDOT:PSS composites for electro-active polymer actuators. *Sensors and Actuators B: Chemical* 194: 59–63. doi: 10.1016/j.snb.2013.12.059.

26 Zhang, H.-B., Yan, Q., Zheng, W.-G. et al. (2011). Tough graphene–polymer microcellular foams for electromagnetic interference shielding. *ACS Applied Materials & Interfaces* 3: 918–924. doi: 10.1021/am200021v.

27 Chen, T., Qiu, J., Zhu, K. et al. (2014). Ultra high permittivity and significantly enhanced electric field induced strain in PEDOT:PSS–RGO@PU intelligent shape-changing electro-active polymers. *RSC Advances* 4: 64061–64067. doi: 10.1039/C4RA10695D.

28 Wang, Y., Wang, W., Xu, R. et al. (2019). Flexible, durable, and thermal conducting thiol-modified rGO-WPU/cotton fabric for robust electromagnetic interference shielding. *Chemical Engineering Journal* 360: 817–828. doi: 10.1016/j.cej.2018.12.045.

29 Wu, H., Fahy, W.P., Kim, S. et al. (2020). Recent developments in polymers/polymer nanocomposites for additive manufacturing. *Progress in Materials Science* 111: 100638. doi: 10.1016/j.pmatsci.2020.100638.

30 Abubakre, O.K., Medupin, R.O., Akintunde, I.B. et al. (2023). Carbon nanotube-reinforced polymer nanocomposites for sustainable biomedical applications: a review. *Journal of Science: Advanced Materials and Devices* 8: 100557. doi: 10.1016/j.jsamd.2023.100557.

31 Lee, S.J., Yoon, S.J., and Jeon, I.-Y. (2022). Graphene/polymer nanocomposites: preparation, mechanical properties, and application. *Polymers* 14: 4733. doi: 10.3390/polym14214733.

32 Sakib-Uz-Zaman, C. and Khondoker, M.A.H. (2023). Polymer-based additive manufacturing for orthotic and prosthetic devices: industry outlook in Canada. *Polymers* 15: 1506. doi: 10.3390/polym15061506.

33 Claes, T. (2021). *Spentys. The manufacturing process of ankle-foot orthosis: traditional vs. 3D printed*. https://www.spentys.com/blog/the-manufacturing-process-of-ankle-foot-orthosis-traditional-vs-3d-printed (accessed 5 July 2022).

34 Jin, Y., Plott, J., Chen, R. et al. (2015). Additive manufacturing of custom orthoses and prostheses – a review. *Procedia CIRP, CIRP 25th Design Conference Innovative Product Creation* 36: 199–204. doi: 10.1016/j.procir.2015.02.125.

35 Kariminejad, M., Tormey, D., Huq, S. et al. (2021). Ultrasound sensors for process monitoring in injection moulding. *Sensors* 21: 5193. doi: 10.3390/s21155193.

36 Song, Y., Gandhi, U., Sekito, T. et al. (2018). CAE method for compression molding of carbon fiber-reinforced thermoplastic composite using bulk materials. *Composites Part A: Applied Science and Manufacturing* 114: 388–397. doi: 10.1016/j.compositesa.2018.09.002.

37 Marino, M., Pattni, S., Greenberg, M. et al. (2015). Access to prosthetic devices in developing countries: pathways and challenges. *2015 IEEE Global Humanitarian Technology Conference (GHTC). Presented at the 2015 IEEE Global Humanitarian Technology Conference (GHTC)*, 45–51. doi: 10.1109/GHTC.2015.7343953.

38 Chinnappan, B.A., Krishnaswamy, M., Xu, H., and Hoque, M.E. (2022). Electrospinning of biomedical nanofibers/nanomembranes: effects of process parameters. *Polymers* 14: 3719. doi: 10.3390/polym14183719.

Index

Please note: Page references in **Bold** refer to separate tables and numbers in *italics* refer to figures.

3D printing 378, *379*, 383–385

a

AA *see* acrylic acid
Acacia 15
acetylation, starch 284
N-acetylglucosamine 13–14
acrylic acid (AA) monomers 64–65
acrylic polymer sensors 247–248
actions of electronic sensors 244–245
active food ingredient encapsulation 299–300, **300**
active packaging 323–324
additive manufacturing (AM) 378, *379*, 383–385
adverse effects, drug delivery systems 173
aerobic bioclogging 279
AEV *see* air entry values
AFPs *see* antifreeze proteins
agarose 13, 205
agarose–PEG diacrylate IPN hydrogels 13
agricultural applications 57–87
  biodegradability 78
  fine-tuning 58–59
  future directions 79–81, *80*
  nutrient release 75–77
  overview 48, **49**
  plant properties 72–75
  preparations and properties **66**
  soil parameter effects 67–71
  soil-plant-atmosphere interactions *60*
  superabsorbent biopolymer composites 58–81
  swelling ratios **66**
  synthesis 64–65, **66**
air entry values (AEV) 70–71
algae 9, 12–13
alginate 12–13
  chemical structure 117, *118*
  electronics applications **222**, 226, 230–231
  food packaging 321
  food technology **293**, 295–296

Applications of Biopolymers in Science, Biotechnology, and Engineering, First Edition.
Edited by Arunima Reghunadhan, Akhina H., and Sabu Thomas.

nanocomposites 230–231
tissue engineering 117–118
alginate–gelatin microcapsules 110–111
aliphatic polycarbonates *34*, 39–40
alkaline phosphatase (ALP) 121, 122
ALP *see* alkaline phosphatase
alteration, biopolymers 9
AM *see* additive manufacturing
amino acid sensing 248
amperometric sensors 245
amylopectin 14
see also starch
amylose 14, 284
see also starch
anaerobic bioclogging 279–280
angiogenesis 109
antibacterial surfaces 201
antifreeze proteins (AFPs) 298
antimicrobial packaging 323–324
applications, overview 45–49, **49**
aramid fiber-reinforced polymer composites (ARPCs) 198, 337, 367–368
ARPCs *see* aramid fiber-reinforced polymer composites
artificial biopolymers, sources 6
L-asparaginase 176
automotive applications, overview 48, **49**

b

bacterial cellulose (BC), electronics applications **222**, 224–225
bacterial nanocellulose (BNC) 140–141
bacterial synthesis 8–9, 122
banana peel powder (BPP) 125
basic fibroblast growth factor (bFGF) 113
BC *see* bacterial cellulose
BCP *see* β-caryophyllene
benzyl starch (BS) 14–15
bFGF *see* basic fibroblast growth factor
bioactive glass nanocomposites 147–148
bioactive glass nanoparticles (nBG) 120, 128
bioadhesives 281–283
biobarriers 280–281
biobased materials 3–4, 32
classification *220*
electronics applications 219–241, 254–256
see also biopolymers
biocementation 275–277
bioceramics 200–201
bioclogging 278–280
biocoatings 283–285
biocompatible polymers
electronic sensors 255
see also bioengineering; biomedical applications; tissue engineering
biocomposites 18–21
biotechnology applications 144–150, **148–149**, 197–199
blood vessel tissue engineering 129
bone tissue engineering 119–123
definition 1–2, 108
electronics applications 228–232
mechanical properties 20
nanocomposites 20–21
oral tissue engineering 126–129
processing techniques 19
skin tissue engineering 123–124
superabsorbent 58–81
biodegradability 78
fine-tuning 58–59
future directions 79–81, *80*
nutrient release 75–77

plant properties 72–75
preparations and properties **66**
soil parameter effects 67–71
soil-plant-atmosphere interactions *60*
swelling ratios **66**
synthesis 64–65, **66**
synthesis 64–65
tissue engineering 119–129
wound healing applications 124–126
bioconcrete 277–278
biodegradability
biomedical applications 91
food packaging 324
packaging materials 352–353
superabsorbent biopolymer composites 78
biodegradable nanoparticles 193–194, **194**
bioelectronic materials 226
bioengineering 89–188
advantages 91, 104–105
bioactive glass composites 120, 128
blood vessels 114–115, 129
bone tissue 109, 114, 118, 119–123
carrageenan 95
cellulose 95
chitin 122
chitosan 94, 111–112, 121, 127
collagen 93–94, 109–110, 127–129
elastin 114–115
gelatin 94, 110–111
gene therapy 173–175
hyaluronic acid 95–96
hybrid films 128–129
oral tissue 126–129
poly(3-hydroxybutyric acid) 122
polyhydroxyalkanoates 93
poly(lactic acid)s 93
polymers used 91–96
silk 95, 112–114
tissue engineering 101–138, 174–175
biofilms 201
biofilters 280–281
biomaterials
definition 32
food packaging 324–325
biomedical applications 89–188, **96**
advantages 91, 104–105
bioactive glass composites 120, 128
biodegradable polymers 91
biopolymers 192–195, **193–194**
biosensors 202, *203*
blood vessels 114–115, 129
bone tissue engineering 109, 114, 118, 119–123
calcium phosphate composites 109, 128
carbon fiber-reinforced polymer composites 197
carrageenan 95
cellulose 95
chitin 122
chitosan 94, 111–112, 121, 127, 195
collagen 93–94, 109–110, 127–129, 194–195
drug delivery 46–47, 93–95, 167–187
elastin 114–115
gelatin 94, 110–111, 195
gels 203–207
glass fiber-reinforced polymer composites 197–198
hyaluronic acid 95–96
hybrid films 128–129
nanocomposites 199–203, *200*
natural fiber-reinforced polymer composites 198–199

natural polymer nanocomposites 256
nerve regeneration and repair 109, 111, 112, 117
oral tissue 126–129
overview 45–47, **49**
poly(3-hydroxybutyric acid) 122
polyhydroxyalkanoates 93, 191–192
poly(lactic acid)s 93
polymers used 91–96
protein-based biopolymers 193–195
silk 95, 112–114
synthetic biopolymers 192–193
tissue engineering 101–138, 174–175
biomolecule delivery 206
bionanocomposites (BNCs) 189–217
cellulose nanofiber composites **333–335**, 338–341, 346
characterization 348–352
chemical modification 340–341
chitin nanofiber composites 341–344
electrospinning 336, 344–345
overview 336
packaging applications 331–360
see also nanofiber composites, packaging applications
physical modification 339–340
renewable 346–348
biopolymer composites *see* biocomposites
biopolymer nanocomposites 20–21
biopolymers 4–12
alteration 9
biomedical applications 192–195, **193–194**
biomedical usage 91–96, **96**
biotechnology applications 139–144, 190–196
blending 9–10
classification 1, 6–8, 32, *33*, 62–63, 90–91, 102–104, *103*, 107–108, *220*, 266–268, *267*, 310–312, *311*
definition 32
drug delivery 175–178
food applications 195–196
food packaging 312–320
gel formation 10–12
IPNs 12–18
key concepts 4–5
plasticization 10
preparation 8–9
properties 43–45, 104, 266–268, *267*
prosthetics 362–365
smart 105–107
sources 8–9, *191*
tissue engineering 108–118
see also specific polymers and composites...
biopolymer thin films, drug delivery 178–179, **180**
biosensors 202, *203*
biosynthesis, polyhydroxyalkanoates 37–38
biotechnology 89–188, 139–165, 189–217
advantages 91
aramid fiber-reinforced polymer composites 198
bioactive glass nanocomposites 147–148
biocomposites 144–150, 197–199
biodegradable polymers 91
biopolymers 139–144, 190–196
biosensors 202, *203*
carbon fiber-reinforced polymer composites 197
cellulose 140–141

chitosan 142–143
clay-based nanocomposites 149–150
food applications 195–196
gelatin 141–142
glass fiber-reinforced polymer composites 197–198
graphene-based nanocomposites 147
mucin 143–144
multifunctional polymer nanocomposites 146–147
nanocellulose 140–141
nanocomposites 144–150, 199–203
nanogels 150–156
natural fiber-reinforced polymer composites 198–199
poly(caprolactone) 144
polymers used 91–96, **96**
see also bioengineering; biomedical applications; drug delivery; medical technology
blending 2–3, 9–10
see also polymer blends
blood vessel formation
collagen scaffolds 109–110
elastin 114–115
overview 129
silk 114
BNC *see* bacterial nanocellulose
bone tissue engineering (BTE) 119–123
alginate 118
bacterially synthesized natural polymers 122
chitin 122
chitosan 121
collagen 109, 120
gelatin 120–121
gels 205
nanocomposites 200–201
poly(3-hydroxybutyrate) 122
poly(hydroxybutyrate-co-hydroxy valerate) 122
silk 114
BPP *see* banana peel powder
brain activity mapping 384
brown algae, alginate 12–13
BS *see* benzyl starch
BTE *see* bone tissue engineering
building blocks
IPNs 12–18
nanoparticles *169*
bulk density, soil 70
β-butyrolactone 37

c

C2C12 cells 205
CAB *see* cellulose acetate butyrate
CAD *see* computer-aided design
calcium bicarbonate biocementation 277
calcium-dependent biocementation 276–277
calcium phosphate ($Ca_3(PO_4)_2$)
biocementation 277
collagen scaffolds 109, 128
cancer therapy, gels 206–207
capacitors 225–226, 229
$Ca_3(PO_4)_2$ *see* calcium phosphate
carbohydrates
food encapsulation 299–300, **300**
see also individual compounds...
carbon fiber-reinforced polymer (CFRP) composites 197, 366–367
carbon nanotubes (CNT)
chitosan-based composites 121
electronics applications **222–223**, 225, 229–230, 231

nanocomposites 229–232
prosthetics 373–374
carboxymethyl cellulose (CMC) 269–270, 284, 298, 324
carboxymethyl chitosan–palladium (CMCS–Pd) 230
carob gum **294**, 295, 297–299
carrageenan 95, **293–294**, 296–298, 321
cartilage repair 114, 205
β-caryophyllene (BCP) 121
β-casein/caseinates 177, 274–275, 321
cell delivery 203–204
cell division 116–117
cellulose
biocoatings 284
bioengineering 95
biotechnology applications 140–141
construction and civil engineering 269–270
drug delivery 178
electronics aplications **221**, 224–225, 229–230
fiber nanocomposites **333–335**, 338–341, 345–346
chemical modification 340–341
electrospinning 344–345
physical modification 339–340
food packaging 314–315, 322–324
IPNs 16
medical applications 95
nanocomposites 229–230
sources and production **8**
cellulose acetate butyrate (CAB) 16
cellulose acetate (CA), food packaging 315
cellulose nanocrystals (CNC) 141, 224–225, 338
cellulose nanofibrils (CNF) 140–141, 224–225, 270, 338
cellulose phosphate 16
cement
chitosan addition 272–273
see also biocementation
CFRP *see* carbon fiber-reinforced polymer
characterization, packaging materials 348–353
chemical fertilizers 59, 75, **76–77**
chemical modification, cellulose fiber nanocomposites 340–341
chemiluminescent films 248
chemiresistive sensors 245, 251, 254–255
chitin
bone tissue engineering 122
chemical structure *112*
food packaging 316
nanofiber composites 341–344
packaging applications 316, 341–344
chitin nanofibers (CHNF/CNF)
hydrogels 206
packaging applications 341–344
chitosan 13–14, 103
bioadhesives 282–283
biocoatings 284–285
bioengineering 94, 111–112, 121, 127
biotechnology applications 142–143
bone tissue engineering 121
cement addition 272–273
chemical structure *112*
construction and civil engineering 272–273
deacetylation 112
drug delivery 178
electronics applications **221–223**, 225–226, 230, 246
food packaging 315–316, 323–324
hydroxyapatite composites 121

medical applications 94, 111–112, 121, 127, 195
muscle regeneration 205
nanocomposites 230
nanoparticles 178, 195
oral tissue engineering 127
polyurethanes 284–285
tissue engineering 111–112, 121, 127
chitosan/disulfiram-loaded hydrogels 207
chitosan/gelatin/PEDOT hybrid scaffolds 111
CHNF *see* chitin nanofibers
chondroitin sulphate-collagen scaffolds 109
chronic wounds
definition 124
see also wound healing
civil engineering 265–290
bioadhesives 281–283
biobarriers 280–281
biocementation 275–277
bioclogging 278–280
biocoatings 283–285
bioconcrete 277–278
biofilters 280–281
biopolymer characteristics 266–268, *267*
biopolymer uses **275**, *276*
cellulose 269–270
chitosan 272–273
gums 273–274
lignin 268–269
protein-based biopolymers 274–275
starch 270–272
classification
biobased materials *220*
biopolymers 1, 6–8, 32, *33*, 62–63, 90–91, 102–104, *103*, 107–108, *220*, 266–268, *267*, 310–312, *311*
electronic sensors 244–245
nanogels 150–152
synthetic biopolymers *33*
clay-based nanocomposites 149–150
CMC *see* carboxymethyl cellulose
CMCS–Pd *see* carboxymethyl chitosan–palladium
CNC *see* cellulose nanocrystals
CNF *see* cellulose nanofibrils; chitin nanofibers
CNT *see* carbon nanotubes
coatings, food packaging 321–323
cobalt merged bioactive glass (CoBG) 120
collagen 16
bioengineering 93–94, 109–110, 120, 126–129
bone tissue engineering 120
chemical structure *110*
dentin 126–127
drug delivery 176–177
electronics applications 227
food coatings 322
hydroxyapatite composites 127–128
medical applications 93–94, 109–110, 126–129, 194–195
nanoparticles 176–177, 194–195
oral tissue engineering 126–129
tissue engineering 109–110
collagen-chondroitin sulphate scaffolds 109
composite nanoparticles *see* nanocomposites
compression molding 380–382
computer-aided design (CAD) 376–378

conducting polymers 247, 368–372
conductometric sensors 245
construction 265–290
  bioadhesives 281–283
  biobarriers 280–281
  biocementation 275–277
  bioclogging 278–280
  biocoatings 283–285
  bioconcrete 277–278
  biofilters 280–281
  biopolymer characteristics 266–268, *267*
  biopolymer uses **275**, *276*
  cellulose 269–270
  chitosan 272–273
  gums 273–274
  lignin 268–269
  protein-based biopolymers 274–275
  starch 270–272
controlled drug release 201–202, 207
controlled nutrient release 75–77
copper oxide **222**
corona modification 339
crosslinkers/crosslinking 64–65, **66**, 152–153
cyanoethyl cellulose 16
cysteine, gluten 7

d

DDS *see* drug delivery systems
deacetylation, chitosan 112
degradation kinetics
  superabsorbent biopolymer composites 78
  see also biodegradability
dental applications 126–129
dental pulp stem cells (DPSC) 46
dentin 126–127
dermis 123
dextran 15, 282
dextran-graft-aniline 205
dielectric materials 225–226, 229–230
dietary fibers 298–299
dietetic products 298–299
diisocyantes, cellulose fiber nanocomposite modification 341
direct poly-condensation, glycolic acid 35–36
disulfide bridges 7
disulfuram (DSF) 207
DPSC *see* dental pulp stem cells
drought 60, 72–75
drug delivery systems (DDS) 46–47, 167–187
  adverse effects 173
  biopolymers 175–178
  biopolymer thin films 178–179, **180**
  β-casein 177
  cellulose 95, 178
  chitosan 94, 178
  collagen 93–94, 176–177
  controlled release 201–202, 207
  EPR effect 172–173
  gelatin 94, 177
  gels 178–179, **180**, 206
  gene therapy 173–175
  insulin 207
  micelles 178
  nanocomposites 201–202
  nanoparticles 168–175, **180**
  natural polymer nanocomposites 256
  nitric oxide 174
  passive 172
  poly(lactic acid)s 93
  self-delivery 172
  silk fibroin 176
  targeting 172–174, 201–202

tissue engineering 174–175
zein 177
DSF *see* disulfuram

e

ECM *see* extracellular matrix
Ecovio 1
edible films, food packaging 321–323
elastin 114–115, *115*
elastin-like polypeptides (ELPs) 114
electrical sensors 245
electroactive polymers (EAPs) 258
electrochemical sensors 244–245
electromagnetic interference (EMI) shielding 226, 230–233
electronics 219–264
alginate **222**, 226, 230–231
bacterial cellulose **222**, 224–225
biocomposites 228–232
carbon nanotubes **222–223**, 225, 229–230, 231
cellulose **221**, 224–225, 229–230
chitosan **221–223**, 225–226, 230, 246
collagen 227, 232
gelatin 227, 232
multiwalled carbon nanotubes **222–223**, 229–230
overview **221–223**
polyaniline **223**
polycarbonate **223**
poly(lactic acid) **221**, **223**, 226–227, 232
polymer blends 233
polypyrrole **223**, 258, 370
polysaccharide nanocomposites 229
polyurethane 368–372
poly(vinylidene fluoride) 251–252
prosthetics 368–372
sensors 243–264
acrylic polymers 247–248
biobased polymers 254–256
biocompatible polymers 255
conducting polymers 247
electroactive polymers 258
EVOH 248–250
flexible polymers 252–253
imprinted 256–258
nanocomposites 250–251
natural polymers 256
polymers used 245–247
poly(vinylidene fluoride) 251–252
prosthetics 368–372
shielding 226, 230–233
silk **223**, 227, 231
electro-responsive polymers (ERPs) 258
electrospinning 336, 344–345
ELPs *see* elastin-like polypeptides
EMI *see* electromagnetic interference
emulsifiers 298
enamel 126
encapsulation, active food ingredients 299–300, **300**
energy storage devices 231
enhanced permeation and retention (EPR) effect 172–173
epidermis 123
EPR effect *see* enhanced permeation and retention effect
ERPs *see* electro-responsive polymers
esterification, cellulose fiber nanocomposites 340–341
ethylene-vinyl alcohol di-block copolymer (EVOH) 248–250
EVOH *see* ethylene-vinyl alcohol di-block copolymer
extracellular matrix (ECM) 109, 192
extraction, gelatin 195

f

fabrication
  prosthetics 375–382
    additive manufacturing 378, *379*
    compression molding 380–382
    conventional processes 376–378
    injection molding 379–380
FC *see* field c3apacity
fenestration sites, drug delivery 172–173
ferroelectric polymers 251–252
fertilizers 59, 75, **76–77**
fibrinogen 16
fibroin 7
  cartilage repair 205
  chemical structure *113*
  drug delivery 176
  insulin administration 207
  tissue engineering 112–114
field capacity (FC), soil 71
films
  drug delivery 178–179, **180**
  food packaging 321–323
  tissue engineering 128–129
flexible polymer sensors 252–253, 368–372
food applications 291–330
  biopolymers 195–196
  pectin **294**, 295–296, 299
  see also food packaging; food technology
food packaging 300–301, 309–330
  active 323–324
  alginate 321
  antimicrobial 323–324
  applications 320–325
  biomaterials 324–325
  biopolymers 312–320
  carboxymethyl cellulose 324
  carrageenan 321
  caseinates 321
  cellulose 314–315, 322–324
  chitosan 315–316, 323–324
  coatings 321–323
  collagen 322
  edible films 321–323
  gelatin 322
  nanoparticles 325–326
  natural polymers 323
  paper 321
  polyhydroxyalkanoates 318–320
  polylactic acid 317–318
  polymer blends 313–314, 316, 319
  polymer properties 310–312
  starch 312–314, 322, 323
  wheat gluten 321, 322, 323
  whey protein 321, 322, 323
  zein 321, 322
food technology 291–308
  alginate **293**, 295–296
  carrageenan **293–294**, 296–298
  dietetic products 298–299
  emulsifiers 298
  encapsulation 299–300, **300**
  gelatin **294**, 297, 299
  gelling agents 295–297
  gels 296–297
  gluten-free formulations 298–299
  guar gum **292**, 295, 297–299
  locust bean gum **294**, 295, 297–299
  pectin **294**, 295–296
  processing aids 297
  stabilizers 298
  thickening agents 295–297
  xanthan gum **292**, 295–296
frameworks, sustainable agriculture *38*
functional biopolymers 105–107

functional food products 298–299
functionalization, cellulose fiber nanocomposites 340–341

g

GA *see* gum arabic
GAG *see* glycosaminoglycan
GAM *see* gene-activated matrix
gamma radiation treatment 340
GE *see* grafting efficiency
gelatin 6–7, 16
  biocoatings 284
  bioengineering 94, 110–111
  biotechnology applications 141–142
  bone tissue engineering 120–121
  chemical structure *111*, *142*
  construction and civil engineering 270–272
  drug delivery 177
  electronics applications 227
  extraction 195
  food coatings 322
  food technology **294**, 297, 299
  medical applications 94, 110–111, 195
  tissue engineering 110–111, 120–121
gelatin/chitosan/PEDOT hybrid scaffolds 111
gelatin methacryloyl (GelMA) hydrogels 205
gelling agents, food 295–297
GelMA *see* gelatin methacryloyl
gels
  alginate 117–118
  biomedical applications 203–207
  biomolecule delivery 206
  cancer therapy 206–207
  cartilage repair 205
  chitosan 112
  drug delivery 178–179, **180**, 206
  food technology 296–297
  formation 10–12
  key concepts 2–3
  polymeric materials **204**
  temperature effects 11
  tissue engineering 203–206
  wound healing 206
  see also nanogels
gene-activated matrix (GAM)-based localized gene delivery 175
gene therapy 173–175
GFRP *see* glass fiber-reinforced polymer
glass fiber-reinforced polymer (GFRP) composites 197–198, 337, 367
glass nanocomposites 147–148
gliadin 7
glucose oxidase sensor (GOD) **222**
gluten
  construction 283
  food packaging 321, 322, 323
  heat treatment 7
  nanofiber composites 344
gluten-free food formulations 298–299
glutenin 7
glycolic acid 35, 93
glycosaminoglycan (GAG) 120
glycosyl connections 6
GOD *see* glucose oxidase sensor
gold nanoparticles 199, 230
gold nanowires 232
GP *see* grafting percentage
grafting efficiency (GE) 64
grafting percentage (GP) 64
graft polymerization 64–65, **66**
graphene-based nanocomposites 147, 374
graphene oxide/cellulose nanocomposites 229–230

guar gum 273, **292**, 295, 297–299
gum arabic (GA) 15
gums, construction and civil engineering 273–274

h

HA *see* hyaluronic acid
heat treatment 7, 339
HEC *see* hydroxyethyl cellulose
heterogeneous composites 3
hexa(methoxymethyl) melamine 282–283
high methylester (HM) pectins 296
HM *see* high methylester
hMSCs *see* human mesenchymal stem cells
HPC *see* hydroxypropyl cellulose
HPG *see* hydroxypropyl guar gum
HPMC *see* hydroxypropyl methyl cellulose
human mesenchymal stem cells (hMSCs) 114, 118, 122–123
human umbilical vein endothelial cells (HUVECs) 110–111
humidity sensors 248
HUVECs *see* human umbilical vein endothelial cells
hyaluronic acid (HA)
  bioengineering 95–96, 115–117
  chemical structure 115–116, *117*
  tissue engineering 115–117
hybrid composites 368
hybrid films 128–129
hybrid IPNs 3, 12
hybrid nanogels 151
hydrazinodeoxy cellulose 16
hydrocolloids 196
  alginate **293**
  carrageenan **293–294**, 296–298
  dietetic products 298–299
  emulsifiers 298
  encapsulation 299–300, **300**
  food technology 291–308, **292–294**
  gelatin **294**, 297, 299
  gelling agents 295–297
  gluten-free food formulations 298–299
  guar gum **292**, 295, 297–299
  locust bean gum **294**, 295, 297–299
  pectin **294**, 295–296, 299
  processing aids 295–297
  stabilizers 298
  thickening agents 295–297
  xanthan gum **292**, 295–296
hydrogels *see* gels
hydrophilic groups, superabsorbent biopolymer composites 59
hydroxyapatite-doped gelatin–silicon microparticles (mSiHA) 121
hydroxyapatite (HA)
  chitosan composites 121
  collagen composites 127–128
  drug delivery systems 201
  oral tissue engineering 127–128
hydroxyethyl cellulose (HEC) 16
hydroxypropyl cellulose (HPC) 16, 300
hydroxypropyl guar (HPG) gum 273
hydroxypropyl methyl cellulose (HPMC) 299, 300
hypodermis 123

i

implants
  carbon fiber-reinforced polymer composites 197
  glass fiber-reinforced polymer composites 197–198
  polymer nanocomposites 199–201

injectable silk fibroin protein hydrogels (iSFH) 207
injection molding 379–380
insulin delivery systems 207
interpenetrating polymer networks (IPNs)
  agarose 13
  alginate 12–13
  benefits 3
  building blocks 12–18
  cellulose 16
  chitosan 13–14
  collagen 16
  dextran 15
  fibrinogen 16
  gelatin 16
  gum arabic 15
  key concepts 3, 12
  polyhydroxyalkanoates 3, 17
  polylactides 17–18
  starch 14–15
intrinsic properties, synthetic biopolymers 43–44
ionic polymer–metal composites (IPMCs) 258
IONPs *see* iron oxide nanoparticles
IPMCs *see* ionic polymer–metal composites
IPNs *see* interpenetrating polymer networks
iron nanoparticles, superparamagnetic 202–203
iron oxide nanoparticles (IONPs) 199, 233
iSFH *see* injectable silk fibroin protein hydrogels
isocyantes, cellulose fiber nanocomposite modification 341

j
j215 biopolymer 13–14

k
Kaltostat 118
kaolin 124–125
Kevlar 198, 367–368

l
LA *see* lactic acid
LAB *see* lactic acid bacteria
lactic acid bacteria (LAB) 7
lactic acid (LA) 7
L-lactide (LPLA) 35
LEDs *see* light-emitting diodes
LFT *see* long fiber-reinforced thermoplastic (LFT)
LFT-D *see* long fiber thermoplastic-direct processes
ligament repair 109, 114–115
light-emitting diodes (LEDs) 224, 227
lignin 268–269
lignosulfonates (LS) 268–269
lithium magnesium silicate hydrate (LMSH) 125–126
LM *see* low methylester
LMSH *see* lithium magnesium silicate hydrate
locust bean gum (LBG) **294**, 295, 297–299
long fiber-reinforced thermoplastic (LFT) 19, 381
long fiber thermoplastic-direct (LFT-D) processes 19
low methylester (LM) pectins 296
LPLA *see* L-lactide
lung tissue repair 114–115

m

magnetic resonance imaging (MRI) 202–203
mass sensitive sensors 245
MC *see* methyl cellulose
mechanical properties
  biopolymer mixtures 20
  packaging materials 349–350
medical devices
  aramid fiber-reinforced polymer composites 198
  glass fiber-reinforced polymer composites 197–198
  prosthetics 361–389
    future challenges and perspectives 382–385
medical imaging, composite nanoparticles 202–203
medical implants
  carbon fiber-reinforced polymer composites 197
  glass fiber-reinforced polymer composites 197–198
  polymer nanocomposites 199–201
medical technology 89–188
  see also biomedical applications
mercerization, cellulose fiber nanocomposites 340
mesenchymal stem cells (MSCs) 114, 118, 122–123
Metabolix PHA 320
metal oxides, electronics applications **222**
2-methyacryloyloxyethyl phosphorylcholine (MPC) 114
methyl cellulose (MC) 269–270, 300
micellar nanogels 152
micellar nanoparticles, drug delivery 178
microencapsulation, food ingredients 299–300, **300**
MIP *see* molecularly imprinted polymers
modified cellulose fiber nanocomposites 338–341
molecularly imprinted polymers (MIP), sensors 256–258
molybdenum disulfide (MoS_2) 225
monoisocyantes, cellulose fiber nanocomposites modification 341
morphology, packaging materials 350–351
MoS_2 *see* molybdenum disulfide
MPC *see* 2-methyacryloyloxyethyl phosphorylcholine
MSCs *see* mesenchymal stem cells
mSiHA *see* hydroxyapatite-doped gelatin–silicon microparticles
mucin, biotechnology 143–144
multifunctional polymer nanocomposites 146–147
multi-responsive nanogels 151
multiwalled carbon nanotubes (MWCNTs), electronics applications **222–223**, 229
muscle regeneration, gels 204–205
MWCNTs *see* multiwalled carbon nanotubes

n

nano biocomposites (NCs) 189
  see also bionanocomposites
nanocellulose (NC) 140–141, **333–335**, 338–341
nanocomposites 20–21
  alginate-based 230–231
  biomedical applications 199–203, *200*

biosensors 202, *203*
biotechnology 144–150, 199–203
bone tissue engineering 200–201
carbon nanotubes 229–232
cellulose-based 229–230
chitosan-based 230
clay-based 149–150
collagen 232
definition 145
drug delivery 201–202
electronics applications 228–232, 250–251
electronic sensors 250–251
food packaging 325–326
gelatin 232
graphene-based 147
medical imaging 202–203
medical implants 199–201
multifunctional 146–147
multiwalled carbon nanotubes 229
poly(lactic acid) 232
polysaccharide-based 229
prosthetics 372–374
protein-based 231–232
recent applications **148–149**
silk 231
see also bionanocomposites
nanofiber composites
electrospinning 336, 344–345
nanocellulose effects **333–335**
packaging applications 331–360
cellulose **333–335**, 338–341, 345–346
chitin 341–344
electrospun 336, 344–345
materials characterization 348–353
modified cellulose fiber nanocomposites 338–341
nanocellulose integration effects **333–335**
nanofiber preparation 336
poly(caprolactone) 347
polyhydroxyalkanoates 347–348
poly(lactic acid) 345–347
poly(vinyl alcohol) 345
properties 336–337
renewable 346–348
requirements 337
zein 344–345
nanofibers, preparation methods 336
nanofibrillated cellulose (NFC) 225
nanogels
advantages 154
applications 153–156
biotechnology 150–156
classification 150–152
cross-linking 152–153
limitations 154
patents 155, **155–156**
preparation 152–153
self-assembly 152
nanoparticles
building blocks *169*
cellulose 178
chitosan 178, 195
collagen 176–177, 194–195
drug delivery 168–175, **180**
food packaging 325–326
gelatin 177
gene therapy 173–174
harmful effects 173
medical imaging 202–203
micelles 178
passive delivery 172
polypeptide 176–178, 193–194, **194**
self-delivery 172
silk fibroin 176

targeting 172–174
zein 177
natural fiber-reinforced polymer composites 198–199
natural fibers, biobased materials 3–4
natural polymers
biobased materials 3–4
electronic sensors 256
food packaging 323
reinforced composites 198–199
see also individual compounds...
nBG *see* bioactive glass nanoparticles
NC *see* nanocellulose
nerve growth factor (NGF) 112
nerve regeneration and repair 109, 111, 112, 117
neural stem cells (NSC) 112
neuroprosthetics 384
NFC *see* nanofibrillated cellulose
NGF *see* nerve growth factor
nitric oxide (NO) 174
nitrogen oxides (NOx) 281
NO *see* nitric oxide
NOx *see* nitrogen oxides
NSC *see* neural stem cells
nutrient release, superabsorbent biopolymers 75–77
Nylon, applications **49**

o

O&P (orthotic and prosthetic) equipment *see* prosthetics
OCP *see* octacalcium phosphate
octacalcium phosphate (OCP) 121
OFETs *see* organic field-effect transistors
OLEDs *see* organic light-emitting diodes
optical sensors 244
oral tissue engineering 126–129
calcium phosphate composites 128
chitosan 127
collagen 126–129
hybrid films 128–129
silver nanoparticle biocomposites 127
organic biopolymers, sources 6
organic field-effect transistors (OFETs) 224, 225, 226
organic light-emitting diodes (OLEDs) 224, 227
orthotic and prosthetic (O&P) equipment *see* prosthetics
osteochondral tissue regeneration 205

p

P3HB *see* poly(3-hydroxybutyrate)
PAA *see* polyacrylic acid
packaging applications
common materials 322
food 300–301, 309–330
active 323–324
alginate 321
antimicrobial 323–324
applications 320–325
biomaterials 324–325
biopolymers 312–320
carboxymethyl cellulose 324
carrageenan 321
caseinates 321
cellulose 314–315, 322–324
chitosan 315–316, 323–324
coatings 321–323
collagen 322
edible films 321–323
gelatin 322
nanoparticles 325–326
natural polymers 323
paper 321
polyhydroxyalkanoates 318–320

polylactic acid 317–318
polymer blends 313–314, 316, 319
polymer properties 310–312
starch 312–314, 322, 323
wheat gluten 321, 322, 323
whey protein 321, 322, 323
zein 321, 322
materials characterization 348–353
nanocellulose effects **333–335**
nanofiber composites 331–360
cellulose **333–335**, 338–341, 345–346
chitin 341–344
electrospun 336, 344–345
modified cellulose fiber nanocomposites 338–341
nanocellulose effects **333–335**
nanofiber preparation 336
poly(caprolactone) 347
polyhydroxyalkanoates 347–348
poly(lactic acid) 345–347
poly(vinyl alcohol) 345
renewable 346–348
requirements 337
zein 344–345
overview 47–48, **49**
paclitaxel 177
PADC *see* poly(alkene dicarboxylate)
PANI *see* polyaniline
paper packaging 321
passive delivery, nanoparticles 172
PAWC *see* plant available water content
PBAT *see* polybutyleneadipate-co-terephthalate
PC *see* polycarbonate
PCL *see* poly(caprolactone)
PDLLA *see* poly DL-lactide
PDMS *see* polydimethoxysilane
PE *see* polyethylene
PEBAX *see* polyether block amides
pectin **294**, 295–296, 299
PEDOT *see* poly(3,4-ethylenedioxythiophene) (PEDOT)
PEDOT NPs *see* poly(3,4-ethylenedioxythiophene) nanoparticles
PEEK *see* polyetheretherketone
PEN *see* poly(ethylene naphthalate)
permanent wilting point (PWC) 71
permeability, packaging materials 348–349
PES *see* polyethersulfone
PET *see* poly(ethylene terephthalate) (PET)
PHB *see* poly(hydroxy butyrate)
PHBV *see* poly(hydroxybutyrate-co-hydroxy valerate)
phosphatidylserine (PS) 120
pH sensors 254
PHV *see* polyhydroxy valerate
physical modification, cellulose fiber nanocomposites 339–340
PI *see* polyimide
piezoelectric sensors 245, 251–252
PLA *see* poly(lactic acid)
plant available water content (PAWC) 71
plant properties, superabsorbent biopolymers 72–75
plasma treatment, cellulose fibers 339
plastixization 10
PLGA *see* poly(lactic-co-glycolic acid)
PLLA *see* poly L-lactide
PMDS *see* polydimethoxysiloxane
PMMA *see* poly(methyl methacrylate)
PNC *see* polymer nanocomposites

poly(3-hydroxybutyrate) (P3HB) 37, **49**, 122
polyacrylic acid (PAA) 248
poly(alkene dicarboxylate) (PADC) *34*, 38
polyamides 6, *34*, 40–41, **49**
poly(α-amino acid) **49**
polyanhydrides *34*, 43
polyaniline (PANI) **223**, 247, 368–369
polybutyleneadipate-co-terephthalate (PBAT) 233
poly(caprolactone) (PCL)
- applications **49**
- biotechnology applications 144
- chitin-chitosan blends 316
- collagen scaffolds 109–110
- food packaging 314, 316, 319
- gelatin scaffolds 111, 120–121
- nanofiber composites 347
- polyhydroxy butyrate blends 319
- starch blends 314

polycarbonate (PC) **223**, 252–253, 363–364
poly(carbonate urethanes) 41–42
poly-condensation, glycolic acid 35–36
polydimethoxysilane (PDMS) 231
polydimethoxysiloxane (PMDS) 365
polydioxanone *34*, 39, **49**
poly(ester amide)s *34*, 40–41, **49**
polyesters 6, 33–38
- applications **49**
- poly(alkene dicarboxylate)s 38
- poly(glycolic acid) 35–37
- poly(hydroxy acid)s 34
- poly(hydroxybutyrate) 37–38
- poly(lactic acid) 34–35, 36–37
- poly(lactic-co-glycolic acid) 36–37
- sources and production **8**

poly(ester urethane) 41–42, **49**
poly(ether amide)s 41
polyether block amides (PEBAX) 41
poly(ether ester) 38–39, **49**
polyetheretherketone (PEEK) 252–253
polyethersulfone (PES) 252–253
poly(ether urethane) 41–42, **49**
poly(3,4-ethylenedioxythiophene) (PEDOT)
- chitosan/gelatin hybrid scaffolds 111
- electronic sensors 258, 371–372
- nanoparticles 205
- prosthetics 371–372

poly(3,4-ethylenedioxythiophene) nanoparticles (PEDOT NPs) 205
poly(ethylene naphthalate) (PEN) 252–253
polyethylene (PE) **8**, 42, 362
poly(ethylene terephthalate) (PET) 253
polyglycolic acid (PGA) *34*, 35–37
- applications **49**
- PLGA synthesis 36–37
- synthesis 35–36

poly(hydroxy acid)s, polyesters 34
polyhydroxyalkanoates (PHAs)
- applications **49**
- bioengineering 93
- biosynthesis 37–38
- biotechnology applications 191–192
- bone tissue engineering 122
- food packaging 318–320
- IPNs 3, 17
- medical applications 93
- nanocomposites 20–21
- nanofiber composites 347–348
- poly(hydroxybutyrate) 37–38
- processing properties 44

synthesis 7–8, 37–38
tissue engineering 122
poly(hydroxybutyrate-co-hydroxy valerate) (PHBV) 7–8
bone tissue engineering 122
nanofiber composites 347–348
poly(hydroxy butyrate) (PHB) 7–8, 17, *34*, 37–38, **49**
food packaging 319–320
nanofiber composites 347–348
PBAT blend 233
polyhydroxy valerate (PHV) 7–8
polyimide (PI), flexible sensors 252–253
poly-ion complexation, gelatin scaffolds 110–111
poly(lactic acid) (PLA) 7
applications **49**
bioengineering 93
electronics applications **221**, **223**, 226–227, 232
food packaging 317–318
intrinsic properties 43–44
IPNs 17–18
medical applications 93
nanocomposites 20–21, 232
nanofiber composites 345–347
PLGA synthesis 36–37
processing properties 44
product properties 45
sources and production **8**
starch blends 313–314
structure *34*
synthesis 34–35
poly(lactic-co-glycolic acid) (PLGA) *34*, 36–37, 93
poly DL-lactide (PDLLA) 317
poly L-lactide (PLLA) 317
poly(lactide)s *see* poly(lactic acid)
polymer-assisted synthesis 196
polymer blends 2–3, 9–10
electronic applications 233
food packaging 313–314, 316, 319
prosthetics 368–372
polymer composite materials (PCM) 145, 145–150, 365–368
see also biocomposites; bionanocomposites; nanofiber composites; polymer nanocomposites
polymer grafting, cellulose fiber nanocomposites 340–341
polymeric biopolymers, definition 2
polymeric micelles, drug delivery 178
polymeric nanoparticles
building blocks *169*
cellulose 178
chitosan 178
collagen 176–177
drug delivery 168–175, **180**
gelatin 177
gene therapy 173–174
harmful effects 173
micelles 178
passive delivery 172
polypeptide 176–178
self-delivery 172
silk fibroin 176
targeting 172–174
zein 177
polymer nanocomposites (PNC) 20–21, 144–150
alginate-based 230–231
biomedical applications 199–203, *200*
biosensors 202, *203*
biotechnology 144–150, 199–203
bone tissue engineering 200–201

carbon nanotubes 229–232
cellulose-based 229–230
chitosan-based 230
clay-based 149–150
collagen 232
definition 145
drug delivery 201–202
electronics applications 228–232, 250–251
electronic sensors 250–251
food packaging 325–326
gelatin 232
graphene-based 147
medical imaging 202–203
medical implants 199–201
multifunctional 146–147
multiwalled carbon nanotubes 229
poly(lactic acid) 232
polysaccharide-based 229
prosthetics 372–374
protein-based 231–232
recent applications **148–149**
silk 231
see also bionanocomposites
poly(methyl methacrylate) (PMMA) 364–365, 374
polynucleotides 104
polyols, synthesis 41–42
poly(ortho-toluidine)–polyurethane composites 370–371
polypeptides
definition 104
drug delivery 176–178
IPNs 3
nanofiber preparation 336
polypropylene (PP) 363
polypyrrole (PPy) **223**, 258, 370
polysaccharides 6, 103–104
bioadhesives 282–283
definition 104
food applications 196
food coatings 322
glycosyl connections 6
IPNs 3
nanocomposite electronics applications 229
superabsorbent biopolymer composites 64–65
see also individual polysaccharides...
poly(siloxane urethane) 41–42
polyurethanes (PU) *34*, 41–42
applications **49**
biocoatings 284
chitosan composites 127, 284
oral tissue engineering 127
prosthetics 364, 368–372
poly(vinyl alcohol) (PVA/PVOH) 42
applications **49**
food packaging 313
gelatin combined scaffolds 121
nanofiber composites 345
prosthetics 377
starch blends 313
poly(vinyl chloride) (PVC) 42
applications **49**
poly(vinylidene fluoride) (PVDF) 251–252
poly(vinyl pyrrolidone) (PVP) 125
porous matrices, bioclogging 278–280
portland cement
chitosan addition 272–273
drawbacks 277
potentiometric sensors 245
PP *see* polypropylene
PPy *see* polypyrrole
pregelatinization, starch 271
preparation, biopolymers 8–9
processing aids, food 297
processing properties, synthetic biopolymers 44

product properties, synthetic biopolymers 44–45
properties
  biopolymers 43–45, 104
  synthetic biopolymers 43–45
prosthetics 361–389
  additive manufacturing 378, *379*, 383–385
  aramid fiber-reinforced 367–368
  carbon-fiber-reinforced 366–367
  carbon nanotubes 373–374
  compression molding 380–382
  conducting polymers 368–372
  fabrication 375–382
  future challenges and perspectives 382–385
  glass fiber-reinforced 367
  graphene 374
  hybrid composites 368
  injection molding 379–380
  polyaniline 368–369
  polycarbonate 363–364
  polydimethoxysiloxane 365
  polyethylene 362
  poly(3,4-ethylenedioxythiophene) 371–372
  polymer blends 368–372
  polymer composites 365–368
  polymer nanocomposites 372–374
  polymers 362–365
  poly(methyl methacrylate) 364–365, 374
  poly(ortho-toluidine) 370–371
  polypropylene 363
  polypyrrole 370
  polyurethane 364, 368–372
protein-based biopolymers
  biomedical applications 193–195
  construction and civil engineering 274–275
  definition 104
  drug delivery 176–178
  edible coatings 322
  food applications 196, 322
  food encapsulation 299–300, **300**, 322
  IPNs 3
  nanofiber preparation 336
  paper packaging 321
  see also individual polymers...
protein-based nanocomposites, electronics applications 231–232
proton conductivity 226
PS *see* phosphatidylserine
PVA *see* poly(vinyl alcohol)
PVC *see* poly(vinyl chloride)
PVDF *see* poly(vinylidene fluoride)
PVOH *see* poly(vinyl alcohol)
PVP *see* poly(vinyl pyrrolidone)
PWC *see* permanent wilting point

q

QDs *see* quantum dots
quantum dots (QDs) 199

r

red algae, agarose 13
renewable energy, in manufacturing 5
renewable fiber composites 346–348
repaglinide (RPG) 178
ring opening polymerization 36, 37
RPG *see* repaglinide

s

SBC *see* superabsorbent biopolymer composites
scaffold tissue engineering 46
SDF-1α *see* stromal-derived factor-1 alpha

SDGs *see* sustainable development goals
seaweed isolates, food coatings 322
self-assembly
  nanofibers 336
  nanogels 152
self-delivery, nanoparticles 172
semi-IPNs 3, 12
  starch-based 14–15
  wound healing 125–126
sensors
  biological 202, *203*
  electronic 243–264
    acrylic polymers 247–248
    actions 244–245
    biobased polymers 254–256
    biocompatible polymers 255
    conducting polymers 247
    electroactive polymers 258
    EVOH 248–250
    flexible polymers 252–253, 368–372
    imprinted 256–258
    nanocomposites 250–251
    natural polymers 256
    polymers used 245–247
    polypyrrole 258, 370
    polyurethane composites 368–372
    poly(vinylidene fluoride) 251–252
    prosthetics 368–372
    types 244–245, 247–253
sericin 7
SF *see* silk fibroin
shear thinning 295–296
shielding, electromagnetic 226, 230–233
silicone polymers 41–42, 365
silk 7
  chemical structure *113*
  drug delivery 176
  electronics applications **223**, 227, 231
  medical applications 95, 112–114
  nanocomposites 231
  tissue engineering 112–114, 205
silk fibroin (SF)
  cartilage repair 205
  chemical structure *113*
  drug delivery 176
  insulin administration 207
  tissue engineering 112–114
silver nanoparticle biocomposites 127, 248
silylation, cellulose fiber nanocomposites 340
siRNA *see* small interfering RNAs
skeletal muscle regeneration 204–205
skin, structure of 123
skin healing 110
skin tissue engineering
  collagen scaffolds 110
  elastin 114–115
  overview 123–124
  silk 113
small interfering RNAs (siRNA) 94
smart biopolymers 105–107
sodium alginate *see* alginate
soil conditioning 58–61, 67–71, 75–77, 278–280
soil parameters, superabsorbent biopolymer composites 67–71
soil-SBC-plant-atmosphere interactions *60*
soil water retention curves (SWRC) 70–71
solid polymer electrolytes 225–226
sonochemical synthesis 336
sources, biopolymers 8–9, *191*
soy protein isolate (SPI) 7, 324

spinal cord injury (SCI) 109
spintronic humidity sensors 248
SPIOs *see* superparamagnetic iron nanoparticles
stabilizers, food 298
starch
  acetylation 284
  benzyl 14–15
  bioadhesives 282–283
  biocoatings 284
  blends 313–314
  construction and civil engineering 270–272
  edible coatings 322
  food packaging 312–314, 322, 323
  IPNs 13–15
  nanocomposites 20–21
  poly(caprolactone) blends 314
  poly(lactic acid) blends 313–314
  poly(vinyl alcohol) blends 313
  pregelatinization 271
  thermoplastic 10
stem cells 114, 116–118
stimuli-responsive biopolymers 105–107
  see also sensors
stimuli-responsive nanogels 151
stromal-derived factor-1 alpha (SDF-1α) 109
superabsorbent biopolymer composites (SBC) 58–81
  biodegradability 78
  fine-tuning 58–59
  future directions 79–81, *80*
  nutrient release 75–77
  plant properties 72–75
  preparations and properties **66**
  soil parameter effects 67–71
  soil-plant-atmosphere interactions *60*
  swelling ratios **66**
  synthesis 64–65, **66**
supercapacitors 225–226, 229
superparamagnetic iron nanoparticles (SPIOs) 202–203
surface imprinted polymers, sensors 256–257
surface radio wave sensors 245
sustainability 5
sustainable agriculture 57–87
  see also agricultural applications
sustainable development goals (SDGs) 57–58
swelling ratios, superabsorbent biopolymer composites **66**
SWRC *see* soil water retention curves
synthetic biopolymers 31–56
  aliphatic polycarbonates 39–40
  applications 45–49, **49**
  biomedical applications 192–193
  classification *33*
  intrinsic properties 43–44
  poly(alkene dicarboxylate) 38
  polyamides 40–41
  polyanhydrides 43
  poly(ester amide)s 40–41
  polyesters 33–38
  poly(ether amide)s 41
  poly(ether ester)s 38–39
  poly(glycolic acid) 35–37
  poly(hydroxy acid)s 34
  poly(hydroxybutyrate) 37–38
  poly(lactic acid) 34–35, 36–37
  poly(lactic-co-glycolic acid) 36–37
  polyurethanes 41–42
  processing properties 44
  product properties 44–45
  properties 43–45
  vinyl polymers 42
synthetic fibers 3–4

t
tannins 283
targeting, drug delivery 172–174, 201–202
temperature, gels 11
template assisted synthesis 336
tendon repair 109, 114–115
2,2,6,6,-tetramethylpiperidine-1-oxy (TEMPO) 341
Tetra Pak 321
thermal properties, packaging materials 351–352
thermoplastic starch (TPS) 10
thickening agents, food 295–297
thin films, drug delivery 178–179, **180**
tissue engineering 46, 101–138, 109, 128
  alginate 117–118
  bacterial synthesis 122
  bioactive glass composites 120, 128
  biopolymer composites 119–129
  biopolymers 108–118
  blood vessels 114–115, 129
  bone 109, 114, 118, 119–123, 200–201, 205
  calcium phosphate composites 109, 128
  cartilage 205
  cell delivery 203–204
  chitin 122
  chitosan 111–112, 121, 127
  collagen 109–110, 120, 127–129
  drug delivery systems 174–175
  elastin 114–115
  gelatin 110–111, 120–121
  gels 203–206
  gene therapy 174–175
  hyaluronic acid 115–117
  hybrid films 128–129
  nanocomposites 200–201
  oral 126–129
  poly(3-hydroxybutyrate) 122
  poly(hydroxybutyrate-co-hydroxy valerate) 122
  silk 112–114
  skeletal muscle 204–205
  skin 113, 114–115, 123–124
  wound healing 109–110, 113, 117–118, 124–126
toxicity, packaging materials 349
TPS *see* thermoplastic starch
transducers 244–245
trifluoroacetyl azobenzene dyes 254–255
type-I collagen 109–110

u
ultrasmall superparamagnetic iron nanoparticles (USPIOs) 202–203
ultraviolet irradiation, cellulose fiber nanocomposites 339
urea-dependent biocementation 276–277
USPIOs *see* ultrasmall superparamagnetic iron nanoparticles

v
vascular tissue engineering
  collagen 109–110
  elastin 114–115
  overview 114–115, 129
  silk 114
vinyl polymers 6, 42, **49**
viral inhibition 95

w
water stress 60, 72–75
WG *see* wheat gluten

wheat gluten (WG)
construction applications 283
food packaging 321, 322, 323
heat treatment 7
whey protein, food packaging 321, 322, 323
wound healing 109–110, 113, 117–118, 124–126, 206
see also tissue engineering
xanthan gum **292**, 295–296

Z

zein 177, 321, 322, 345

Printed and bound by CPI Group (UK) Ltd, Croydon, CR0 4YY

17/09/2026

14964252-0007